Die Grundlehren der mathematischen Wissenschaften

in Einzeldarstellungen
mit besonderer Berücksichtigung
der Anwendungsgebiete

Band 126

O. Lehto · K. I. Virtanen

Quasiconformal Mappings in the Plane

Translated from the German
by K. W. Lucas

With 15 Figures

Second Edition

Springer-Verlag
Berlin Heidelberg New York 1973

O. Lehto · K. I. Virtanen
University of Helsinki/Finland
Department of Mathematics

Translator
K. W. Lucas
Aberystwyth/Great Britain

AMS Subject Classifications (1970)
30 A 60

Title of the Original Edition
Quasikonforme Abbildungen
(Grdl. d. math. Wiss. Bd. 126) 1965

ISBN-13: 978-3-642-65515-9 e-ISBN-13: 978-3-642-65513-5
DOI: 10.1007/978-3-642-65513-5

O. Lehto · K. I. Virtanen

Quasiconformal Mappings in the Plane

Translated from the German
by K. W. Lucas

With 15 Figures

Second Edition

Springer-Verlag
New York Heidelberg Berlin 1973

O. Lehto · K. J. Virtanen
University of Helsinki/Finland
Department of Mathematics

Translator
K. W. Lucas
Aberystwyth/Great Britain

Geschäftsführende Herausgeber
B. Eckmann
Eidgenössische Technische Hochschule Zürich

B. L. van der Waerden
Mathematisches Institut der Universität Zürich

AMS Subject Classifications (1970)
30 A 60

Title of the Original Edition
Quasikonforme Abbildungen
(Grdlg. d. math. Wiss. Bd. 126) 1965

ISBN-13: 978-3-642-65515-9 e-ISBN-13: 978-3-642-65513-5
DOI: 10.1007/978-3-642-65513-5

Preface

The present text is a fairly direct translation of the German edition "Quasikonforme Abbildungen" published in 1965. During the past decade the theory of quasiconformal mappings in the plane has remained relatively stable. We felt, therefore, that major changes were not necessarily required in the text. In view of the recent progress in the higher-dimensional theory we found it preferable to indicate the two-dimensional case in the title.

Our sincere thanks are due to K. W. Lucas, who did the major part of the translation work. In shaping the final form of the text with him we received many valuable suggestions from A. J. Lohwater.

We are indebted to Anja Aaltonen and Pentti Dyyster for the preparation of the manuscript, and to Timo Erkama and Tuomas Sorvali for the careful reading and correction of the proofs. Finally, we should like to express our thanks to Springer-Verlag for their friendly cooperation in the production of this volume.

Helsinki, April 1973

Olli Lehto · K. I. Virtanen

Contents

List of special symbols VIII
Introduction 1

I. Geometric Definition of a Quasiconformal Mapping

Introduction to Chapter I 4

§ 1. Topological Properties of Plane Sets 5
§ 2. Conformal Mappings of Plane Domains 13
§ 3. Definition of a Quasiconformal Mapping. 16
§ 4. Conformal Module and Extremal Length 19
§ 5. Two Basic Properties of Quasiconformal Mappings 28
§ 6. Module of a Ring Domain 30
§ 7. Characterization of Quasiconformality with the Help of Ring Domains 38
§ 8. Extension Theorems for Quasiconformal Mappings 41
§ 9. Local Characterization of Quasiconformality 47

II. Distortion Theorems for Quasiconformal Mappings

Introduction to Chapter II 52

§ 1. Ring Domains with Extremal Module. 53
§ 2. Module of Grötzsch's Extremal Domain 59
§ 3. Distortion under a Bounded Quasiconformal Mapping of a Disc. . . 63
§ 4. Order of Continuity of Quasiconformal Mappings 68
§ 5. Convergence Theorems for Quasiconformal Mappings 71
§ 6. Boundary Values of a Quasiconformal Mapping 79
§ 7. Quasisymmetric Functions. 88
§ 8. Quasiconformal Continuation 96
§ 9. Circular Dilatation 105

III. Auxiliary Results from Real Analysis

Introduction to Chapter III 109

§ 1. Measure and Integral 110
§ 2. Absolute Continuity 117
§ 3. Differentiability of Mappings of Plane Domains 127
§ 4. Module of a Family of Arcs or Curves. 132
§ 5. Approximation of Measurable Functions 136
§ 6. Functions with L^p-derivatives 143
§ 7. Hilbert Transformation 154

IV. Analytic Characterization of a Quasiconformal Mapping

Introduction to Chapter IV 161

§ 1. Analytic Properties of a Quasiconformal Mapping 162
§ 2. Analytic Definition of Quasiconformality 166
§ 3. Variants of the Geometric Definition 170
§ 4. Characterization of Quasiconformality with the Help of the Circular Dilatation 177
§ 5. Complex Dilatation 182

V. Quasiconformal Mappings with Prescribed Complex Dilatation

Introduction to Chapter V 190

§ 1. Existence Theorem 191
§ 2. Local Dilatation Measures 195
§ 3. Removable Point Sets 199
§ 4. Approximation of a Quasiconformal Mapping 207
§ 5. Application of the Hilbert Transformation to Quasiconformal Mappings 211
§ 6. Conformality at a Point 219
§ 7. Regularity of a Mapping with Prescribed Complex Dilatation . . . 233

VI. Quasiconformal Functions

Introduction to Chapter VI 239

§ 1. Geometric Characterization of a Quasiconformal Function 240
§ 2. Analytic Characterization of a Quasiconformal Function 246

Bibliography 249

Index 253

List of special symbols

C^n, C^∞, C_0^∞ 139
(C, α) oriented arc 8
$D(z)$ dilatation quotient 17, 49
$E(p_1, p_2, \ldots)$ Cantor set 125
$F(z)$ maximal dilatation at a point 47
$f \circ g$ composed mapping
$f * \vartheta$ 139
$\|f\|_p$ 135
Fr A boundary of A 8
$H(z)$ circular dilatation 105
$J(z)$ Jacobian 9
$K(G)$ maximal dilatation 16
$k(z_1, z_2)$ spherical distance 5
l linear measure 22, 110, 116
l_ϱ 21
$L^p(A)$ 135
m area measure 22, 110
m_ϱ 21, 132
$M(\mathscr{C})$ module of a family of arcs or curves 133
$M(Q)$ module of a quadrilateral 15
$N\{E_n\}$ kernel of a sequence of sets 76
$O(z - z_0)$, $o(z - z_0)$ 49
$p(K)$ 215
$Q(z_1, z_2, z_3, z_4)$ quadrilateral 14
s_a distance between a-sides 22
$S\omega$ Hilbert transform 157
$T_D\omega$, $T\omega$ 154, 155
w_z, $w_{\bar{z}}$ complex derivatives 49
$\partial_\alpha w$ derivative in the direction α 17
∂G positively oriented boundary 8
$\lambda(K)$ 81
μ, μ^* measure, outer measure 110, 111
μ_α α-dimensional measure 115
$\mu(r)$ module of Grötzsch's extremal domain 53
μ_f 118
μ_f' 120
φ_K 63

Introduction

The theory of quasiconformal mappings in the plane is closely connected with the theory of analytic functions of one complex variable. All the standard definitions of quasiconformality are based on direct generalizations of certain characteristic properties of conformal mappings, and several fundamental theorems on analytic functions remain valid for quasiconformal mappings at least in a modified form.

However, there are aspects of the modern theory of quasiconformal mappings which do not have analogues in complex analysis. One difference is striking: while conformality implies very strong regularity properties, a quasiconformal mapping need not even be differentiable. A considerable part of the theory of quasiconformal mappings consists, therefore, of problems which must be treated by the methods of measure and integration theory.

In order to arrive at the concept of a quasiconformal mapping we must define a dilatation measure for a homeomorphism. This can be done in many ways. If the mapping w is regular, i.e. continuously differentiable together with its inverse, the dilatation quotient of w at a point z_0 is equal to the ratio of the upper and lower limits of the expression $|w(z) - w(z_0)|/|z - z_0|$ as $z \to z_0$.

A regular homeomorphism is said to be quasiconformal if its dilatation quotient is bounded. This classical definition of quasiconformality cannot be immediately extended to non-regular homeomorphisms. For performing the generalization there are two standard methods, called the analytic and the geometric definitions of quasiconformality. In the former, one assumes a kind of absolute continuity, which implies differentiability at almost all points, and then requires the boundedness of the dilatation quotient almost everywhere. The geometric definition, which forms the starting point in this book, rests on a concept of dilatation defined by means of the conformal module of quadrilaterals.

The above three definitions date from different times. Grötzsch [2] introduced the regular quasiconformal mappings in 1928 and gave impetus to a research directed first to problems of function-theoretic character. On one hand, many analogues between conformal and quasiconformal mappings were discovered. On the other hand, it was soon noticed, primarily thanks to the investigations of Teichmüller, that quasiconformal mappings provide an important tool in studies concerning analytic functions and Riemann surfaces.

The analytic definition of quasiconformality is contained in a paper of Morrey [1] in 1938. He studied homeomorphic solutions of a Beltrami system which is a generalization of the Cauchy-Riemann equations. These solutions agree with mappings quasiconformal in the present terminology, but the connection was not noticed until almost twenty years later.

In the meantime, the geometric definition was introduced. This was done at the beginning of the fifties by Pfluger [1] and Ahlfors [1]. Since the definition is of global character and assumes no a priori differentiability, the local behaviour of the mappings became the target of intensive investigation. The studies of Mori, Yûjôbô, Bers, and others soon gave the result that the geometric definition implies differentiability almost everywhere and is equivalent to the analytic definition. This meant a very satisfactory unification of the general theory.

The contents of this book center around the above definitions. Keeping to this main theme as systematically as possible, we have not separately treated the class of quasiconformal mappings introduced by Lavrentieff [1] in 1935.[1] Neither have we dealt more closely with the many applications to complex analysis. To these we include Teichmüller's problem of moduli[2] whose presentation would require many concepts and tools outside the general theory of quasiconformal mappings.

As mentioned above, we have chosen the geometric definition for the starting point in our representation. This decision, which largely determines the composition of the book, is due to the fact that in our opinion the geometric definition provides the most direct approach to a large part of the theory.

The results emanating directly from the geometric definition are contained mainly in Chapters I and II. Here we have resorted to some results of function theory which are not necessarily needed but simplify the presentation; for instance, in proving some theorems we assume the corresponding result for conformal mappings to be known.

---

[1] An exposition of the theory of these mappings until around 1950 is given in the monograph of Volkovyskij [1].

[2] Introductions to this theory have been given by Bers [3] and Ahlfors [4].

In Chapter III, we interrupt the treatment of quasiconformal mappings and present the theorems of real analysis needed in the rest of the book. They are used in Chapter IV for formulating the analytic definition and proving its equivalence to the geometric one. Parallel application of these definitions provides new methods for developing the theory in Chapters IV and V.

In the first five chapters of the book the quasiconformal mappings are homeomorphisms between domains in the plane. In Chapter VI, the notion of quasiconformality is generalized to the case in which the mapping is only locally homeomorphic up to isolated points.

Every chapter begins with an introduction describing its contents. The chapters are divided into sections which consist of numbered subsections. The references, such as III.2.1, are made with respect to this threefold division. In references within a chapter, however, the first number is omitted.

In the text we have primarily referred to papers which we utilize directly. A more extensive bibliography can be found in the monograph of Künzi [1].

The reader is assumed to be acquainted with the fundamentals of general topology, complex analysis, and measure and integration theory. Beyond these, the auxiliary theorems are proved unless we have been able to give precise references to easily accessible literature. However, we have permitted a few exceptions from this rule in constructing examples which are not essential for developing the general theory.

I. Geometric Definition of a Quasiconformal Mapping

Introduction to Chapter I

The quasiconformal mappings studied in this monograph are either homeomorphisms between plane domains (Chapters I—V) or, except for isolated points, locally homeomorphic mappings of plane domains into the plane (Chapter VI). It is therefore natural to begin the presentation of the theory with some general remarks on the topological properties of plane sets.

This will be carried out in § 1. It will be assumed that the reader is acquainted with the elementary concepts of general topology, while we have tried to avoid applying advanced topological concepts. Such well-known but non-trivial properties of the plane as contained in the theorem on the invariance of domains, the Jordan curve theorem, and the orientation theorem, are mentioned without proof. Detailed proofs can be found for example in Newman's book [1] on the topology of plane sets.

Some basic properties of one-to-one conformal mappings are collected in § 2. By applying the Riemann mapping theorem and the theorem on boundary correspondence, we introduce the concept of the conformal module of a quadrilateral.

The definition of a quasiconformal homeomorphism is given in § 3 with the help of the conformal module of quadrilaterals. Following the historical development we have already here made some remarks on so-called regular quasiconformal mappings, that is, those which together with their inverses are continuously differentiable.

From the definition of quasiconformality it is evident that we need methods of estimating the module of a quadrilateral characterized by geometric properties. To deal with different problems of this sort in a uniform way, we introduce the concept of extremal length. This will not only serve as a technical expedient, but will also make possible the definition of the module of a quadrilateral without the use of conformal mapping.

The method of extremal length is discussed in § 4, where some properties of the conformal module, fundamental for the further presentation, are derived. These will be immediately applied in § 5, whose contents can be regarded as a motivation for the definition of quasiconformality given in § 3.

In § 6 we introduce the conformal module of a ring domain and, in close analogy with the discussion of § 4, prove some of its important properties. In § 7 we then show that quasiconformality can also be defined by means of the modules of ring domains.

The last two sections are closely connected in method with the preceding ones. In § 8, using the module estimations already obtained, we prove some extension theorems for quasiconformal mappings. Finally, in § 9 we show that quasiconformal mappings can also be locally characterized with the help of small quadrilaterals.

§ 1. *Topological Properties of Plane Sets*

1.1. The plane. By *plane* we shall, throughout this book, understand the extended complex plane (the Riemann sphere). The usual euclidean plane will be called the *finite plane.*

Besides the euclidean metric we shall use the spherical metric in the plane. Two finite points z_1 and z_2 have the spherical distance

$$k(z_1, z_2) = \arc\tan \left| \frac{z_1 - z_2}{1 + \bar{z}_1 z_2} \right| ,$$

where $0 \leqq k(z_1, z_2) \leqq \pi/2$. For $z_2 = \infty$ we have $k(z_1, \infty) = \arc\tan |1/z_1|$.

The spherical metric determines a topology in every plane set A. A set which is open in this topology is called *open in A*. The concepts of "open" (i.e. open in the plane) and "open in A" coincide for subsets of A only if A itself is open. The same holds if the word "open" is replaced everywhere by "closed". On the other hand, the property of a subset of A of being compact or connected is independent of whether the topology of the plane or that of A is considered.

If one considers a mapping $f : A \to A'$ of a plane set A into a plane set A', then in general A and A' are conceived as being subspaces of the plane provided with the above topology. However, the continuity of f does not depend on whether f is taken to be a mapping into the plane or into the space A' (cf. also the theorem on the invariance of open sets in 1.2).

1.2. Homeomorphisms. A one-to-one mapping f of a set A onto a set A' is called a homeomorphism if f and its inverse mapping $f^{-1} : A' \to A$ are both continuous. In certain special cases the continuity of f^{-1} is a consequence of one-to-oneness and of the continuity of f. This holds in all Hausdorff spaces if A is compact, since the image of every closed subset of A is then closed. For plane sets A the continuity of f^{-1} also follows if A is open; because of the many later applications we formulate this result as a lemma.

Lemma 1.1. *Every one-to-one and continuous mapping of an open set of the plane onto a plane set is a homeomorphism.*

This result can be deduced immediately from the following property of the plane (Newman [1], p. 122):

Theorem on the invariance of open sets. *If f is a one-to-one continuous mapping of an open set G of the plane onto the plane set G', then G' is also an open set.*

Since the continuous image of a connected set is connected, the above theorem holds if "open set" is replaced by "domain" (*theorem on the invariance of domains*).

1.3. Separation theorems. The word "line" appears in this book in three different meanings. A line, unqualified, is compact and contains one point at infinity, a finite line consists only of finite points, and an extended line is formed from a finite line by adding the points $+\infty$ and $-\infty$. In contrast to the other sets considered here the extended line is not regarded as a subset of the plane. Where no misunderstanding is possible the word "line" will be used in all three meanings.

A line segment (or an interval) is a connected subset of a finite line. Unless otherwise stated, a line segment is always assumed to be bounded. As usual a line segment is called closed or open depending on whether it contains its endpoints or not.

A *Jordan curve* is a set which is homeomorphic to a circle and a *Jordan arc* is the topological image of a line segment. A Jordan arc C is called open or closed depending on whether it is homeomorphic to an open or a closed segment. By a parametric representation of C we understand a homeomorphism $z : I \to C$, where I is a bounded or unbounded segment. A closed arc always has two endpoints, while an open arc C has endpoints only when its closure $\bar{C}$ is a closed arc.

If a Jordan arc or Jordan curve C consists of a finite number of line segments, it is called a polygonal arc or closed polygon, respectively.

A set A of the plane Ω separates the plane sets A_1 and A_2 if A_1 and A_2 lie in different components of the complement $-A = \Omega - A$ of A. If $A_1, A_2 \subset E \subset \Omega$, then one defines separation in E by A in a corresponding way.

A Jordan arc has a connected complement, while Jordan curves possess the following fundamental separation property (Newman [1], p. 115).

Jordan curve theorem. *The complement of a plane Jordan curve C consists of two disjoint domains, which both have C as boundary.*

The part of the Jordan curve theorem which tells us that the complement of a Jordan curve is not connected is contained in the following more general result (Newman [1], p. 117) as a special case.

Lemma 1.2. *If F_1 and F_2 are two continua whose intersection is not connected, then the complement of $F_1 \cup F_2$ is not connected.*[3]

This result can be generalized in the following way:

Lemma 1.2′. *Let F_i, $i = 1, \ldots, n$, be continua of which any three have no common points. If $F_1 \cap F_2$ is not connected, then the complement of the union of the sets F_i is not connected.*

Proof: For $n = 2$ the result follows from Lemma 1.2. We assume that the assertion is true for $n = m$ and show that it then holds for $n = m + 1$ also. Let $F_1, F_2, \ldots, F_{m+1}$ be sets which satisfy the hypotheses of the lemma. Then the complement of the set

$$F = \bigcup_{i=1}^{m} F_i$$

is not connected.

If $F \cap F_{m+1} = \emptyset$, the set $-(F \cup F_{m+1})$ contains points of every component of $-F$. Then $-(F \cup F_{m+1})$ cannot be connected.

If, on the other hand, $F \cap F_{m+1} \neq \emptyset$, there is a k, $1 \leqq k \leqq m$, such that $F_k' = F_k \cup F_{m+1}$ is a continuum. Thus it is enough to show that $F_1 \cap F_2$ is not connected if F_k is replaced by F_k'. For $k \neq 1,2$ this is clear. If say $k = 1$, then $(F_1 \cap F_2) \cap (F_{m+1} \cap F_2) = \emptyset$, and therefore $F_1' \cap F_2 = (F_1 \cap F_2) \cup (F_{m+1} \cap F_2)$ is not connected.

There is also a result which is in a certain sense converse to Lemma 1.2 (Newman [1], p. 112).

Lemma 1.3. *Let F_1 and F_2 be closed sets with a connected intersection. Two points which are separated neither by F_1 nor by F_2 are not separated by $F_1 \cup F_2$.*

---

[3] It should be noted that the empty set $\emptyset$ is connected.

The following way of carrying out a separation will also find application later (Newman [1], p. 142).

Lemma 1.4. *Let G be a domain with a connected complement and F a closed subset of G. Then F and the complement of G can be separated by a closed polygon.*

1.4. Orientation. For a given Jordan arc C let us consider all homeomorphisms f of segments of the x-axis onto C. Two such homeomorphisms f_1 and f_2 are called equivalent if $f_2^{-1}(f_1(x))$ increases with increasing x. By means of this equivalence relation the homeomorphisms considered are divided into two classes, α and β, called the *orientations* of C. The pairs $C^+ = (C, \alpha)$ and $C^- = (C, \beta)$ are called *oriented arcs.*

The orientation of a Jordan curve C can be defined analogously. We consider all homeomorphisms of a circle $K = \{r\, e^{i\varphi} \mid 0 \leqq \varphi \leqq 2\pi\}$ onto C and say that the homeomorphisms f_1 and f_2 belong to the same orientation of C if $f_2^{-1}(f_1(\varphi))$ increases with increasing φ. The pair formed by a Jordan curve and one of its two orientations is called an *oriented curve.*

The orientation of a Jordan curv eC can also be defined by means of a sequence of three (or more) points on C: by the orientation p_1, p_2, p_3 we understand the class of those homeomorphisms f for which $\varphi_1 = \arg f^{-1}(p_1)$, $\varphi_2 = \arg f^{-1}(p_2)$, $\varphi_3 = \arg f^{-1}(p_3)$ is an increasing sequence for the branch of the argument between φ_1 and $\varphi_1 + 2\pi$.

Let C be a Jordan curve and G_1 and G_2 the disjoint domains bounded by C. We want to define when the orientation of C is *positive or negative with respect to* G_1. For this purpose we choose a linear conformal mapping t such that $t(G_1)$ is a bounded domain containing the origin. Let $f: K \to C$ be a representative of the orientation α. As the argument of a point z of K increases from 0 to 2π, each continuous branch of $\arg t(f(z))$ changes by either 2π or -2π. In the first case the orientation α is called positive, in the second negative, with respect to G_1. It is easy to see that this definition does not depend on the choice of the mapping t and that the orientation which is positive with respect to G_1 is negative with respect to G_2.

For a *Jordan domain* G, i.e. a domain whose boundary is a Jordan curve, we denote by ∂G the positively oriented (with respect to G) boundary curve. On the other hand, for the boundary as a point set we shall use the notation $\operatorname{Fr} G$.

1.5. Sense-preserving homeomorphisms. Let $w: \overline{D} \to A$ be a homeomorphism of the closure $\overline{D}$ of a Jordan domain D onto the plane set A. It follows from the invariance of domains that A is the closure of a Jordan domain D' and that the restriction of w to the boundary of D maps this curve topologically onto Fr D'.[4]

If $f : K \to \text{Fr } D$ is a homeomorphism, then so is the composite mapping $w \circ f : K \to \text{Fr } D'$. If f_1 and f_2 belong to the same orientation of Fr D, then the same is true of $w \circ f_1$ and $w \circ f_2$ relative to Fr D'. The homeomorphism w thus induces a mapping of the orientations of Fr D onto those of Fr D'. If the positive orientations with respect to D and D' are transformed onto one another by this mapping, we say that w preserves the orientation of the boundary of D. If the positive orientation with respect to D is given by three points of Fr D in the form p_1, p_2, p_3, it follows from the definition that w preserves the orientation of the boundary of D if and only if $w(p_1), w(p_2), w(p_3)$ is positive with respect to D'.

More generally, we now consider a homeomorphism $w : A \to A'$ where A and A' are arbitrary point sets of the plane. The mapping w is called *sense-preserving* if w preserves the orientation of the boundary of every Jordan domain D such that $\overline{D} \subset A$. However, in all cases which concern us it is sufficient to consider a single Jordan domain D. In fact we have the following result (cf. Newman [1], p. 197):

Orientation theorem. *Let G be either a plane domain or the closure of a Jordan domain and w a homeomorphism of G onto a plane set G'. If there is a Jordan domain D, $\overline{D} \subset G$, such that w preserves the orientation of the boundary of D, then w is sense-preserving.*

It should be noted that the inverse of a sense-preserving homeomorphism is sense-preserving, as is also a composite mapping of sense-preserving homeomorphisms.

1.6. Regular points of a mapping. As above, let G be either a plane domain or the closure of a Jordan domain and $w : G \to G'$ a homeomorphism. We suppose that w is differentiable at an interior point z of G, i.e. that the real and imaginary parts of w are differentiable at z. Here w is called differentiable at infinity if the mapping $\tilde{w}$, $\tilde{w}(z) = w(1/z)$, is differentiable at the origin, and differentiability at a point where $w = \infty$ means that $1/w$ is differentiable there. From I.9 onwards the word differentiable refers only to finite points and finite functions.

The Jacobian of w at the point z will be denoted by $J_w(z)$ or $J(z)$. If $z = \infty$ or $w(z) = \infty$, then we only define whether $J_w(z)$ is positive,

[4] A mapping and its restriction will be denoted in the same way, when no misunderstanding is possible.

negative or zero in such a way that $J_w(z)$ is allotted the sign of the Jacobian of $\tilde{w}$ at the origin or that of $J_{1/w}(z)$, respectively.

We say that z is a *regular* point of $w : G \to G'$ (or that w is regular at the point z) if z lies in the interior of G, w is differentiable at z, and $J_w(z)$ does not vanish.

Let z be a regular point where $J_w(z) > 0$. A simple calculation shows that z possesses a disc neighbourhood D, $\bar{D} \subset G$, such that w preserves the orientation of the boundary of D. It follows from the orientation theorem that w is sense-preserving.

If, on the other hand, $J_w(z) < 0$, then one can find a disc D such that w does not preserve the orientation of the boundary of D. In this case w cannot be sense-preserving.

To sum up, we have:

If the homeomorphism $w : G \to G'$ possesses a regular point z where $J_w(z) > 0$, then w is sense-preserving. Conversely, the Jacobian of a sense-preserving homeomorphism is positive at every regular point.

1.7. Connectivity of a domain. A plane domain G is called *simply connected* if its complement is connected. If the complement of G consists of n components $(1 < n < \infty)$, then G is called *n-tuply connected.*

Connectivity is a topological invariant; on the other hand, two n-tuply connected domains can always be mapped topologically onto each other if $n > 1$ (Newman [1], p. 157). The same is true for two simply connected domains if neither is the whole plane, which is naturally homeomorphic only with itself.

Let F be a component of the complement of a domain G. Being a component of a closed set, F itself is closed. Its complement is a simply connected domain (Newman [1], p. 78).

The boundary of an n-tuply connected domain G (which is not the whole plane) consists of n components. First of all it is clear that every component F of the complement of G contains a component of the boundary of G. The number of boundary components is thus at least n. To prove that it is exactly n, we must show that $\bar{G} \cap F = \overline{(-F)} \cap F$ is connected. If the contrary were true we could apply Lemma 1.2 to the sets F and $\overline{(-F)}$, since $\overline{(-F)}$ is connected according to the above. We would then come to the contradiction that the complement of $F \cup \overline{(-F)}$, i.e. the empty set, is disconnected.

Let G be a simply connected domain. A Jordan arc $C \subset G$ with endpoints on $\operatorname{Fr} G$ separates G into two simply connected domains both having each point of C as a boundary point (Newman [1], pp. 118 and 145). If, besides, G is a Jordan domain, then $\overline{C}$ separates in $\overline{G}$ the open subarcs of $\operatorname{Fr} G$ which are bounded by the endpoints of C (Newman [1], p. 119).

1.8. Boundary behaviour of topological mappings. Let $w : G \to G'$ be a homeomorphism and $A \subset G$ a set, all of whose limit points belong to the boundary of G. Then the limit points of the image set $w(A)$ lie on the boundary of G'. Let us consider all sets $A \subset G$ whose limit points belong to a given set E of boundary points of G. The set E' of limit points of all sets $w(A)$ is called the set of limit points of E under the mapping w. We say that E and E' *correspond to one another* under the mapping w if each of the sets E and E' is the set of limit points of the other.

Under a homeomorphism $w : G \to G'$ between two n-tuply connected domains the n boundary components of G and G' correspond pairwise. To prove this we consider a boundary component E' of G'. According to 1.7, the complement of that component of $-G'$ which contains E' is a simply connected domain. By Lemma 1.4, there exists a Jordan curve C' in G' which separates E' from the other boundary components of G'. The preimage $C = w^{-1}(C')$ separates the plane into two domains D_1 and D_2, both of which contain only whole boundary components of G. From Lemma 1.3 we conclude, by setting $F_1 = C$, $F_2 = -G$, that the sets $G \cap D_i$, $i = 1, 2$, are connected, and therefore their images are also connected. Thus, if there exists a sequence of points converging to a boundary component $E \subset D_1$ of G such that the image sequence tends to E', then $w(G \cap D_1)$ lies in the same domain bounded by C' as E'. It follows that no sequence converging to E has an image with limit points on any of the boundary components of G' other than E'.

1.9. Free boundary arcs. An open Jordan arc (Jordan curve) C on the boundary of the domain G is called a *free boundary arc* (*free boundary curve*) of G if the following two conditions are fulfilled:

1° The set $C \cap \overline{(\operatorname{Fr} G - C)}$ is empty.
2° If E denotes that boundary component of G which contains C, then $E - C$ is connected.

A closed boundary arc is called free if it is a subarc of an open free boundary arc.

Condition 2°, often omitted in the literature, excludes the so-called multiple boundary points. In the case of a Jordan curve C condition 2°

is not needed, since it follows from 1°. It also appears from the definition that a subarc of a free boundary arc or free boundary curve is also a free boundary arc of G.

Every point of a free boundary arc C of G is a limit point both of G and of $-\bar{G}$. We see this in the following way: In the first place one may obviously suppose that C is open and possesses two endpoints. Let E be the boundary component of G which contains C. Since the set $E-C$ is connected, all components of its complement are simply connected domains (Newman [1], p. 144). Now G is contained in one of these; this we denote by D. Since C lies in D and joins two boundary points of D, then, in view of 1.7, $D-C$ consists of two domains which both have C as a boundary arc. Our assertion follows from this.

The following result gives information about the possibility of joining points of domains in the neighbourhood of a free boundary arc.

Lemma 1.5. *Let a be a point of a free boundary arc of the domain G and U a given neighbourhood of a. Then we have:*

1° *There exists a disc neighbourhood V of a, $\bar{V} \subset U$, such that every pair of points of $G \cap V$ can be joined by a Jordan arc in $G \cap U$.*

2° *If V has the property described in 1°, and U_1 is a disc neighbourhood of a contained in V, then there exists a neighbourhood V_1 of a, $V_1 \subset U_1$, such that every pair of points in $(V - \bar{U}_1) \cap G$ can be joined by a Jordan arc lying in $(U - \bar{V}_1) \cap G$.*

Proof: Since U can be made smaller, we may suppose without loss of generality that U is a disc and that the intersection of $\bar{U}$ with $\operatorname{Fr} G$ consists of points of the free boundary arc only. Let C be an open boundary arc of G lying in U and containing the point a. Let $V \subset U$ then be chosen as a disc neighbourhood of a such that $\bar{V} \cap (\operatorname{Fr} G - C) = \emptyset$.

Let E denote the component of $\operatorname{Fr} G$ which contains the point a. To prove assertion 1°, we apply Lemma 1.3 to the closed sets $F_1 = E$, $F_2 = (E - C) \cup \operatorname{Fr} U$. According to the definition of a free boundary arc the set $E - C = F_1 \cap F_2$ is connected. Since neither E nor $(E - C) \cup \operatorname{Fr} U$ separates a pair of points lying in $G \cap V$, such points must by Lemma 1.3 belong to the same component of the complement of the set $F_1 \cup F_2 = E \cup \operatorname{Fr} U$. The first assertion of the lemma follows from this, since all boundary components of G other than E lie outside U.

To prove assertion 2° we consider a closed subarc C_1 of the given free boundary arc such that C_1 lies in U_1 and contains the point a. Let

$V_1 \subset U_1$ be such a disc neighbourhood of a that $\overline{V}_1 \cap (\mathrm{Fr}\, G - C_1) = \emptyset$. If one sets $F_1 = E \cup \mathrm{Fr}\, U$, $F_2 = \mathrm{Fr}\, V_1 \cup C_1$, then 2° follows as did 1° above by application of Lemma 1.3.

By the aid of Lemma 1.5, it is possible to prove the validity of the following result concerning the approach to a point on a free boundary arc; for the details we refer to Behnke-Sommer [1], p. 359.

Lemma 1.6. *Let a be a point of a free boundary arc of the domain G and z_n a sequence of points in G which converges to a. Then there exists a Jordan arc C passing through the points z_n with $C - \{a\} \subset G$ and with an endpoint at a.*

§ 2. *Conformal Mappings of Plane Domains*

2.1. The Riemann mapping theorem. In § 3 we shall define quasiconformal mappings as direct generalizations of conformal mappings. It is advisable therefore to collect together some fundamental results from the theory of conformal mappings. We remark that, as before, domains of the extended plane are to be considered. The conformality of w at the point ∞ and at a point z_0 where $w(z_0) = \infty$ means that $\tilde{w}$, $\tilde{w}(z) = w(1/z)$, is conformal at the origin and $1/w$ at the point z_0, respectively.

As is known, there are no non-constant bounded functions analytic in the whole finite plane. Thus the finite plane is not conformally equivalent to any bounded domain. Simply connected domains fall, therefore, into at least three conformal equivalence classes which have as representatives the following canonical domains: 1° the whole plane, 2° the finite plane, and 3° the unit disc.

On the other hand, every simply connected domain belongs to one of the above classes, in view of the following classical theorem:

The Riemann mapping theorem. *Every simply connected domain of the plane can be mapped conformally onto one of the canonical domains 1°–3°. A conformal mapping onto the finite plane exists if and only if the complement of the domain consists of a single point.*

A Jordan domain is simply connected and has infinitely many boundary points; it can therefore be mapped conformally onto the unit disc. Thus all Jordan domains are conformally equivalent.

2.2. Boundary behaviour of conformal mappings. The question of the boundary behaviour of conformal mappings between domains bounded by Jordan curves is solved by the following theorem (see e.g. Behnke-Sommer [1], p. 371).

Theorem on correspondence of boundaries. *Let G and G' be n-tuply connected domains which both have n Jordan curves as boundary components. Then every conformal mapping $w: G \to G'$ can be extended to a homeomorphism of $\overline{G}$ onto $\overline{G}'$.*

In fact we have the following still more general result: Let G and G' be domains, C and C' their free boundary arcs or free boundary curves, and $w: G \to G'$ a conformal mapping under which C and C' correspond. Then w can be extended to a homeomorphism of $G \cup C$ onto $G' \cup C'$. For our first applications, however, the above simpler theorem on boundary correspondence is sufficient, and the more general result will be proved in 8.2 even for quasiconformal mappings.

The conformal mapping w of G extended to $\overline{G}$ will often also be called conformal and denoted by w. However, if we speak of a conformal mapping without mention of the set to be mapped we always mean a mapping of a domain.

A conformal mapping $w: G \to G'$ is sense-preserving because $J_w(z) = |w'(z)|^2$, and in the case of two Jordan domains G and G' so is the extended mapping $w: \overline{G} \to \overline{G}'$ (cf. 1.6). If the points p_1, p_2, p_3 lie on the boundary curve of G and if p_1, p_2, p_3 is the positive orientation of the boundary with respect to G, then $w(p_1), w(p_2), w(p_3)$ is the positive orientation of the boundary of G' with respect to G'. Conversely, with the above notation, we have:

If G and G' are Jordan domains and the orientations p_1, p_2, p_3 and q_1, q_2, q_3 are both positive relative to G and G' respectively, then there is a uniquely determined conformal mapping $w: G \to G'$ which carries the boundary points p_i into the boundary points $w(p_i) = q_i$.

2.3. Quadrilaterals and their mappings. A *quadrilateral* consists of a Jordan domain Q and a sequence z_1, z_2, z_3, z_4 of boundary points of Q. The points z_i are called the *vertices* of the quadrilateral. In the following we shall consider only quadrilaterals $Q(z_1, z_2, z_3, z_4)$ whose sequence of vertices agrees with the positive orientation with respect to Q.

For the sake of simplicity it is often adivisable to interpret a quadrilateral as a point set; for example, we speak of interior points and of the closure of a quadrilateral. If no misunderstanding is possible, we use the same notation for a quadrilateral and its corresponding domain.

The vertices of a quadrilateral $Q(z_1, z_2, z_3, z_4)$ divide its boundary into four Jordan arcs, the *sides* of the quadrilateral. The arcs $\widehat{z_1 z_2}$ and $\widehat{z_3 z_4}$ are called the *a-sides* and the other two arcs, the *b-sides* of Q.

By a *homeomorphism of the quadrilateral* $Q(z_1, z_2, z_3, z_4)$ onto the quadrilateral $Q'(w_1, w_2, w_3, w_4)$ we understand a topological mapping $w : \overline{Q} \to \overline{Q}'$ which carries the points z_i into the points $w_i = w(z_i)$. If the restriction of w to Q is conformal, then w is called a *conformal mapping* of $Q(z_1, z_2, z_3, z_4)$ onto $Q'(w_1, w_2, w_3, w_4)$. It is not in general possible to map given quadrilaterals onto one another conformally, since the images of three boundary points determine the mapping uniquely. All quadrilaterals are therefore divided into several conformal equivalence classes.

2.4. Conformal module of a quadrilateral. It follows from the Riemann mapping theorem that every quadrilateral $Q(z_1, z_2, z_3, z_4)$ can be mapped onto a quadrilateral $Q'\,(-1/k,\ -1,\ 1,\ 1/k)$ where $0 < k < 1$ and Q' is the upper half plane. From the classical theory of the elliptic integral we obtain further that the function w,

$$w(z) = \int_0^z \frac{d\zeta}{\sqrt{(1-\zeta^2)\,(1-k^2\zeta^2)}}\,,$$

maps the quadrilateral $Q'\,(-1/k,\ -1,\ 1,\ 1/k)$ conformally onto a quadrilateral which consists of a rectangle and its corners. Where no misunderstanding is possible, we call such a quadrilateral simply a rectangle.

By combining the above mappings, we can map an arbitrary quadrilateral conformally onto a rectangle. Such a mapping will be called the *canonical* mapping of the quadrilateral and the corresponding rectangle the canonical rectangle of the quadrilateral.

Every conformal equivalence class of quadrilaterals thus contains rectangles and all similar rectangles obviously belong to the same class. Conversely it follows from the reflection principle that every conformal mapping between two rectangles is a similarity transformation.

All canonical rectangles of a given quadrilateral Q have therefore the same ratio of sides $a/b = M(Q)$, where a denotes the length of the a-sides and b the length of the b-sides. The number $M(Q)$ is called the *(conformal) module* of Q. From the above it follows that two quadrilaterals are conformally equivalent if and only if they possess the same module.

If one carries out a rearrangement of the vertices of Q, then the module behaves as follows:

$$M\big(Q(z_1, z_2, z_3, z_4)\big) = M\big(Q(z_3, z_4, z_1, z_2)\big) = 1/M\big(Q(z_2, z_3, z_4, z_1)\big)\,.$$

§ 3. *Definition of a Quasiconformal Mapping*

3.1. Dilatation of a quadrilateral under a homeomorphism. Let G be a domain and w a sense-preserving homeomorphism of G. A quadrilateral $Q = Q(z_1, z_2, z_3, z_4)$ whose closure is contained in G is mapped by w onto a quadrilateral Q'. The ratio $M(Q')/M(Q)$ of the modules of Q' and Q is called the *dilatation* of Q under the mapping w.

The number

$$K(G) = \sup_{\bar{Q} \subset G} \frac{M(Q')}{M(Q)}$$

is called the *maximal dilatation* of w in the domain G. Since the dilatations of $Q(z_1, z_2, z_3, z_4)$ and $Q(z_2, z_3, z_4, z_1)$ are reciprocal numbers, $K(G)$ is at least 1.

If w is conformal, then Q and Q' have the same module. The dilatation of Q is then always 1 and consequently $K(G) = 1$. On the other hand, we shall show in 5.1 that the maximal dilatation of a non-conformal, sense-preserving homeomorphism is always greater than 1. Thus $K(G)$ can be regarded as a measure of how much the mapping deviates from conformality in G.

3.2. Quasiconformal mapping. Quasiconformal mappings can now be defined as follows:

Definition. *A sense-preserving homeomorphism w of the domain G is called quasiconformal if its maximal dilatation $K(G)$ is finite. If $K(G) \leqq K < \infty$, then w will be called K-quasiconformal.*

Following the terminology used in the conformal case, we also call a quasiconformal homeomorphism a quasiconformal mapping.

The requirement that a quasiconformal mapping be sense-preserving entails certain formal simplifications. On the other hand, most of the following theorems also hold for an "anti-quasiconformal" mapping, i.e. a quasiconformal mapping followed by a reflection.

It follows from the definition that a conformal mapping is 1-quasiconformal. As we have already mentioned, we shall prove in 5.1 that conversely every 1-quasiconformal mapping is conformal.

As will be shown later, quasiconformal mappings permit many other characterizations which are non-trivially equivalent to the above definition. These we state as theorems. The above definition, due to Pfluger [1] and Ahlfors [1], is according to the former's suggestion called the *geometric definition*.

The following properties of quasiconformal mappings follow immediately from the definition:

The inverse of a K-quasiconformal mapping is K-quasiconformal. The composition of a K_1-quasiconformal and K_2-quasiconformal mapping is $K_1 K_2$-quasiconformal.

3.3. Regular quasiconformal mappings. We consider as an example a class of mappings introduced by Grötzsch [2] in 1928. It is shown in 3.4 that these are quasiconformal according to the above definition. Here we shall call them regular quasiconformal. The term "quasiconformal" formerly referred to these mappings, but with isolated singularities often permitted.

To define the class of Grötzsch mappings, we consider a sense-preserving homeomorphism w of the finite domain G of the $z = x + i\,y$-plane onto the finite domain G'. We suppose that w is regular throughout G and that the partial derivatives w_x and w_y are continuous in G. Then the directional derivative

$$\partial_\alpha w(z) = \lim_{r \to 0} \frac{w(z + r\,e^{i\alpha}) - w(z)}{r\,e^{i\alpha}} = e^{-i\alpha}\left(w_x(z)\cos\alpha + w_y(z)\sin\alpha\right)$$

exists at every point $z \in G$, and the *dilatation quotient* D,

$$D(z) = \frac{\max_\alpha |\partial_\alpha w(z)|}{\min_\alpha |\partial_\alpha w(z)|}, \tag{3.1}$$

is bounded in every compact subset of G. If D is bounded in the whole domain G, we say that w is a *regular quasiconformal mapping* of G. If

$$\sup_{z \in G} D(z) \leqq K,$$

then w is called a *regular K-quasiconformal mapping*.

If w is conformal, $\partial_\alpha w(z)$ does not depend on the direction α. Then w is a regular quasiconformal mapping with the dilatation quotient $D(z) = 1$ at every point $z \in G$, i.e. w is a regular 1-quasiconformal mapping.

The composition $w_3 = w_1 \circ w_2$ of two regular quasiconformal mappings is regular quasiconformal and $D_3(z) \leqq D_1(w_2(z))\,D_2(z)$, where D_i denotes the dilatation quotient of w_i. Equality always holds when either w_1 or w_2 is conformal; thus in the first case $D_3(z) = D_2(z)$, in the second $D_3(z) = D_1(w_2(z))$.

Since D is conformally invariant, we can define $D(z)$ at $z = \infty$ or at a finite point $z = z_0$ where $w(z_0) = \infty$ by means of an auxiliary linear transformation. The definition of a regular quasiconformal mapping can thus be extended to the general case where G and G' are arbitrary domains of the plane. The restriction that the partial derivatives be continuous must be replaced by the requirement that the composition of w and the auxiliary mapping have continuous derivatives (cf. 1.6).

3.4. Grötzsch's inequality. Let $\alpha = \alpha_1$ be a direction in which the derivative $|\partial_\alpha w(z)|$ of a regular quasiconformal mapping w takes its greatest value at the point z. The image of an infinitesimal square with one corner at z and sides in the directions α_1 and $\alpha_1 + \pi/2$ is then a rectangle whose module is $D(z)$. This indicates a connection between the dilatation quotient and the dilatation of a quadrilateral.

As a result of this type, we shall now prove that *a regular K-quasiconformal mapping is also K-quasiconformal in the sense of Definition* 3.2. This assertion is equivalent to the following inequality due to Grötzsch [2]:

Theorem 3.1. *Between the maximal dilatation and the dilatation quotient of a regular quasiconformal mapping we have the inequality*

$$K(G) \leqq \sup_{z \in G} D(z) . \tag{3.2}$$

Proof: We have to prove that the dilatation of an arbitrary quadrilateral Q, $\overline{Q} \subset G$, is at most equal to $\sup_z D(z) = K$. To this end we map the quadrilateral Q and its image quadrilateral Q' conformally onto their canonical rectangles $R = R(0, M, M + i, i)$ and $R' = R'(0, M', M' + i, i)$. Here M and M' are, by definition, the modules of Q and Q', respectively. Thus we must prove that

$$M' \leqq K M . \tag{3.3}$$

We denote by $w: R \to R'$ the composition of the original regular quasiconformal mapping and the two conformal mappings. Since the dilatation quotient is invariant with respect to conformal mappings, we have the inequality $\max_\alpha |\partial_\alpha w(z)| \leqq K \min_\alpha |\partial_\alpha w(z)|$ at each interior point of R. It follows from this (with $z = x + i y$) that

$$\begin{aligned} J_w(z) &= \max_\alpha |\partial_\alpha w(z)| \min_\alpha |\partial_\alpha w(z)| \\ &\geqq \frac{1}{K} (\max_\alpha |\partial_\alpha w(z)|)^2 \geqq \frac{1}{K} |w_x(z)|^2 . \end{aligned}$$

Thus for the area of R' we have[5]

$$M' = \iint_R J_w(z)\, dx\, dy \geqq \frac{1}{K} \int_0^1 dy \int_0^M |w_x(x + i y)|^2\, dx . \tag{3.4}$$

[5] For a rigorous proof of (3.4) one must either carry out a simple limiting process or apply the theorem of Fubini quoted in III.1.5.

Application of Schwarz's inequality yields

$$\int_0^M |w_x(x+i\,y)|^2\,dx \geqq \frac{1}{M}\left(\int_0^M |w_x(x+i\,y)|\,dx\right)^2.$$

The integral on the right is $\geqq M'$ since it expresses the length of a curve which joins the b-sides of R'. If we set this estimation in (3.4) we obtain $M' \geqq M'^2/(K\,M)$ which is the same as (3.3).

As a completion of Theorem 3.1 we shall show later (Theorem 9.3) that, conversely, every K-quasiconformal mapping satisfies the inequality $D(z) \leqq K(G)$ at any regular point z. It follows that inequality (3.2) is in fact always an equality.

Let it be mentioned at this point that the general quasiconformal mappings defined in 3.2 have, because of their greater generality, the following advantage over the regular mappings: As will be shown in I.5.2, the uniform convergence of a sequence of K-quasiconformal mappings to a homeomorphism implies that the limit mapping is also K-quasiconformal. However, the limit mapping need not be regular K-quasiconformal even when the mappings of the sequence are.

Remark. For the affine mapping w, $w(z) = K\,x + i\,y$, $K \geqq 1$, we have $D(z) = K$ for every z. Thus, according to Theorem 3.1, w is K-quasiconformal. From this we conclude: For every pair of quadrilaterals Q and Q' with $M(Q') = K\,M(Q)$, there exists a homeomorphism of Q onto Q' which is K-quasiconformal in the interior of Q. In fact, we need only map Q and Q' canonically onto the rectangles $R\,(0, M, M+i, i)$ and $R'\,(0, K\,M, K\,M+i, i)$, $M = M(Q)$, and map R onto R' by the above affine mapping w.

§ 4. Conformal Module and Extremal Length

4.1. Preparatory remarks towards a new characterization of the module. The definition of a quasiconformal mapping given in 3.2 is based on the concept of the module, which itself is defined by means of a conformal (i.e. 1-quasiconformal) mapping. It is therefore important to define the module of a quadrilateral without the application of conformal mappings. This we shall now demonstrate. Moreover, the new characterization is of great practical significance in the development of the theory of quasiconformal mappings, since it offers us methods of estimating the module.

To better understand this characterization, we make some introductory remarks. Let Q be a quadrilateral and f its canonical (conformal) mapping onto the rectangle $R = R(0, a, a + ib, ib)$ of the $w = u + iv$-plane. With the possible exception of the point at infinity, we have $J_f(z) = |f'(z)|^2$ everywhere in the interior of Q. For the area of R we obtain therefore the expression

$$\iint_Q |f'(z)|^2 \, d\sigma = a\, b\,, \tag{4.1}$$

where $d\sigma = d\sigma_z = dx\,dy$ denotes the area element of the $z = x + iy$-plane.

In order to obtain the module $M(Q) = a/b$, we write the height b of R in a suitable integral form. For this purpose we consider the vertical segments $C'_u = \{u + iv \mid 0 < v < b\}$ lying in R and their preimages $f^{-1}(C'_u) = C_u$.

According to the theorem in 2.2 on boundary correspondence under conformal mapping, each of the open analytic Jordan arcs[6] C_u has two endpoints each lying on one a-side of Q. The length of the image segment C'_u of C_u is b, and we obtain by integrating

$$\int_{C_u} |f'(z)|\,|dz| = \int_{C'_u} |dw| = b\,. \tag{4.2}$$

We now consider a more general class $\mathscr{C}_a$ of open Jordan arcs C with the following properties:

1. Every $C \in \mathscr{C}_a$ lies in the interior of Q and has one endpoint on each a-side of Q.
2. $\mathscr{C}_a$ contains all the arcs C_u.
3. $\mathscr{C}_a$ consists of arcs sufficiently regular for all subsequent integrations.

Condition 3 is deliberately indefinite: In Chapters I and II we can limit ourselves to such regular families $\mathscr{C}_a$ that the validity of condition 3 can be verified by the use of elementary calculus. In III.4 the discussion is carried out under more general hypotheses; see also the footnote on page 2.

The image $f(C) = C'$ of an arc $C \in \mathscr{C}_a$ is an open Jordan arc which joins the a-sides of R and whose length is thus at least equal to the height b of R. Consequently, by means of the same integration as in (4.2), one obtains for $C \in \mathscr{C}_a$

$$\int_C |f'(z)|\,|dz| \geqq b\,, \tag{4.3}$$

where equality holds for all the arcs C_u.

[6] A curve is called analytic if it is the image of a circle under a conformal mapping of a neighbourhood of this circle. Correspondingly, a closed arc is called analytic if it is the image of a closed segment I under a conformal mapping of a neighbourhood of I. An open arc is called analytic if all its closed subarcs are analytic.

Since $M(Q) = a\,b/b^2$ we obtain from (4.1) and (4.3) the following preliminary result: *The module of a quadrilateral Q admits the representation*

$$M(Q) = \frac{\iint\limits_Q |f'|^2\, d\sigma}{\left(\inf\limits_{C \in \mathscr{C}_a} \int\limits_C |f'|\,|dz|\right)^2}, \tag{4.4}$$

where f' is the derivative of the canonical mapping of Q and the family $\mathscr{C}_a$ possesses the above properties 1—3.

4.2. Characterization of the module without conformal mapping. Besides the derivative $|f'|$ we now consider more general non-negative functions ϱ defined in the interior of Q; let $\mathscr{P}$ denote a family of such functions. We suppose that $\mathscr{P}$ contains the derivative $|f'|$ and that the functions $\varrho \in \mathscr{P}$ are not too discontinuous for the integrations to be carried out below[7].

With every function $\varrho \in \mathscr{P}$ we associate the function ϱ_1, $\varrho_1(f(z)) = \varrho(z)/|f'(z)|$, defined in the canonical rectangle. By means of an integration corresponding to (4.1) and (4.2) one obtains first

$$\iint\limits_Q \varrho^2\, d\sigma = \int\limits_0^a du \int\limits_0^b (\varrho_1\,(u + i\,v))^2\, dv\,,$$

and for the curvilinear integrals

$$\inf_{C \in \mathscr{C}_a} \int\limits_C \varrho|dz| \leqq \int\limits_{C_u} \varrho|dz| = \int\limits_0^b \varrho_1\,(u + i\,v)\, dv\,.$$

Hence, by Schwarz's inequality,

$$\iint\limits_Q \varrho^2\, d\sigma \geqq \int\limits_0^a \frac{1}{b}\left(\int\limits_0^b \varrho_1\,(u + i\,v)\, dv\right)^2 du \geqq \frac{a}{b}\left(\inf_{C \in \mathscr{C}_a} \int\limits_C \varrho|dz|\right)^2,$$

where $a/b = M(Q)$. Because of (4.4) equality holds here if $\varrho = |f'|$.

Using the notations

$$\int\limits_C \varrho|dz| = l_\varrho(C)\,, \qquad \iint\limits_Q \varrho^2\, d\sigma = m_\varrho(Q)\,, \tag{4.5}$$

we thus obtain the desired characterization of the module in the following form (Ahlfors-Beurling [1]):

---

[7] As was mentioned in connection with the curve family $\mathscr{C}_a$, we are not concerned here with carrying through the discussion under the most general conditions, this being inessential from the point of view of the applications in Chapters I and II. In III.4 we shall return to this question and then only suppose that the arcs C are locally rectifiable and the functions ϱ Borel-measurable.

Theorem 4.1. *The module of a quadrilateral admits the representation*[8]

$$M(Q) = \inf_{\varrho \in \mathcal{P}} \frac{m_\varrho(Q)}{\left(\inf\limits_{C \in \mathcal{C}_a} l_\varrho(C)\right)^2} . \tag{4.6}$$

The infimum is attained for $\varrho = |f'|$, *where* f *is a canonical mapping of* Q.

The reciprocal value of the expression on the right in (4.6) is called the *extremal length* of the curve family $\mathcal{C}_a$.

In view of Theorem 4.1 the definition of quasiconformality can be expressed without application of the theory of conformal mappings. The conformal mappings could rather be defined as 1-quasiconformal mappings (cf. Theorem 5.1); for these one could then prove the Riemann mapping theorem and the theorem on boundary correspondence, which are in any case of crucial importance for the theory of quasiconformal mappings. However, in order to derive the most important properties of quasiconformal mappings as quickly as possible, we have preferred not to start from a definition of quasiconformality based directly on (4.6).

In a euclidean space of dimension three or more there is no theorem corresponding to the Riemann mapping theorem. Although we shall only deal with the case of the plane in this book, it should be remarked that a definition of the module similar to Theorem 4.1 makes it possible to carry over certain parts of the theory of quasiconformal mappings into higher dimensions.

4.3. Module estimation by means of euclidean length and area. If $\varrho = 1$ in (4.5), then $l_\varrho(C)$ is the euclidean length $l(C)$ and $m_\varrho(Q)$ the euclidean area $m(Q)$. We call

$$s_a = s_a(Q) = \inf_{C \in \mathcal{C}_a} l(C)$$

the *distance between the a-sides of* Q. The dependence of s_a on the choice of the family $\mathcal{C}_a$ has no meaning here since the estimations below hold for every $\mathcal{C}_a$ which satisfies the conditions listed in 4.1. The distance s_b between the b-sides is defined in a corresponding way. Because $M(Q(z_2, z_3, z_4, z_1)) = 1/M(Q(z_1, z_2, z_3, z_4))$ one obtains from (4.6) the following result, which later will be applied repeatedly.

Rengel's inequality. *The module of a quadrilateral* Q *satisfies the double inequality*

$$\frac{(s_b(Q))^2}{m(Q)} \leqq M(Q) \leqq \frac{m(Q)}{(s_a(Q))^2} .$$

Equality holds in both cases when and only when Q *is a rectangle.*

---

[8] Naturally we only allow here functions ϱ for which the numerator and denominator in (4.6) are not simultaneously zero or infinite.

As to the last statement, equality is clearly true in the case of a rectangle. In order to prove the necessity part, we consider the inverse $f^{-1} = g$ of the canonical mapping of Q onto $R(0, M(Q), M(Q) + i, i)$. Then

$$M(Q)\,(s_a(Q))^2 = M(Q) \inf_{C \in \mathscr{C}_a} \left(\int_{f(C)} |g'|\,|dw|\right)^2 \leqq \int_0^{M(Q)} \left(\int_0^1 |g'(u + i\,v)|\,dv\right)^2 du$$

and

$$m(Q) = \int_0^{M(Q)} du \int_0^1 |g'(u + i\,v)|^2\,dv\,.$$

From $M(Q)\,(s_a(Q))^2 = m(Q)$ it thus follows that

$$\int_0^1 |g'(u + i\,v)|^2\,dv = \left(\int_0^1 |g'(u + i\,v)|\,dv\right)^2 = \inf_{C \in \mathscr{C}_a} \left(\int_{f(C)} |g'|\,|dw|\right)^2$$

for every u. For the first equation to be possible, i.e. for Schwarz's inequality to become an equation, $|g'|$ must be independent of v. From the second relation one then concludes that g' is a constant. This means that g is a similarity transformation, and Q is thus a rectangle.

In the same way one can show that the equation $M(Q) = (s_b(Q))^2/m(Q)$ can only hold for a rectangle.

In order to estimate the module of a quadrilateral with the help of Rengel's inequality one needs a lower bound for s_a (or s_b). From an upper bound for s_a we cannot draw any inference about $M(Q)$ if, apart from this, we know only the area $m(Q)$. In fact, we can construct quadrilaterals with arbitrary given module and area for which s_a is arbitrarily small.

4.4. Module estimation by means of two euclidean lengths. In cases when a lower bound for s_a and an upper bound for s_b are known the lemma below is often useful, especially if there is no information about the area $m(Q)$.

Lemma 4.1. *The module of a quadrilateral Q satisfies the inequality*

$$M(Q) \leqq \pi \frac{1 + 2\log(1 + 2\,s_a/s_b)}{(\log(1 + 2\,s_a/s_b))^2}\,. \tag{4.7}$$

Note that this also yields a lower bound for $M(Q)$ if s_a and s_b are interchanged and $M(Q)$ is replaced by $1/M(Q)$.

To prove (4.7) we consider an arc C_0 which lies in the interior of Q and joins the b-sides of Q. We may assume that the length $l(C_0) = s_b$. For $l(C_0)$ can be chosen arbitrarily near to s_b, and we obtain the lemma by proceeding to the limit as soon as the proof is carried out with $l(C_0)$ instead of s_b.

The proof depends on the application of Theorem 4.1. To construct a suitable mass function ϱ we denote by z_0 the point which divides the arc C_0 into two parts of length $s_b/2$, and set

$$\varrho(z) = \begin{cases} \dfrac{2}{s_b} & \text{for} \quad |z - z_0| \leqq \dfrac{1}{2} s_b\,, \\[2ex] \dfrac{1}{|z - z_0|} & \text{for} \quad \dfrac{1}{2} s_b < |z - z_0| \leqq s_a + \dfrac{1}{2} s_b\,, \\[2ex] 0 & \text{for} \quad |z - z_0| > s_a + \dfrac{1}{2} s_b\,. \end{cases}$$

The function ϱ is continuous in the whole plane with the exception of the circle $|z - z_0| = s_a + \frac{1}{2} s_b$. From this it obviously follows that ϱ can be considered a member of the family $\mathcal{P}$ in (4.6). Thus we have

$$M(Q) \leqq \frac{m_\varrho(Q)}{\left(\inf\limits_{C \in \mathscr{C}_a} l_\varrho(C)\right)^2}\,, \tag{4.8}$$

where $\mathscr{C}_a$ is a family of arcs joining the a-sides of Q and possessing the properties 1—3 listed in 4.1. For $m_\varrho(Q)$ one immediately obtains

$$m_\varrho(Q) \leqq \pi \left(1 + 2 \log \left(1 + \frac{2\, s_a}{s_b}\right)\right). \tag{4.9}$$

In order to estimate $l_\varrho(C)$ for an arbitrary $C \in \mathscr{C}_a$ we notice first that C and C_0 have a common point (or a common limit point if C and C_0 go into the same vertex). This follows from the remark made at the end of 1.7 since C joins the a-sides and C_0 the b-sides of Q. It follows from the choice of the point z_0 that C_0 lies in the disc $|z - z_0| < s_b/2$. The arc C therefore possesses common points with $|z - z_0| \leqq s_b/2$. Since the length of C is at least s_a and ϱ is a decreasing function of $|z - z_0|$ we obtain a minorant for $l_\varrho(C)$ if we integrate ϱ over a segment of length s_a joining the circles $|z - z_0| = s_b/2$ and $|z - z_0| = s_a + s_b/2$. Thus for every $C \in \mathscr{C}_a$ we have the estimation

$$l_\varrho(C) \geqq \int\limits_{s_b/2}^{s_a + s_b/2} \frac{dr}{r} = \log\left(1 + \frac{2\, s_a}{s_b}\right). \tag{4.10}$$

The assertion (4.7) follows from (4.8), (4.9) and (4.10), and the lemma is proved.

4.5. Degenerate quadrilaterals. The majorant in (4.7) tends to zero as $s_b \to 0$. Thus if Q_n, $n = 1, 2, 3, \ldots$, is a sequence of quadrilaterals for which

$$\lim_{n \to \infty} s_b(Q_n) = 0\,, \qquad \liminf_{n \to \infty} s_a(Q_n) > 0\,, \tag{4.11}$$

then

$$\lim_{n \to \infty} M(Q_n) = 0 .$$

We now introduce the concept of a *degenerate quadrilateral* $Q(z_1, z_1, z_2, z_3)$ which is in fact a triangle, or in the case $z_2 = z_3$ a lune, with one or two double corners, respectively. On the other hand, the coincidence of three or four corners is not allowed. The double corner z_1 is called the a-side of the degenerate quadrilateral Q.

It can be shown without difficulty that for a degenerate quadrilateral $Q(z_1, z_1, z_2, z_3)$ the quantity s_b vanishes. In view of (4.11) it is therefore meaningful to define

$$M(Q(z_1, z_1, z_2, z_3)) = 0 , \qquad M(Q(z_1, z_2, z_2, z_3)) = \infty . \tag{4.12}$$

4.6. Superadditivity and monotonicity of the module. As a first application of Rengel's inequality, we shall prove a lemma which can be interpreted either as a superadditivity relation or as a monotonicity property of the module.

Lemma 4.2. *Let Q_n, $n = 1, 2, \ldots$, be quadrilaterals without common interior points and with closures contained in the closure of a quadrilateral Q. If the a-sides of every Q_n are contained one in each a-side of Q, then*

$$\sum_n M(Q_n) \leqq M(Q) . \tag{4.13}$$

If Q is a rectangle, then equality holds here if and only if all Q_n are rectangles and $\sum m(Q_n) = m(Q)$.

Proof: Due to conformal invariance of the module, we may suppose that Q is the rectangle $R(0, M(Q), M(Q) + i, i)$. Then $s_a(Q_n) \geqq 1$ and therefore, according to Rengel's inequality,

$$\sum M(Q_n) \leqq \sum m(Q_n) \leqq m(Q) = M(Q) .$$

For equality to hold here, Rengel's inequality itself must become an equality for every Q_n. By 4.2 this happens if and only if Q_n is a rectangle. If, in addition, $\sum m(Q_n) = m(Q)$, then indeed $\sum M(Q_n) = M(Q)$.

It should be noted that this lemma obviously also holds if one or more of the quadrilaterals Q, Q_n are degenerate.

If one thinks of the quadrilateral Q as being divided up by Jordan arcs connecting its a-sides, then (4.13) can be seen as a superadditivity property of the module. On the other hand, in the case of a single quadrilateral Q_n, it can be interpreted as a monotonicity property.

4.7. Continuity of the module. As a second application of Rengel's inequality, we prove two theorems on the convergence of the module of a sequence of quadrilaterals. The first of these theorems will be fully proved. The proof of the second more general result will be reduced to a function theoretical result, since its complete proof with the means developed here would be rather complicated. For our applications the first, simpler theorem is sufficient, and the more general theorem will never be required.

Let us consider a sequence Q_n of quadrilaterals with sides a_1^n, b_1^n, a_2^n, and b_2^n. We say that the sequence Q_n, $n = 1, 2, \ldots$, *converges* to the quadrilateral Q *from inside* if, first, $\overline{Q}_n \subset \overline{Q}$ for every n and second, to every $\varepsilon > 0$ there corresponds an n_ε such that for $n \geqq n_\varepsilon$ every point of a_i^n and b_i^n $(i = 1, 2)$ has a spherical distance $< \varepsilon$ (cf. 1.1) from the corresponding sides a_i and b_i of Q.

Lemma 4.3. *If the sequence Q_n of quadrilaterals converges to the quadrilateral Q from inside, then*

$$\lim_{n \to \infty} M(Q_n) = M(Q) . \tag{4.14}$$

Proof: We note first that the canonical mapping f of Q onto the rectangle $R = R\left(0, M(Q), M(Q) + i, i\right)$ is uniformly continuous in $\overline{Q}$ since $\overline{Q}$ is compact. That is, for $z_1, z_2 \in \overline{Q}$, we have

$$\lim_{\delta \to 0} \sup_{k(z_1, z_2) < \delta} k\left(f(z_1), f(z_2)\right) = 0 .$$

It follows that the images $f(Q_n) = Q_n'$ of the Q_n converge to R from inside, not only in the spherical metric, but also in the euclidean, since for bounded sets the two definitions of convergence coincide.

Thus to every ε, $0 < \varepsilon < \min\left(1/2, M(Q)/2\right)$, there corresponds an n_ε such that for $n \geqq n_\varepsilon$ every side of Q_n' lies in a closed strip of width ε, with the corresponding side of R in its boundary. Hence $s_a(Q_n') \geqq 1 - 2\varepsilon$, $s_b(Q_n') \geqq M(Q) - 2\varepsilon$. Because $m(Q_n') \leqq M(Q)$, Rengel's inequality yields

$$\frac{(M(Q) - 2\varepsilon)^2}{M(Q)} \leqq M(Q_n') \leqq \frac{M(Q)}{(1 - 2\varepsilon)^2} . \tag{4.15}$$

Since the canonical mapping is conformal, we have $M(Q_n') = M(Q_n)$, and the lemma is proved.

Remark. It is not difficult to show that Lemma 4.3 remains true if Q is a degenerate quadrilateral.

4.8. Approximation to a quadrilateral from inside. In view of later application of Lemma 4.3 we construct, for a given quadrilateral Q, a particular sequence of quadrilaterals converging to Q from inside. This is done by first mapping Q conformally by a function f onto a quadrilateral consisting of the unit disc $|z| < 1$ and the vertices z_1, z_2, z_3, z_4. After this we consider the quadrilateral Q_n' consisting of the disc $|z| < 1 - 1/n$ and the vertices $(1 - 1/n)\ z_i$, $i = 1, 2, 3, 4$. Since the inverse of f is uniformly continuous in $|z| \leqq 1$ the preimages $Q_n = f^{-1}(Q_n')$ converge to Q from inside. The boundary of every Q_n is an analytic curve; such quadrilaterals will in future be termed *analytic*. We have thus obtained the following result.

Lemma 4.4. *Every quadrilateral Q is the limit of a sequence of analytic quadrilaterals having closures in Q and converging to Q from inside.*

4.9. Generalized continuity theorem. For the sake of completeness, we also prove a more general convergence theorem, which contains Lemma 4.3 as a special case. In fact we can free ourselves of the requirement that the convergence $Q_n \to Q$ takes place from inside.

We again consider a sequence of quadrilaterals Q_n and use the same notation as above. By definition *the sequence Q_n converges to the quadrilateral Q* if to every $\varepsilon > 0$ there corresponds an n_ε such that for $n \geqq n_\varepsilon$, every point of a_i^n, b_i^n, $i = 1, 2$, and every interior point of Q_n has a spherical distance of at most ε from a_i, b_i and Q, respectively.

Theorem on the convergence of the module. *If the sequence of quadrilaterals Q_n converges to a quadrilateral Q, then*

$$\lim_{n \to \infty} M(Q_n) = M(Q)\,.$$

Proof[9]: We first construct a sequence of Jordan domains $G_1 \supset G_2 \supset \cdots \supset \overline{Q}$, which converge monotonically to Q and whose boundaries converge not only in the sense of our definition, but also in the following stronger sense of Fréchet: To every $\varepsilon > 0$ there exists an n_ε such that for $n \geqq n_\varepsilon$ one can define a homeomorphism between the boundaries of G_n and Q such that every point and its image have a spherical distance less than ε. The sequence G_n can be constructed as follows: We map the exterior of Q conformally onto a disc D, choose a sequence of discs D_n converging to D from inside, and then take the exterior of the preimage of D_n as G_n.

The proof now rests on the following function theoretical convergence theorem of Radó (for the proof, see for example Golusin [1], p. 50):

Let $G_1 \supset G_2 \supset \cdots \supset \overline{Q}$ be a sequence of Jordan domains such that the boundary of G_n converges to that of Q in the sense of Fréchet. Further

---

[9] For the idea of this proof we are indebted to E. Reich.

let f_n and f be conformal mappings of a Jordan domain R onto G_n and Q, respectively, which are normalized at a point $z_0 \in R$ by the conditions $f_n(z_0) = f(z_0)$, $\arg f_n'(z_0) = \arg f'(z_0)$. Then $f_n(z) \to f(z)$ uniformly in $\overline{R}$.

Let us apply this theorem as follows: Let R be the canonical rectangle of Q, f the inverse of the canonical mapping of Q onto R, and f_n the conformal mapping of the domain R onto G_n, normalized at a point $z_0 \in R$ as above. From the uniform convergence $f_n \to f$ and the uniform continuity of f^{-1} it follows that the sequence of preimages $f_n^{-1}(Q)$ converges to the rectangle R from inside. Thus to every $\varepsilon > 0$ there corresponds an N such that every point of a side of $f_N^{-1}(Q) = Q'$ is at a distance $< \varepsilon/2$ from the corresponding side of R.

We consider now the original sequence Q_n converging to Q. From some n_1 onwards, every Q_n is contained in $\overline{G}_N$ and thus $f_N^{-1}(Q_n) = Q_n'$ is contained in $\overline{R}$. From the uniform continuity of f_N^{-1} it follows that the sequence Q_n', $n = n_1, n_1 + 1, \ldots$, converges to the quadrilateral Q'. Hence, there exists an n_ε such that for $n \geqq n_\varepsilon$, the points of each side of Q_n' lie at a distance $< \varepsilon/2$ from the corresponding side of Q'. It follows that each side of Q_n' lies in a strip of width ε whose boundary contains the corresponding side of R. Hence $M(Q_n') = M(Q_n)$ satisfies inequality (4.15), and the theorem on the convergence of the module is proved.

§ 5. *Two Basic Properties of Quasiconformal Mappings*

5.1. 1-quasiconformal mappings. We now turn back to quasiconformal mappings. As a first application of the module estimations obtained in § 4, we prove two theorems (Ahlfors [1]) already referred to in § 3. The first (Theorem 5.1) is the converse of the result mentioned in 3.2, i.e. that a conformal mapping is 1-quasiconformal. The second (Theorem 5.2) states that no non-quasiconformal homeomorphism can arise as the limit of a uniformly convergent sequence of K-quasiconformal mappings. We want to prove these theorems first since they justify the definition of quasiconformality given in 3.2.

Theorem 5.1. *A 1-quasiconformal mapping $G \to G'$ is conformal.*

Proof: Since every point of G lies in the interior of a quadrilateral Q, $\overline{Q} \subset G$, it is enough to demonstrate the conformality of the given 1-quasiconformal mapping in such a quadrilateral Q. For this purpose we map Q and its image quadrilateral Q' conformally onto their canonical rectangles R and R', respectively. For a 1-quasiconformal mapping, $M(Q) = M(Q')$ by definition. We can therefore choose the rectangles R and R' to be congruent, with the same corners $0, M, M + i, i$, where $M = M(Q)$.

From the original 1-quasiconformal mapping $Q \to Q'$ and the conformal mappings $R \to Q$ and $Q' \to R'$ we obtain a composite mapping $w : R \to R'$, which is also 1-quasiconformal. We shall show that w reduces to the identity mapping, which proves the theorem.

We choose an arbitrary interior point $z_0 = x_0 + i\, y_0$ of R and divide R into two rectangles $R_1 = R_1\,(0, x_0, x_0 + i, i)$ and $R_2 = R_2\,(x_0, M, M + i, x_0 + i)$. If $R_1' = w(R_1)$, $R_2' = w(R_2)$ denote the image quadrilaterals, then, by the 1-quasiconformality of the original mapping,

$$M(R_1') + M(R_2') = M(R_1) + M(R_2) = x_0 + (M - x_0) = M(R') .$$

However, according to Lemma 4.2, we have $M(R_1') + M(R_2') = M(R')$ only if R_1 is a rectangle. Its base, the real part of $w(z_0)$, must equal x_0 since $M(R_1') = M(R_1) = x_0$.

In the same way we conclude that the imaginary part of $w(z_0)$ equals y_0. Thus an arbitrary point z_0 remains invariant under the mapping w.

5.2. Limit of a sequence of quasiconformal mappings. The result mentioned at the beginning of 5.1 concerning a uniformly convergent sequence of quasiconformal mappings can be easily proved with the help of Lemmas 4.3 and 4.4.

Theorem 5.2. *If the K-quasiconformal mappings w_n of the domain G converge uniformly to a homeomorphism w of G in every compact subset of G, then w is K-quasiconformal in G.*

Proof: We must show that the inequality $M\big(w(Q)\big) \leqq K\, M(Q)$ holds for an arbitrary quadrilateral Q, $\overline{Q} \subset G$. To do this we construct a sequence of quadrilaterals Q_n converging to Q and having the properties specified in Lemma 4.4. From a certain k onwards, $w_k(Q_n)$ is contained in $w(Q)$, and we obtain $w_n(Q_n) \subset w(Q)$ by selecting a suitable subsequence. Since the mappings w_n converge uniformly in $\overline{Q}$ to the mapping w, and this is uniformly continuous, the quadrilaterals $w_n(Q_n)$ converge to $w(Q)$ from inside. Thus by Lemma 4.3, $\lim M\big(w_n(Q_n)\big) = M\big(w(Q)\big)$ and $\lim M(Q_n) = M(Q)$. This implies the assertion $M\big(w(Q)\big) \leqq K\, M(Q)$, since the inequality $M(w_n(Q_n)) \leqq K\, M(Q_n)$ holds for every n.

Theorem 5.2 will be strengthened in II.5, where we shall free ourselves from the condition that the limit function w be a homeomorphism.

5.3. Further inferences from the continuity of the module. In this connection we prove two consequences of Lemmas 4.3 and 4.4, which will be needed later.

The former result states that we may restrict ourselves to analytic quadrilaterals in the definition of quasiconformality. Besides its intrinsic interest, this result is important in certain applications.

Theorem 5.3. *Let w be a sense-preserving homeomorphism of a domain G which satisfies the condition $M(w(Q)) \leqq K\,M(Q)$ for every analytic quadrilateral Q, $\overline{Q} \subset G$. Then w is a K-quasiconformal mapping of G.*

Proof: We choose an arbitrary quadrilateral Q, $\overline{Q} \subset G$, and construct a sequence of analytic quadrilaterals Q_n possessing the properties mentioned in Lemma 4.4. Since w is uniformly continuous in $\overline{Q}$, the quadrilaterals $w(Q_n)$ converge to $w(Q)$ from inside. By Lemma 4.3 we have therefore $\lim M(Q_n) = M(Q)$ and $\lim M(w(Q_n)) = M(w(Q))$. Since the inequality $M(w(Q_n)) \leqq K\,M(Q_n)$ holds for all n, the dilatation of Q is at most K, and the theorem is proved.

The second result is concerned with a homeomorphism of the quadrilateral Q (see 2.3) which is K-quasiconformal in the interior of Q. Such a homeomorphism will be called a *K-quasiconformal mapping* of Q. For these mappings we have the following result which will be often used below:

Lemma 5.1. *If a quadrilateral Q is mapped K-quasiconformally onto a quadrilateral Q', then $M(Q') \leqq K\,M(Q)$.*

To prove this we need only replace the arbitrary quadrilateral of the proof of Theorem 5.3 by the given quadrilateral Q, and the proof holds without further alteration.

§ 6. *Module of a Ring Domain*

6.1. Definition of the module of a ring domain. In this section it will be shown that doubly connected domains, which we call *ring domains*, could be chosen instead of quadrilaterals as a starting point for the theory of quasiconformal mappings. This depends on the fact that the conformal equivalence class of a ring domain can be expressed by means of a single real module in the same way as that of a quadrilateral.

The definition of the module of a ring domain is based on the following mapping theorem: *Every ring domain B can be mapped conformally* *onto an annulus*[10] $\{z|0 \leqq r_1 < |z| < r_2 \leqq \infty\}$. Such a mapping is called a *canonical mapping* and the corresponding annulus a *canonical image* of B.

If $r_1 > 0$ and $r_2 < \infty$ for one canonical image of B, then the ratio of the radii r_2/r_1 is the same for all canonical images of B. The number

$$M(B) = \log \frac{r_2}{r_1}$$

which then determines the conformal equivalence class of B, is called the *module* of B. If, on the other hand, $r_1 = 0$ or $r_2 = \infty$ for any canonical image, then we define $M(B) = \infty$. This happens if and only if at least one boundary component of B consists of a single point.

The ring domains with infinite module fall into two conformal equivalence classes, according as only one or both of the boundary components are points. In the first case, either $r_1 = 0$, $r_2 < \infty$ or $r_1 > 0$, $r_2 = \infty$[11], in the second $r_1 = 0$, $r_2 = \infty$.

6.2. Direct characterization of the module. In this section we prove some basic properties of the module of a ring domain. These show a great similarity with the results on the module of a quadrilateral obtained in § 4. By a method which is completely analogous to the one used in 4.1—4.2, we show first that the module of a ring domain can also be defined without the use of conformal mapping.

Let B be a ring domain, $(-B)_1$ and $(-B)_2$ the components of the complement of B, and f the canonical mapping of B onto $\{w|r_1 < |w| < r_2\}$. Let the family of the preimages of the circles $\{w|\,|w| = r\}$, $r_1 < r < r_2$, be denoted by $\mathscr{C}_1$. Every curve $C \in \mathscr{C}_1$ separates the sets $(-B)_1$ and $(-B)_2$.

Let $\mathscr{C} \supset \mathscr{C}_1$ be a family of Jordan curves which separate the sets $(-B)_1$ and $(-B)_2$. Further, let us suppose that all curves in $\mathscr{C}$ are sufficiently regular to admit all following integrations (cf. the footnote on p. 21).

Denote by $\mathscr{P}$ a family of non-negative functions in B which are not too discontinuous for the following integrations; we also suppose that $\mathscr{P}$ contains the function $|f'/f|$.

[10] In our terminology an annulus is a domain bounded by two concentric circles which may degenerate into points.

[11] These cases are equivalent since the points ∞ and 0 can be interchanged by means of an inversion.

We associate with any function $\varrho \in \mathcal{P}$ the function ϱ_1, $\varrho_1(f(z)) = \varrho(z)|f(z)/f'(z)|$. By integrating and using Schwarz's inequality we then obtain

$$m_\varrho(B) = \iint_B \varrho^2\, d\sigma = \int_{r_1}^{r_2} \frac{dr}{r} \int_0^{2\pi} \varrho_1^2(r\, e^{i\varphi})\, d\varphi \geqq \frac{1}{2\pi} \int_{r_1}^{r_2} \frac{dr}{r} \left(\int_0^{2\pi} \varrho_1(r\, e^{i\varphi})\, d\varphi \right)^2.$$

If $C_1 \in \mathcal{C}_1$, i.e. $f(C_1)$ is a circle $|w| = r$, then

$$\int_0^{2\pi} \varrho_1(r\, e^{i\varphi})\, d\varphi = \int_{C_1} \varrho |dz| \geqq \inf_{C \in \mathcal{C}_1} l_\varrho(C) \geqq \inf_{C \in \mathcal{C}} l_\varrho(C)\,.$$

It follows from the above that, in the case $M(B) < \infty$,

$$m_\varrho(B) \geqq \frac{1}{2\pi} \int_{r_1}^{r_2} \left(\inf_{C \in \mathcal{C}} l_\varrho(C) \right)^2 \frac{dr}{r}$$

$$= \frac{1}{2\pi} \log \frac{r_2}{r_1} \left(\inf_{C \in \mathcal{C}} l_\varrho(C) \right)^2 = \frac{1}{2\pi} M(B) \left(\inf_{C \in \mathcal{C}} l_\varrho(C) \right)^2,$$

where equality holds for $\varrho = |f'/f|$. In the case $M(B) = \infty$ we have $m_\varrho(B) = \infty$ if $\inf l_\varrho(C) > 0$.

Thus we have obtained the following result corresponding to Theorem 4.1 (cf. the footnote on p. 22).

Theorem 6.1. *The module of a ring domain B is*

$$M(B) = \inf_{\varrho \in \mathcal{P}} \frac{2\pi\, m_\varrho(B)}{\left(\inf_{C \in \mathcal{C}} l_\varrho(C) \right)^2}\,, \tag{6.1}$$

where the infimum is attained for $\varrho = |f'/f|$.

The module of the ring domain B can also be characterized in another way. To see this, denote by $\mathcal{C}_2$ the family of the preimages $f^{-1}(C)$ of the segments $C = \{w | r_1 < |w| < r_2,\ \arg w = \varphi\}$. Let $\mathcal{C}^* \supset \mathcal{C}_2$ be a family of (sufficiently regular) Jordan arcs lying in B and joining the components of $-B$, i.e. having a limit point on each of $(-B)_1$ and $(-B)_2$. Let $\mathcal{P}$ be a family of functions with the same properties as the above one. Then, in the above notation, if $\varrho \in \mathcal{P}$ we obtain

$$m_\varrho(B) = \int_0^{2\pi} d\varphi \int_{r_1}^{r_2} \varrho_1^2(r\, e^{i\varphi}) \frac{dr}{r} \geqq \frac{1}{\log \frac{r_2}{r_1}} \int_0^{2\pi} d\varphi \left(\int_{r_1}^{r_2} \varrho_1(r\, e^{i\varphi}) \frac{dr}{r} \right)^2,$$

and

$$l_\varrho(C) = \int_{r_1}^{r_2} \varrho_1(r\, e^{i\varphi}) \frac{dr}{r} \geqq \inf_{C \in \mathcal{C}^*} l_\varrho(C)\,.$$

Hence,

$$M(B) = 2\pi \sup_{\varrho \in \mathcal{P}} \frac{\left(\inf\limits_{C \in \mathscr{C}^*} l_\varrho(C)\right)^2}{m_\varrho(B)}, \tag{6.2}$$

where, as before, the supremum is attained for $\varrho = |f'/f|$.

6.3. Analogue of Rengel's inequality. If the ring domain B separates the points 0 and ∞, then these points will also be separated by every curve of the family $\mathscr{C}$. Consequently, we have for $C \in \mathscr{C}$,

$$2\pi = \left|\int_C d \arg z\right| \leqq \int_C \frac{|dz|}{|z|}.$$

Setting $\varrho(z) = 1/|z|$ in (6.1) we obtain:

The module of a ring domain B which separates the points 0 and ∞ satisfies the inequality

$$M(B) \leqq \frac{1}{2\pi} \iint_B \frac{d\sigma}{|z|^2}. \tag{6.3}$$

If B is an annulus of centre $z = 0$, then equality holds in (6.3). To examine a necessary condition for equality, we consider the inverse $f^{-1} = g$ of the canonical mapping of B onto $r_1 < |w| < r_2$. If equality holds in (6.3), then

$$\log \frac{r_2}{r_1} = M(B) = \frac{1}{2\pi} \iint_B \frac{d\sigma}{|z|^2} = \frac{1}{2\pi} \int_{r_1}^{r_2} dr \int_0^{2\pi} \left|\frac{g'(w)}{g(w)}\right|^2 r \, d\varphi,$$

where $w = r e^{i\varphi}$. On the other hand, for constant $|w| = r$ we have

$$\int_0^{2\pi} \frac{g'(w)}{g(w)} w \, d\varphi = \int_{|w|=r} d \arg g(w) = \pm 2\pi. \tag{6.4}$$

It follows that, in the case $M(B) < \infty$, the inequality

$$\frac{1}{2\pi r} \left|\int_0^{2\pi} \frac{g'(w)}{g(w)} w \, d\varphi\right|^2 \leqq \int_0^{2\pi} \left|\frac{g'(w)}{g(w)}\right|^2 r \, d\varphi$$

must become an equality for every r, $r_1 < r < r_2$. This is possible only if $w\, g'(w)/g(w)$ has a constant value on every circle $|w| = r$ (cf. 4.3). By (6.4) this value must be either $+1$ or -1. Then either $g(w)/w$ or $w\, g(w)$ is a constant, and we have shown:

Equality holds in (6.3) *if and only if either B is an annulus with centre* 0 *or $M(B)$ is infinite.*

6.4. Module estimation in the spherical metric. In the case of the spherical metric $\varrho(z) = (1 + |z|^2)^{-1}$ (cf. 1.1) we always have $m_\varrho(B) \leqq \pi$. It follows from (6.1) that

$$M(B) \leqq \frac{2\pi^2}{k_0^2}, \tag{6.5}$$

where k_0 is the greatest lower bound of the spherical length of curves which separate $(-B)_1$ and $(-B)_2$.

From (6.5) we obtain the following result, which will play an important role in II.5:

Lemma 6.1. *If the ring domain B separates the pair of points a_1, b_1 from the pair a_2, b_2 and if the spherical distance satisfies $k(a_i, b_i) \geqq \delta > 0$, $i = 1, 2$, then*

$$M(B) \leqq \frac{\pi^2}{2\delta^2}.$$

Proof: If a curve C separates the components of the complement of B, then C also separates the points a_1, b_1 from the points a_2, b_2. Since $k(a_i, b_i) \geqq \delta$, elementary geometrical considerations show that $k_0 \geqq 2\delta$, and the assertion follows from (6.5)[12].

6.5. Degenerating ring domains. It was shown in 4.5 that the module of a quadrilateral tends to zero under a limiting process in which the distance between the b-sides tends to 0 while the distance between the a-sides remains greater than a fixed positive bound. The corresponding situation for ring domains is as follows: The distance between the boundary components tends to 0, while their diameters remain bounded away from 0. Then the module tends to 0, as the following lemma shows.

Lemma 6.2. *Let B be a ring domain whose boundary components have spherical diameters $> \delta$ and a mutual spherical distance $< \varepsilon$. Then, for $\varepsilon < \delta$, we have*

$$M(B) \leqq \frac{\pi^2}{\log \dfrac{\tan(\delta/2)}{\tan(\varepsilon/2)}}. \tag{6.6}$$

Proof: Let $z_i \in (-B)_i$, $i = 1, 2$, be two points with $k(z_1, z_2) = \varepsilon$. By a linear transformation which corresponds to a rotation of the Riemann sphere we can carry the points z_i into the points $\pm \tan(\varepsilon/2)$.

Both components of the complement of the annulus

$$A = \{z \mid \tan(\varepsilon/2) < |z| < \tan(\delta/2)\}$$

[12] The inequality $k_0 \geqq \delta$, which is sufficient for our applications, is immediate.

then contain points both of $(-B)_1$ and $(-B)_2$. It follows that every curve C which separates $(-B)_1$ and $(-B)_2$ has points in both components of the complement of A. If $\varrho(z) = 1/|z|$ for $z \in A$ and $\varrho(z) = 0$ otherwise, then

$$l_\varrho(C) \geqq 2 \log \frac{\tan(\delta/2)}{\tan(\varepsilon/2)} .$$

On the other hand, we have

$$m_\varrho(B) \leqq 2\pi \log \frac{\tan(\delta/2)}{\tan(\varepsilon/2)} ,$$

and (6.6) follows from Theorem 6.1.

6.6. Superadditivity of the module. The theorem on superadditivity for the module of a ring domain which corresponds to Lemma 4.2 for quadrilaterals reads as follows:

Lemma 6.3. *Let $B, B_1, B_2, \ldots$ be ring domains such that $B_1, B_2, \ldots$ are disjoint subdomains of B. If every B_n separates $(-B)_1$ and $(-B)_2$, then*

$$\sum_n M(B_n) \leqq M(B) . \tag{6.7}$$

If B is an annulus with finite module then equality holds if and only if every B_n is an annulus and $\sum m(B_n) = m(B)$.

Proof: Without loss of generality we may suppose B to be the annulus $0 < r_1 < |z| < r_2 < \infty$. Then every $M(B_n)$ is finite (cf. 6.1), and by (6.3) we have

$$M(B_n) \leqq \frac{1}{2\pi} \iint_{B_n} \frac{d\sigma}{|z|^2} ,$$

where equality holds if and only if B_n is an annulus of centre 0. Hence,

$$\sum_n M(B_n) \leqq \frac{1}{2\pi} \iint_{\cup B_n} \frac{d\sigma}{|z|^2} \leqq \frac{1}{2\pi} \iint_{B} \frac{d\sigma}{|z|^2} = M(B) .$$

In the case of a single ring domain B_n inequality (6.7) can be interpreted as a monotonicity property of the module.

6.7. Continuity of the module. To formulate a convergence theorem corresponding to Lemma 4.3 we introduce the following definition: A sequence of ring domains B_n converges from inside to a ring domain B if $B_n \subset B$, $n = 1, 2, \ldots$, and to every $\varepsilon > 0$ there corresponds an n_ε such that for $n \geqq n_\varepsilon$ every point of $(-B_n)_i$, $i = 1, 2$, lies within a spherical distance $< \varepsilon$ of the set $(-B)_i$. With this definition of convergence the following continuity property of the module can easily be proved:

Lemma 6.4. *If B_n is a sequence of ring domains which converges to a ring domain B from inside, then*

$$\lim_{n\to\infty} M(B_n) = M(B) .$$

For the proof we need only remark that from some n onwards the complement of B_n lies outside an arbitrary given compact subset of B. The images of the ring domains B_n in a canonical image B_0 of B converge therefore to B_0 from inside and from some n onwards contain any given annulus whose closure lies in B_0. The lemma then follows from the monotonicity of the module.

A ring domain will henceforth be called analytic if its boundary consists of two analytic curves. *If B is an arbitrary ring domain then there exists a sequence of analytic ring domains B_n, $\overline{B}_n \subset B$, which converge to B from inside.* This sequence can be constructed by choosing a sequence of annuli which converges to the canonical image of B from inside. Their preimages in B then possess the required properties.

6.8. Relations between the modules of ring domains and quadrilaterals. The close relationship between the modules of quadrilaterals and ring domains will be evident if we cut an annulus $B = \{z | 0 < r_1 < |z| < r_2 < \infty\}$ along the segment $r_1 < x < r_2$ of the real axis. The logarithm function then maps the resulting simply connected domain conformally onto a rectangle with the ratio of sides

$$\frac{1}{2\pi}\log\frac{r_2}{r_1} = \frac{1}{2\pi} M(B) .$$

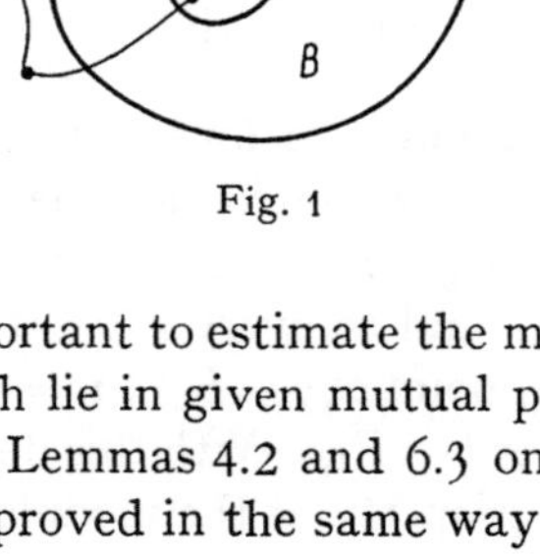

Fig. 1

For the sequel it is important to estimate the modules of quadrilaterals and ring domains which lie in given mutual positions. The following result brings to mind Lemmas 4.2 and 6.3 on superadditivity of the module, and it will be proved in the same way (Fig. 1).

Lemma 6.5. *Let Q_n, $n = 1, 2, \ldots$ be quadrilaterals having no common interior points. If B is a ring domain which separates the a-sides of every Q_n, then*

$$\Sigma\, M(Q_n) \leqq \frac{2\pi}{M(B)} \,. \tag{6.8}$$

If B is an annulus, then equality holds if and only if the a-sides of each Q_n lie on the boundary of B, the b-sides are radial segments, and $\Sigma\, m(Q_n) = m(B)$.

Proof: In proving (6.8) there is no loss of generality if we suppose that B is an analytic ring domain and that the a-sides of each Q_n lie outside $\overline{B}$ (cf. 6.7). We shall estimate the module of Q_n with the help of Theorem 4.1. To this end we define a function ϱ by

$$\varrho(z) = \begin{cases} |f'(z)/f(z)| \,, & z \in B \,, \\ 0 \,, & z \notin B \,, \end{cases}$$

where f is the canonical mapping of B onto the annulus $r_1 < |w| < r_2$. Let $\mathscr{C}$ be the family of open analytic arcs C which join the a-sides of Q_n. Since every $C \in \mathscr{C}$ cuts the boundary of B in only a finite number of points, ϱ is continuous on C with the possible exception of finitely many points. The function ϱ can therefore be considered as an element of the family $\mathscr{P}$ appearing in Theorem 4.1.

Since B separates the a-sides of Q_n, every $C \in \mathscr{C}$ joins the components of $-B$. It follows that

$$l_\varrho(C) = \int\limits_C \varrho\, |dz| \geqq \int\limits_{C \cap B} \left|\frac{f'}{f}\right| |dz| \geqq \int\limits_{r_1}^{r_2} \frac{dr}{r} = M(B) \,.$$

Hence by Theorem 4.1 we have

$$M(Q_n) \leqq \frac{1}{(M(B))^2} \iint\limits_{Q_n \cap B} \left|\frac{f'}{f}\right|^2 d\sigma \,. \tag{6.9}$$

Since

$$\iint\limits_B \left|\frac{f'}{f}\right|^2 d\sigma = \int\limits_{r_1}^{r_2}\int\limits_0^{2\pi} \frac{dr\, d\varphi}{r} = 2\pi\, M(B) \,,$$

it follows from (6.9) that

$$\Sigma\, M(Q_n) \leqq \frac{1}{(M(B))^2} \iint\limits_{\cup Q_n \cap B} \left|\frac{f'}{f}\right|^2 d\sigma \leqq \frac{2\pi}{M(B)} \,,$$

which is the required inequality (6.8).

If B is an annulus and the a- and b-sides of Q_n satisfy the property mentioned in the second part of the lemma, then

$$M(Q_n) = \frac{2\pi\, m(Q_n)}{m(B)\, M(B)} .$$

In this case equality holds in (6.8) if $\sum m(Q_n) = m(B)$.

Remark. The lemma also holds for degenerate quadrilaterals as well as in the case $M(B) = \infty$.

§ 7. *Characterization of Quasiconformality with the Help of Ring Domains*

7.1. Quasiconformal mappings of ring domains. The result obtained in § 6 on the modules of ring domains will now be applied to quasiconformal mappings. We prove two theorems, which taken together state that under a homeomorphism w of a domain G the maximal dilatation

$$\sup_{\bar{B}\subset G} \frac{M(w(B))}{M(B)}$$

of ring domains B agrees with the maximal dilatation $K(G)$ of quadrilaterals.

Theorem 7.1. *If there exists a K-quasiconformal mapping w of the ring domain B onto the ring domain B', then*

$$\frac{1}{K} M(B) \leqq M(B') \leqq K\, M(B) .$$

Proof: It is enough to prove only one of these inequalities, e.g. $M(B') \leqq K\, M(B)$, since the same proof, applied to the inverse of w, gives the other inequality.

Since no assumptions were made about the boundary behaviour of the mapping w, we first approximate B' by analytic ring domains. By 6.7, if $\varepsilon > 0$ is given, we can construct an analytic ring domain B'_ε, $\bar{B}'_\varepsilon \subset B'$, which separates the boundary components of B' and whose module satisfies the inequality $M(B'_\varepsilon) > M(B') - \varepsilon$ (or $M(B'_\varepsilon) > 1/\varepsilon$ if $M(B') = \infty$). The preimage B_ε of B'_ε is a ring domain in B bounded by two Jordan curves, and it follows from the monotonicity of the module that $M(B_\varepsilon) < M(B)$. If we can demonstrate the validity of the relation $M(B'_\varepsilon) \leqq K\, M(B_\varepsilon)$, then we shall obtain our theorem by letting $\varepsilon \to 0$.

Thus from now on we may assume that the boundaries of B and B' consist of Jordan curves and that the mapping w is a homeomorphism of $\bar{B}$ onto $\bar{B}'$. Since the canonical mapping of B can then be topologically extended to $\bar{B}$ by 2.2, we may further restrict ourselves to the case when B is the annulus $\{z|0 < r_1 < |z| < r_2 < \infty\}$.

We now make use of Lemma 6.5 by cutting the annulus B along the segments $r_1 < x < r_2$ and $-r_2 < x < -r_1$ of the real axis. Then B is divided into two domains Q_1 and Q_2 which can be thought of as quadrilaterals with the cuts as their b-sides. Let the image quadrilaterals of Q_1 and Q_2 in B' be Q_1' and Q_2', respectively. It then follows from Lemma 6.5 that

$$M(Q_1') + M(Q_2') \leqq \frac{2\pi}{M(B')}, \qquad M(Q_1) + M(Q_2) = \frac{2\pi}{M(B)}.$$

Since w is K-quasiconformal, we obtain

$$M(B') \leqq \frac{2\pi}{M(Q_1') + M(Q_2')} \leqq \frac{2\pi K}{M(Q_1) + M(Q_2)} = K\,M(B),$$

and the theorem is proved.

7.2. Sufficient module condition for quasiconformality. Theorem 7.1 has a converse as we shall now show[13]:

Theorem 7.2. *Let w be a sense-preserving homeomorphism of a domain G. If the inequality $M(w(B)) \leqq K\,M(B)$ holds for every ring domain B, $\bar{B} \subset G$, then w is a K-quasiconformal mapping.*

Proof: We have to show that the dilatation of an arbitrary quadrilateral Q, $\bar{Q} \subset G$, is at most K. Since the modules of ring domains are invariant under conformal mapping, we may replace Q and $w(Q)$ by their canonical rectangles $R\,(0, M, M+i, i)$ and $R'\,(0, M', M'+i, i)$, respectively.

We denote by R_n' the rectangle lying in R' whose sides are at a constant distance $1/n$ from the corresponding sides of R' (where $2/n < \min(1, M')$). Let R_n be the preimage of R_n' in R. As $n \to \infty$, the quadrilaterals R_n converge to R from inside, and it follows from Lemma 4.3 that

$$\lim_{n\to\infty} M(R_n) = M\,. \tag{7.1}$$

Let R_n' be divided by vertical lines into n^3 rectangles R_{nk}' with horizontal a-sides, each rectangle having the same width

$$(M' - 2/n)/n^3\,.$$

---

[13] Gehring-Väisälä [1]; their proof is based on the analytic definition given in IV.2. The above direct proof is due to Reich [1].

If R_{nk} denotes the preimage of R'_{nk}, then by the superadditivity of the module (Lemma 4.2),

$$\sum_{k=1}^{n^3} M(R_{nk}) \leqq M(R_n) .$$

Thus there exists an R_{np} for which

$$M(R_{np}) \leqq \frac{M(R_n)}{n^3} . \tag{7.2}$$

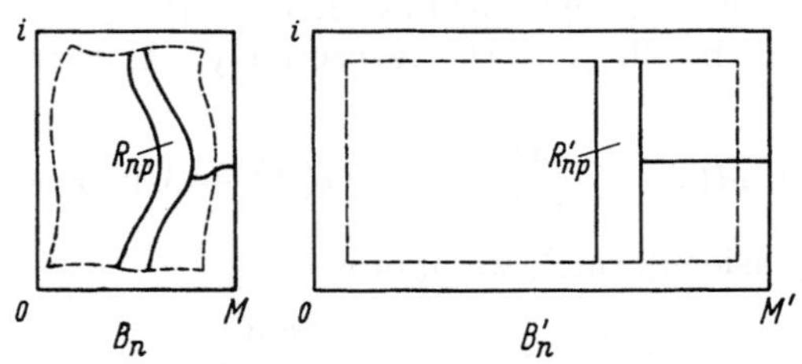

Fig. 2

We now construct the ring domain B'_n represented in Fig. 2, where the unbroken segments constitute its boundary. Let the preimage of B'_n be B_n. Since the b-sides of R_{np} lie one on each boundary component of B_n, it follows from Lemma 6.5 that

$$\frac{1}{M(R_{np})} \leqq \frac{2\pi}{M(B_n)} .$$

Thus from (7.2) we obtain

$$M(B_n) \leqq \frac{2\pi}{n^3} M(R_n) . \tag{7.3}$$

We have now to find a lower bound for $M(B'_n)$. To this end we make use of equation (6.2). Choose $n^2 > M'$ and set

$$\varrho(z) = \frac{n^3}{M' - \frac{2}{n}}$$

if z belongs to R'_{np} or the vertical strip containing R'_{np}, and $\varrho(z) = n$ at all other points of B'_n. Then $l_\varrho(C) \geqq 1$ for any arc C joining the boundary components of B'_n. Thus (6.2) implies

$$M(B'_n) \geqq \frac{2\pi}{m_\varrho(B'_n)} > \frac{2\pi}{n^3\left(\frac{M'}{n} + \frac{1}{M' - \frac{2}{n}}\right)} .$$

Finally, in view of (7.3), we deduce that

$$\frac{M(B_n')}{M(B_n)} > \frac{1}{\left(\frac{M'}{n} + \frac{1}{M' - \frac{2}{n}}\right) M(R_n)} .$$

By (7.1) the expression on the right hand side tends to M'/M as $n \to \infty$, while $M(B_n')/M(B_n)$ is at most K by the hypothesis. It follows that $M' \leqq K\,M$, which proves Theorem 7.2.

If we combine the conclusions of Theorems 7.1 and 7.2 we obtain the aforementioned characterization of K-quasiconformality:

A sense-preserving homeomorphism w of a domain G is K-quasiconformal in G if and only if the module condition $M(w(B)) \leqq K\,M(B)$ holds for every ring domain B, $\overline{B} \subset G$.

§ 8. Extension Theorems for Quasiconformal Mappings

8.1. Isolated boundary points. In § 2 we mentioned two theorems on the extension of conformal mappings: the theorem on correspondence of boundaries and the reflection principle. A third classical result of this type is the theorem on isolated singularities. In this section we shall show that all three theorems also hold for quasiconformal mappings.

A boundary point z_0 of the domain G is called isolated if there exists a neighbourhood of z_0 which consists entirely of points of G, apart from z_0 itself. The set $G \cup \{z_0\}$ is then a domain.

Let w be a K-quasiconformal mapping of G. The isolated boundary point z_0 is called a *removable singularity* of w if w can be extended to a K-quasiconformal mapping of $G \cup \{z_0\}$. For $K = 1$, i.e. for the conformal case, this is known to be always true, and we now show that the same holds for all K-quasiconformal mappings.

Theorem 8.1. *The isolated boundary points of a domain G are removable singularities of every K-quasiconformal mapping of G.*

Proof: Let z_0 be an isolated boundary point of G and $w: G \to G'$ a K-quasiconformal mapping. The proof falls into two parts: First we prove the existence of a homeomorphic extension of w to $G \cup \{z_0\}$ and then show that the continued mapping is K-quasiconformal.

To prove the first part, we consider a neighbourhood D of z_0, $\overline{D} - \{z_0\} \subset G$, bounded by a circle C. The module of the ring domain $D_0 = D - \{z_0\}$

is infinite. By Theorem 7.1, the module of the image domain $w(D_0)$ is also infinite. It follows (cf. 6.1) that one component of the complement of $w(D_0)$ consists of a single point w_0. The other component contains the image $w(C)$ of C.

Since the boundary components of D_0 and $w(D_0)$ correspond pairwise (cf. 1.8) and C is mapped onto $w(C)$, z_0 and w_0 must be corresponding boundary components. This means that every sequence converging to z_0 must have as its image a sequence converging to w_0.

If we set $w(z_0) = w_0$, then w will be extended to a one-to-one continuous mapping of $G \cup \{z_0\}$. By Lemma 1.1 such a mapping is a homeomorphism, and the first part of the proof is completed.

To prove that the extended mapping is K-quasiconformal, we consider a quadrilateral Q whose closure lies in $G \cup \{z_0\}$. Since w is K-quasiconformal in G, we have $M(w(Q)) \leqq K\, M(Q)$, if $z_0 \notin \overline{Q}$. By Lemma 5.1 this condition is also satisfied if z_0 is a boundary point of Q. Finally, if z_0 is an interior point of Q, we divide the canonical rectangle $R' = R'(0, M(w(Q)), M(w(Q)) + i, i)$ of $w(Q)$ into two rectangles R_1' and R_2' by a vertical line going through the image point of w_0. The a-sides of R_1' and R_2' lie on those of R'. The preimages R_i in Q of the rectangles R_i' have z_0 as a boundary point, and so

$$M(R_i') \leqq K\, M(R_i)\,, \qquad i = 1, 2\,.$$

On the other hand, by Lemma 4.2, we have

$$M(R_1) + M(R_2) \leqq M(Q)\,, \qquad M(R_1') + M(R_2') = M(w(Q))\,,$$

so that $M(w(Q)) \leqq K\, M(Q)$ in this case, too. Thus the extension of w to $G \cup \{z_0\}$ is K-quasiconformal, and the theorem is proved.

Remark. From this theorem, which will be generalized in V.3.5–7, we conclude in particular: *Quasiconformal and conformal mappings divide simply connected domains into the same equivalence classes* (cf. 2.1).

8.2. Extension to free boundary arcs. As a second problem we investigate the boundary behaviour of quasiconformal mappings between domains with a free boundary arc. The following result contains the theorem on boundary correspondence mentioned in 2.2.

Theorem 8.2. *Let G and G' be two domains, C and C' free boundary arcs or curves of G and G', respectively, and $w : G \to G'$ a quasiconformal mapping. If C and C' correspond under the mapping w, then w can be extended to a homeomorphism of $G \cup C$ onto $G' \cup C'$.*

Proof: Let z_0 be an arbitrary point of C; we may assume that $z_0 \neq \infty$. We show first that the set of limit points of z_0 under the mapping w

consists of a single point $w_0 \in C'$, i.e., if $z_n \in G$, $z_n \to z_0$, then $w(z_n) \to w_0$. Otherwise there would exist two sequences z_{in}, $i = 1, 2$, $z_{in} \to z_0$, in G, such that $w(z_{in}) \to w_i$, $w_1 \neq w_2$. We derive a contradiction.

Since w_1 and w_2 lie on C', Lemma 1.6 tells us that there exist two Jordan arcs $b_i' \subset G'$, $i = 1, 2$, such that b_i' goes through the points $w(z_{in})$ and has w_i as endpoint. Since $w_1 \neq w_2$, we may assume that b_1' and b_2' have a positive mutual distance d. Let the preimages of b_i' in G be denoted by b_i (Fig. 3).

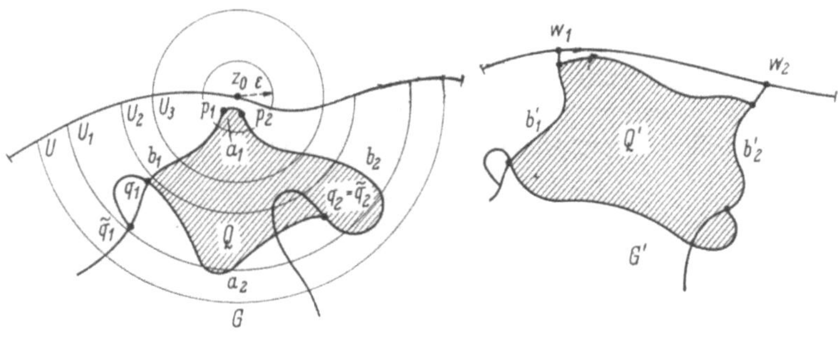

Fig. 3

Let U be such a small disc neighbourhood of z_0 that $U \cap \operatorname{Fr} G \subset C$ and $C - U \neq \emptyset$. We may assume without loss of generality that $w(G \cap U)$ lies in a disc of radius R.

By Lemma 1.5 there are discs $U_3 \subset U_2 \subset U_1 \subset U$, $z_0 \in U_3$, $b_i - \overline{U}_2 \neq \emptyset$, $i = 1, 2$, such that every two points of $G \cap (U_1 - \overline{U}_2)$ can be joined in $G \cap (U - \overline{U}_3)$. Let the distance between $- U_3$ and z_0 be denoted by δ.

Now let ε, $0 < \varepsilon < \delta$, be given. Since $z_{in} \to z_0$ and b_i contains the points z_{in}, Lemma 1.5 says that b_1 and b_2 can be joined in the set $G \cap \{z \mid |z - z_0| < \varepsilon\}$ by a closed Jordan arc a_1. We may assume that a_1 has only its endpoints p_1 and p_2 in common with the arcs b_1 and b_2.

Next we choose points $\tilde{q}_i \in b_i \cap (U_1 - \overline{U}_2)$ such that the subarcs $\tilde{b}_i$ of b_i, bounded by p_i and $\tilde{q}_i$, lie in U. By the above, $\tilde{q}_1$ and $\tilde{q}_2$ can be joined in $G \cap (U - \overline{U}_3)$ by a closed Jordan arc. By changing over to a subarc we further ensure that this arc a_2 has only its endpoints q_i in common with $\tilde{b}_i$.

Together with the subarcs $\widehat{p_i q_i}$ of b_i, a_1 and a_2 bound a Jordan domain Q lying in U. We show that Q lies in G. Otherwise there exist boundary points of G in Q, since we certainly have $G \cap Q \neq \emptyset$. On the other hand,

$U \cap \operatorname{Fr} G \subset C$, and so $Q \cap C \neq \emptyset$. This is impossible, since $C - \overline{Q} \supset C - U \neq \emptyset$ and $\operatorname{Fr} Q \cap C = \emptyset$. Thus $Q \subset G$.

Now let Q be thought of as a quadrilateral with a_1 and a_2 as its a-sides. By the above argument, the annulus $\varepsilon < |z - z_0| < \delta$ separates the a-sides of Q. From Lemma 6.5 we therefore conclude that

$$M(Q) \leqq \frac{2\pi}{\log \frac{\delta}{\varepsilon}} . \tag{8.1}$$

Thus Q may be constructed so that its module is arbitrarily small.

For the image quadrilateral Q' we have $m(Q') \leqq \pi R^2$, since $w(G \cap U)$ lies in a disc of radius R. As s_b is at least the distance d between b_1' and b_2' it follows from Rengel's inequality that

$$M(Q') \geqq \frac{d^2}{\pi R^2} .$$

Hence $M(Q')$ is greater than a fixed positive bound. This with (8.1) contradicts the quasiconformality of w.

The set of limit points of $z_0 \in C$ under the mapping w therefore consists of a single point $w_0 \in C'$. We now define $w(z_0) = w_0$, which extends w to C. It is clear that the inverse mapping w^{-1} can be extended in the same way to C' and that the extended mapping is the inverse of w.

To prove that the extended mapping w is a homeomorphism of $G \cup C$ onto $G' \cup C'$, we now have only to show that w is continuous at an arbitrary point $z_0 \in C$. Otherwise there exists a neighbourhood $U' = \{w \mid |w - w_0| < \varepsilon\}$ of $w(z_0)$ such that for no neighbourhood $U_n = \{z \mid |z - z_0| < 1/n\}$, $n = 1, 2, \ldots$, of z_0, does the image of $(G \cup C) \cap U_n$ lie in U'. Then there is also a point $z_n \in G \cap U_n$ such that $|w(z_n) - w(z_0)| > \varepsilon/2$, which gives a contradiction as $n \to \infty$.

8.3. Removability of an analytic arc. The maximal dilatation of a topological mapping w of the domain G was defined in 3.1 as

$$K(G) = \sup_{\overline{Q} \subset G} \frac{M(w(Q))}{M(Q)} .$$

We use the same definition in the case when G is any open set.

We shall now show that the maximal dilatation of a topological mapping of a domain G remains unchanged if a closed analytic arc $C \subset G$ is removed (Ahlfors [1]). This result will later turn out to be a special case of Theorem V.3.2. Nevertheless, we wish to prove it here, not only because it permits a proof methodologically connected with this chapter, but above all because it is needed in proving the reflection principle, which is often used in Chapter II.

Theorem 8.3. *Let w be a homeomorphism of a domain G and C a closed analytic arc which lies in G with the possible exception of its end points. For the maximal dilatation of w we then have*

$$K(G) = K\,(G - C)\,.$$

Proof: In any case we have $K(G) \geqq K\,(G - C)$, and we may therefore assume that $K\,(G - C) = K < \infty$. Let Q, $\overline{Q} \subset G$, be a quadrilateral whose canonical mapping has a conformal extension to a domain containing $\overline{Q}$. By continuity of the module it is sufficient to show that the dilatation of Q under w is at most K. Since every side of Q is a closed analytic arc, $\overline{Q} \cap C$ is a finite union of disjoint closed analytic arcs.

Using the canonical mappings we pass to the case when Q and its image Q' are rectangles. The problem can then be stated as follows: Let w be a homeomorphism of the rectangle $R\,(0, M, M + i, i)$ of the $z = x + i\,y$ — plane onto the rectangle $R'\,(0, M', M' + i, i)$. Let $C_1, \ldots, C_n$ be closed disjoint analytic arcs in $\overline{R}$ and let the maximal dilatation of w in $R - \bigcup C_i$ be at most K. We have to show that $M' \leqq K\,M$.

If a C_i is not a horizontal segment, it can be divided into a finite number of arcs which intersect any horizontal line at most once. Thus we can divide R by horizontal lines into rectangles $R_h(i\,y_{h-1}, M + i\,y_{h-1}, M + i\,y_h, i\,y_h)$ where $h = 1, \ldots, m$, and $y_0 = 0 < y_1 < \cdots < y_m = 1$, so that the following conditions are fulfilled: 1° For every h and i, the set $R_h \cap C_i$ is a finite union of arcs joining the horizontal sides of R_h. 2° For every h, the sum of the diameters of the components of $w\,(R_h \cap (\cup\, C_i))$ is smaller than a previously given $\varepsilon > 0$.

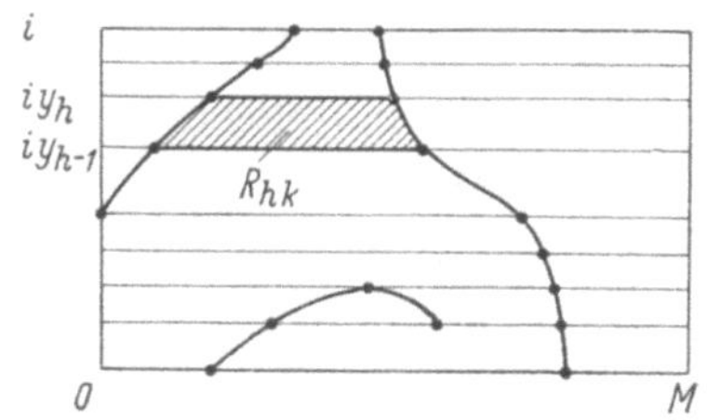

Fig. 4

The arcs C_i divide R_h into a finite number of (possibly degenerate) quadrilaterals R_{hk} (Fig. 4), whose horizontal sides we take as a-sides. Denote the image quadrilaterals of R_h and R_{hk} by R'_h and R'_{hk}. If s_{hk} is the distance between the b-sides of R'_{hk}, then it follows from condition 2° above that

$$\sum_k s_{hk} > M' - \varepsilon . \tag{8.2}$$

On the other hand, applying Rengel's inequality to R'_{hk} we obtain

$$M(R'_{hk}) \geqq \frac{s^2_{hk}}{m(R'_{hk})} .$$

Thus, by Schwarz's inequality, we deduce from (8.2) that

$$\sum_k M(R'_{hk}) \geqq \frac{\left(\sum_k s_{hk}\right)^2}{\sum_k m(R'_{hk})} > \frac{(M' - \varepsilon)^2}{m(R'_h)} . \tag{8.3}$$

By Lemma 5.1, $M(R'_{hk}) \leqq K\, M(R_{hk})$ since no points of the arcs $C_1, \ldots, C_n$ lie in the interior of R_{hk}. Thus in view of Lemma 4.2 on the superadditivity of the module we have

$$\sum_k M(R'_{hk}) \leqq K\, M(R_h) = \frac{K\, M}{y_h - y_{h-1}} .$$

Hence by (8.3)

$$\frac{y_h - y_{h-1}}{K\, M} < \frac{m(R'_h)}{(M' - \varepsilon)^2} .$$

Summing over h we obtain

$$\frac{1}{K\, M} < \frac{M'}{(M' - \varepsilon)^2} .$$

Since ε was arbitrary the required inequality $M' \leqq K\, M$ follows.

8.4. Reflection principle. We now generalize the reflection principle for conformal mappings to quasiconformal mappings. This principle refers to domains which have a circular arc as a free boundary arc. By circular arc we mean here either an open arc of a circle or a full circle or their images under a linear transformation.

We say that the domain G can be reflected in the circular arc C if G satisfies the following conditions:

1° C is a free boundary arc of G.

2° G has no common points with the domain G^* which is obtained from G by reflection in C.

If G permits a reflection in C, then $G \cup C \cup G^*$ is obviously a domain.

Let G and G' be two domains which permit reflections in circular arcs C and C', respectively, and suppose that $w : G \to G'$ is a K-quasiconformal mapping under which the boundary arcs C and C' correspond. Then w can be continued to $G \cup C \cup G^*$ as follows:

By Theorem 8.2 there exists a homeomorphic extension of w to $G \cup C$. In order to continue w to G^* we denote by z^* the reflection of $z \in G$ in C and by $w(z)^*$ the reflection of $w(z)$ in C'. If we set

$$w(z^*) = w(z)^* \,, \tag{8.4}$$

then the extended mapping $w : G \cup C \cup G^* \to G' \cup C' \cup G'^*$ is a sense-preserving homeomorphism. For every quadrilateral $Q^*(z_1^*, z_2^*, z_3^*, z_4^*)$ in G^* and its reflection Q in G we have

$$M(Q(z_1^*, z_2^*, z_3^*, z_4^*)) = M(Q(z_4, z_3, z_2, z_1)) \,,$$

and a corresponding relation holds between reflected quadrilaterals in G'^* and G'. The maximal dilatation of the continued mapping w is therefore at most K in $G \cup G^*$. By Theorem 8.3 the continuation is K-quasiconformal in $G \cup C \cup G^*$.

Reflection principle. *Let G and G' be two domains which can be reflected in circular arcs C and C', respectively, and let $w : G \to G'$ be a K-quasiconformal mapping under which the boundary arcs C and C' correspond. Then the continued mapping $w : G \cup C \cup G^* \to G' \cup C' \cup G'^*$ satisfying* (8.4) *is K-quasiconformal.*

§ 9. Local Characterization of Quasiconformality

9.1. Maximal dilatation at a point. The definition of a quasiconformal mapping takes account of all quadrilaterals and has in this sense a global nature. We now show that, analogously to the classical case of regular mappings, our general definition can be given in a local form.

Let w be a sense-preserving homeomorphism of the domain G. Let $U_z \subset G$ denote a neighbourhood of $z \in G$ and $K(U_z)$, as before, the maximal dilatation of w in U_z. The number

$$F_w(z) = F(z) = \inf_{U_z} K(U_z)$$

is called *the maximal dilatation of the mapping w at the point z.*

It follows immediately from this definition that $F(z)$ is a conformal invariant and that the maximal dilatation of the inverse mapping at the image point of z is also $F(z)$. The number $F(z)$ is finite if and only if the mapping is quasiconformal in a neighbourhood of z.

9.2. Local and global maximal dilatation. We shall show that there exist points z in G where $F(z)$ is arbitrarily near to the maximal dilatation $K(G)$. This result can also be interpreted to mean that the dilatation of small quadrilaterals is enough to determine $K(G)$.

Lemma 9.1. *The maximal dilatation of a sense-preserving homeomorphism of the domain G satisfies*

$$K(G) = \sup_{z \in G} F(z) . \tag{9.1}$$

Proof: By the definition of $F(z)$ we have $\sup_z F(z) \leqq K(G)$. It is therefore sufficient to show that the closure of any Jordan domain D, $\overline{D} \subset G$, contains a point z at which

$$F(z) \geqq K(D) . \tag{9.2}$$

Because of the conformal invariance of the module we may suppose further that D is a square.

Divide D into four congruent open squares. By Theorem 8.3, w has the maximal dilatation $K(D)$ in at least one of these squares. Denote such a square by D_1. Carrying on this division process, after n steps, $n = 2, 3, \ldots$, we obtain a square D_n, one of the four congruent subsquares of D_{n-1}, such that w has the maximal dilatation $K(D)$ in D_n.

The squares D_n converge to a point $z_0 \in \overline{D}$. The required inequality (9.2) is true for $z = z_0$, since every neighbourhood of z_0 contains squares D_n and $K(D_n) = K(D)$ holds for every n.

From Lemma 9.1 we immediately obtain the following local characterization of quasiconformality.

Theorem 9.1. *A sense-preserving homeomorphism of a domain G is K-quasiconformal if and only if*

$$F(z) \leqq K$$

at every point $z \in G$.

9.3. Semicontinuity of the local maximal dilatation. Let w be a sense-preserving homeomorphism of the domain G, z a point of G, and ε an arbitrary positive number. In the case $F(z) < \infty$ there exists a neighbourhood U_z of z such that $F(z) > K(U_z) - \varepsilon$. Then

$$F(z) > F(\zeta) - \varepsilon \tag{9.3}$$

for every $\zeta \in U_z$. This implies that F is upper semicontinuous.

In fact, a stronger result holds. By Theorem 8.1 we have $K(U_z) = K(U_z - \{z\})$. Hence by Lemma 9.1 there exists in every U_z a point

$\zeta \neq z$ such that $F(\zeta) > K(U_z) - \varepsilon$. A fortiori

$$F(z) < F(\zeta) + \varepsilon$$

for such a ζ. Together with (9.3) this yields the following result, which is obviously also true when $F(z) = \infty$.

Theorem 9.2. *A sense-preserving homeomorphism of a domain G satisfies*

$$F(z) = \limsup_{\zeta \to z} F(\zeta)$$

at every point $z \in G$.

9.4. Dilatation quotient at a regular point. Let G be a domain of the $z = x + i\,y$-plane, $w : G \to G'$ a sense-preserving homeomorphism and $z_0 \in G$ a finite point at which w is differentiable. Besides the partial derivatives w_x and w_y we shall also use the *complex derivatives*

$$w_z(z) = \frac{1}{2}\left(w_x(z) - i\,w_y(z)\right), \qquad w_{\bar z}(z) = \frac{1}{2}\left(w_x(z) + i\,w_y(z)\right)$$

since these often simplify our notation. The differentiability of w at z_0 then means that a representation

$$w(z) = w(z_0) + w_z(z_0)\,(z - z_0) + w_{\bar z}(z_0)\,(\bar z - \bar z_0) + o\,(z - z_0)$$

exists[14].

Expressed in terms of the complex derivatives the derivative of w in a direction α can be written in the form (cf. 3.3)

$$\partial_\alpha w(z_0) = w_z(z_0) + w_{\bar z}(z_0)\,e^{-2i\alpha},$$

and for the Jacobian we obtain

$$J(z_0) = |w_z(z_0)|^2 - |w_{\bar z}(z_0)|^2\,. \tag{9.4}$$

Since w is sense-preserving, $J(z_0) \geqq 0$ (cf. 1.6), and so $|w_z(z_0)| \geqq |w_{\bar z}(z_0)|$. Therefore

$$\max_\alpha |\partial_\alpha w(z_0)| = |w_z(z_0)| + |w_{\bar z}(z_0)|\,,$$

$$\min_\alpha |\partial_\alpha w(z_0)| = |w_z(z_0)| - |w_{\bar z}(z_0)|\,. \tag{9.5}$$

We now suppose that z_0 is a regular point of w, i.e. that w is differentiable at z_0 and $J(z_0) > 0$. By (9.4) and (9.5) we then have $\min_\alpha |\partial_\alpha w(z_0)| > 0$. The dilatation quotient

$$D(z) = \frac{\max_\alpha |\partial_\alpha w(z)|}{\min_\alpha |\partial_\alpha w(z)|} = \frac{|w_z(z)| + |w_{\bar z}(z)|}{|w_z(z)| - |w_{\bar z}(z)|}\,,$$

[14] Here as usual $o\,(z - z_0)$ denotes an expression which, when divided by $z - z_0$, tends to zero as $z \to z_0$. We shall also use the notation $O\,(z - z_0)$ for an expression which, when divided by $z - z_0$, is bounded in a neighbourhood of $z = z_0$.

introduced in 3.3 for regular quasiconformal mappings, can therefore be defined also for w at the regular point z_0. In the case $z_0 = \infty$ or $w(z_0) = \infty$, $D(z_0)$ can be defined by means of an auxiliary linear mapping (cf. 3.3).

9.5. General dilatation condition. Not only $F(z)$ but also $D(z)$ can be viewed as a local measure of dilatation at a regular point z. The following *general dilatation condition* gives information about the mutual relationship between these two quantities.

Theorem 9.3. *If the quasiconformal mapping w is differentiable at the point z_0, then*

$$\max_\alpha |\partial_\alpha w(z_0)| \leqq F(z_0) \min_\alpha |\partial_\alpha w(z_0)| . \tag{9.6}$$

Proof: Using translations and rotations we need consider only the situation in which $z_0 = 0$, $w(0) = 0$, $w_z(0) = |w_z(0)|$, and $w_{\bar{z}}(0) = |w_{\bar{z}}(0)|$. Then

$$w(z) = u(z) + i\, v(z) = |w_z(0)|\, z + |w_{\bar{z}}(0)|\, \bar{z} + o(z) . \tag{9.7}$$

We choose $\delta > 0$ so small that the closure of the square $R_\delta\big((-1-i)\,\delta,\ (1-i)\,\delta,\ (1+i)\,\delta,\ (-1+i)\,\delta\big)$ lies in the domain of w. It follows from (9.7) that the b-sides of the image R'_δ of R_δ are at a mutual distance

$$s_b = 2\,\delta\big(|w_z(0)| + |w_{\bar{z}}(0)|\big) + o(\delta)$$

and that R'_δ has the area

$$m(R'_\delta) = 4\,\delta^2\big(|w_z(0)|^2 - |w_{\bar{z}}(0)|^2\big) + o(\delta^2) .$$

By Rengel's inequality we have

$$M(R'_\delta) \geqq \frac{4\,\delta^2(|w_z(0)| + |w_{\bar{z}}(0)|)^2 + o(\delta^2)}{4\,\delta^2(|w_z(0)|^2 - |w_{\bar{z}}(0)|^2) + o(\delta^2)} . \tag{9.8}$$

On the other hand, since $M(R_\delta) = 1$, we have by Lemma 5.1

$$M(R'_\delta) = \frac{M(R'_\delta)}{M(R_\delta)} \leqq K(R_\delta) .$$

Thus by (9.8)

$$\big(|w_z(0)| + |w_{\bar{z}}(0)|\big)^2 + o(\delta)/\delta \leqq K(R_\delta)\ \big(|w_z(0)|^2 - |w_{\bar{z}}(0)|^2 + o(\delta)/\delta\big) .$$

The inequality (9.6) follows from this as $\delta \to 0$, in view of (9.5) and the limiting relation

$$\lim_{\delta \to 0} K(R_\delta) = F(0) .$$

Remark. We see immediately from the above proof that Theorem 9.3 can be modified as follows: *If the sense-preserving homeomorphism w*

of the domain G satisfies the dilatation condition $M(w(R)) \leqq K\, M(R)$ *for every square* R, $\overline{R} \subset G$, *then*

$$\max_{\alpha} |\partial_\alpha w(z)| \leqq K \min_{\alpha} |\partial_\alpha w(z)|$$

at every point $z \in G$ *at which* w *is differentiable.* It should be mentioned that every square must indeed be taken into consideration, although, because of the initial transformations, we have only considered squares whose sides are parallel to the coordinate axes.

9.6. Local maximal dilatation at a regular point. Let the homeomorphism w be differentiable at a point z and quasiconformal in a neighbourhood of z. Then $F(z) < \infty$, and in the case $\min_\alpha |\partial_\alpha w(z)| = 0$ it follows from (9.6) that the derivatives $\partial_\alpha w(z)$ vanish for every α. Expressed in terms of complex derivatives this means that $w_z(z)$ is always equal to zero if $|w_z(z)| = |w_{\bar{z}}(z)|$. From this we conclude:

If the quasiconformal mapping w *is differentiable at the point* z, *then the three conditions* $J(z) > 0$, $\min_\alpha |\partial_\alpha w(z)| > 0$ *and* $w_z(z) \neq 0$ *are equivalent.*

Theorem 9.3 therefore yields the following corollary:

A quasiconformal mapping satisfies

$$D(z) \leqq F(z) \tag{9.9}$$

at every regular point.

If w is a regular quasiconformal mapping of G, then

$$K(U_z) \leqq \sup_{\zeta \in U_z} D(\zeta)$$

for every neighbourhood $U_z \subset G$ of z, by Theorem 3.1. Since $D(\zeta)$ varies continuously with ζ, it follows that $F(z) \leqq D(z)$. In view of (9.9) we have proved the following extension of Theorem 3.1:

Theorem 9.4. *A regular quasiconformal mapping of the domain G satisfies* $D(z) = F(z)$ *for every* $z \in G$.

II. Distortion Theorems for Quasiconformal Mappings

Introduction to Chapter II

Methodologically Chapter II is closely connected with Chapter I. Here too the geometric definition of quasiconformality given in I.3.2 is used as a basis, and the characterization of modules by means of extremal length plays an important role.

Most of the questions to be considered lead back to the following *distortion problem,* which is therefore of crucial importance in the discussion:

Let $\mathcal{W}$ be a family of quasiconformal mappings w of the domain G, and z_1, z_2 two points in G. Find an upper bound for the maximal distance between the image points $w(z_1)$ and $w(z_2)$, $w \in \mathcal{W}$.

In the simplest case G is the unit disc $|z| < 1$, z_1 is the origin, and $\mathcal{W}$ comprises all K-quasiconformal mappings normalized by the condition $w(0) = 0$ and mapping the unit disc onto itself. The distortion problem is then equivalent to an extremal problem for ring domains. Its solution, *Grötzsch's extremal domain,* will be repeatedly utilized below.

In the first two sections we investigate Grötzsch's extremal domain and some other closely related ring domains with extremal modules. The results obtained will help us to deal with the above-mentioned distortion problem, which will be taken up in § 3. This problem is connected with the question of the uniform continuity of a family of K-quasiconformal mappings, discussed in § 4. The results achieved will allow us to prove important convergence theorems for quasiconformal mappings in § 5.

In § 6 we deal with the boundary value problem for quasiconformal mappings between Jordan domains. The derivation of a necessary condition is connected with the above distortion problem, while the existence of a solution will be proved by means of a direct construction. The boundary condition is analyzed in greater detail in § 7.

In close connection with the results of §§ 6—7, we shall investigate in § 8 conditions under which a quasiconformal mapping can be continued quasiconformally over the boundary of its domain.

Various characterizations will be given of boundary arcs or curves allowing such an extension.

Finally, in § 9 we define a local dilatation measure, the so-called circular dilatation, which coincides at regular points with the dilatation quotient defined previously. We investigate the relation between the circular dilatation and the local maximal dilatation by appealing to prior results in this chapter.

§ 1. Ring Domains with Extremal Module

1.1. Grötzsch's module theorem. In this section we deal with extremal problems of the following type: *Among all ring domains which separate two given closed sets E_1 and E_2, $E_1 \cap E_2 = \emptyset$, find one whose module has the greatest value.*

If E_1 and E_2 are connected, the extremal problem is not difficult to solve. In fact, one component of the complement of $E_1 \cup E_2$ is then doubly connected[15]. It follows from the monotonicity of the module (cf. the remark at the end of I.6.6) that this ring domain is extremal.

In the next simplest case E_1 is a continuum and E_2 consists of two points not separated by E_1. Because of the conformal invariance of the module we may then suppose that E_1 is the circle $|z| = 1$ and that the points of E_2 are $z = 0$ and $z = r$, $0 < r < 1$.

It seems almost evident that the extremal problem, normalized as above, has as solution the ring domain whose boundary consists of the circle $|z| = 1$ and the segment $0 \leqq x \leqq r$ of the real axis. This domain $B(r)$ is called *Grötzsch's extremal domain* (Fig. 5). Its module $M(B(r)) = \mu(r)$ will be often used later.

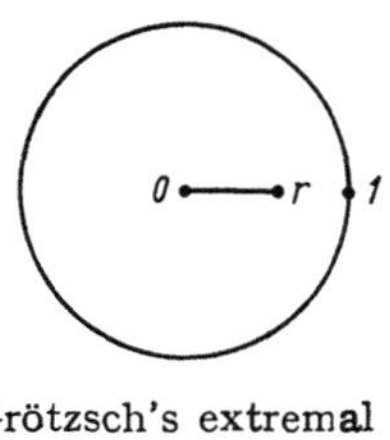

Fig. 5. Grötzsch's extremal domain

In the following theorem we show that $B(r)$ in fact possesses the required extremal property (Grötzsch [1]).

[15] This assertion will not be proved here, since we are only interested in special cases.

Grötzsch's module theorem. *If the ring domain B separates the points 0 and r from the circle $|z| = 1$, then*

$$M(B) \leqq \mu(r) .$$

Proof: We can construct a canonical mapping f of $B(r)$ onto the annulus $1 < |w| < e^{\mu(r)}$ by mapping the upper half of $B(r)$ conformally onto that of the annulus and then extending the map to the whole domain $B(r)$ by reflection in the real axis. From this construction we see that the function

$$\varrho = \left|\frac{f'}{f}\right|$$

is symmetric with respect to the real axis. Therefore, ϱ can be continued to a function[16] which is single-valued and continuous in the unit disc punctured at 0 and r.

We now utilize the characterization of the module given in I.6.2. Let $\mathscr{C}$ be the family of all analytic curves C which separate the two components of the complement of B. Setting the above function ϱ in formula (6.1) of I.6.2 we deduce that

$$M(B) \leqq \frac{4\pi^2 \mu(r)}{\left(\inf\limits_{C \in \mathscr{C}} l_\varrho(C)\right)^2}, \tag{1.1}$$

since $m_\varrho(B) \leqq m_\varrho(B(r)) = 2\pi\mu(r)$.

We now consider a curve $C \in \mathscr{C}$. Since C separates the points 0 and r from the circle $|z| = 1$, C can be divided into two subarcs C_1 and C_2 which both have one endpoint on each of the segments $-1 < x < 0$ and $r < x < 1$. If C_i lies in the closure of the upper half of the unit disc, then its ϱ-length is at least π. One can see this immediately by considering the image curve of C_i in the canonical image of $B(r)$. If C_i has points in the lower half of the unit disc, then, by the symmetry of ϱ, every subarc of C_i lying in the lower half-plane has the same ϱ-length as its reflection in the real axis. Hence $l_\varrho(C_i) \geqq \pi$ in this case also. Because $l_\varrho(C) = l_\varrho(C_1) + l_\varrho(C_2)$, the inequality $M(B) \leqq \mu(r)$ follows from (1.1).

In § 2 the properties of the function μ will be investigated in detail. Here we remark only that $\mu(r)$ decreases monotonically with increasing r; this follows from the monotonicity of the module (cf. I.6.6).

1.2. Teichmüller's extremal domain. We now consider our extremal problem in the case when both E_1 and E_2 have only two points a_1, b_1

[16] It should be mentioned that ϱ is independent of the choice of the canonical mapping. For, using the above f we can write all canonical mappings of $B(r)$ in the form cf or c/f, c being a constant.

and a_2, b_2, respectively. Among all ring domains separating a_1, b_1 from a_2, b_2 we have to find one with the greatest module.

It should at this point be remarked that for many applications a rather rough upper bound for the extremal module is sufficient. Such a bound is contained in Lemma I.6.1, on which the theory of normal families in § 5 is essentially based.

Here we shall solve the extremal problem in the case when all four points a_i, b_i lie on the same line or circle in the order a_1, b_1, a_2, b_2.[17] By a linear transformation these points can be carried into the points $-r_1, 0, r_2, \infty$ of the real axis. We call the plane, slit along the real axis from $-r_1$ to 0 and r_2 to ∞, *Teichmüller's extremal domain* (Fig. 6), and show that this domain is a solution of our extremal problem. This follows from the next theorem, where μ denotes the function introduced in 1.1.

Fig. 6- Teichmüller's extremal domain

Theorem 1.1. *If the ring domain B separates the points 0 and $-r_1 < 0$ from the points $r_2 > 0$ and ∞, then*

$$M(B) \leqq 2\mu\left(\sqrt{\frac{r_1}{r_1 + r_2}}\right). \tag{1.2}$$

Equality holds for Teichmüller's extremal domain.

Proof: We show first that equality holds in (1.2) for Teichmüller's extremal domain. To this end, note that the boundary components of Teichmüller's extremal domain are reflections of one another in the circle $|z + r_1| = \sqrt{r_1(r_1 + r_2)}$. This circle divides the domain into two subdomains which are conformally equivalent to the Grötzsch extremal domain $B\left(\sqrt{r_1/(r_1 + r_2)}\right)$. We map one of these subdomains conformally onto its canonical annulus. By a reflection the mapping can be extended to a canonical map f of Teichmüller's extremal domain. Therefore, by Lemma I.6.3, this domain has the module

$$2\,M\left(B\left(\sqrt{r_1/(r_1+r_2)}\right)\right) = 2\,\mu\left(\sqrt{r_1/(r_1 + r_2)}\right),$$

and the second half of the theorem is proved.

To complete the proof we notice first that the function $\varrho = |f'/f|$ is symmetric with respect to the real axis. It can therefore be extended to a function which is single-valued and continuous in the whole plane

[17] For the general case see Schiffer [1].

Ω with the exception of the points $-r_1, 0, r_2, \infty$. From this we deduce in the same way as in 1.1 that the ϱ-length of an analytic curve which separates $-r_1$, 0 from r_2, ∞ is at least 2π. For a ring domain B which separates these pairs of points we therefore obtain from Theorem I.6.1

$$M(B) \leqq \frac{1}{2\pi} m_\varrho(B) \leqq \frac{1}{2\pi} m_\varrho(\Omega) = \frac{1}{2\pi} \iint\limits_{\Omega} \left|\frac{f'}{f}\right|^2 d\sigma = 2\mu\left(\sqrt{\frac{r_1}{r_1 + r_2}}\right),$$

and Theorem 1.1 is proved.

1.3. Teichmüller's module theorem. Our extremal problem is more difficult, if the points a_i, b_i do not lie on the same line or circle. We prove here the following generalization of Theorem 1.1 (Teichmüller [1]), which in general does not give the best possible bound for $M(B)$.

Teichmüller's module theorem. *If the ring domain B separates the points 0 and z_1 from z_2 and ∞, then*

$$M(B) \leqq 2\mu\left(\sqrt{\frac{|z_1|}{|z_1| + |z_2|}}\right).$$

Proof: Let $(-B)_2$ denote the component of the complement of B which contains the points z_2 and ∞. The complement G of $(-B)_2$ is a simply connected domain (see I.1.7). If ζ is the conformal mapping of G onto the unit disc for which $\zeta(0) = 0$ and $\zeta(z_1) = \zeta_1 > 0$, then the function w,

$$w(z) = \frac{-4|z_2|\,\zeta(z)}{(1 - \zeta(z))^2}$$

maps G conformally onto the plane slit along the positive real axis from $|z_2|$ to ∞. The ring domain B is mapped by w onto a ring domain B' which separates the points $|z_2|$ and ∞ from $w(0) = 0$ and $w(z_1) = -4|z_2|\,\zeta_1/(1 - \zeta_1)^2 < 0$. Hence, by Theorem 1.1,

$$M(B) = M(B') \leqq 2\mu\left(\sqrt{\frac{-w(z_1)}{-w(z_1) + |z_2|}}\right). \tag{1.3}$$

Since μ is a decreasing function of r, we must now show that $-w(z_1) \geqq |z_1|$. For this purpose we appeal to two results from the theory of univalent conformal mappings. The *Koebe distortion formula* (see for example Behnke-Sommer [1], p. 394) applied to the inverse ζ^{-1} of ζ yields

$$|z_1| = |\zeta^{-1}(\zeta_1)| \leqq \frac{|(\zeta^{-1})'(0)|\,\zeta_1}{(1 - \zeta_1)^2}. \tag{1.4}$$

Further, by Koebe's "one-quarter theorem",

$$|(\zeta^{-1})'(0)| \leqq 4|z_2|. \tag{1.5}$$

From (1.4) and (1.5) we obtain the required inequality

$$- w(z_1) = \frac{4|z_2|\zeta_1}{(1-\zeta_1)^2} \geqq |z_1| ,$$

and Teichmüller's module theorem is proved.

1.4. Modification of Teichmüller's module theorem. From Teichmüller's module theorem we now deduce a module estimation which is suitable for some of our applications. To this end the ring domains considered will be doubled by means of a square root operation. The result takes a simpler form for ring domains which separate a pair of points z_1 and z_2 from 0 and ∞.

Theorem 1.2. *If the ring domain B separates z_1 and z_2 from 0 and ∞, then*

$$M(B) \leqq \mu\left(\frac{|\sqrt{z_1} - \sqrt{z_2}|}{\sqrt{2(|z_1| + |z_2|)}}\right), \tag{1.6}$$

where $\sqrt{z_1}$ and $\sqrt{z_2}$ belong to the same branch of the square root, single-valued in B.

Proof: Let $(-B)_1$ and $(-B)_2 \ni 0$ be the components of the complement of B and G the complement of $(-B)_2$. Since G is simply connected and does not contain the points 0 and ∞, there are two single valued analytic branches of the square root in G; one will be denoted by $\sqrt{\ }$ and the other by $-\sqrt{\ }$. The two functions give conformal maps of G onto non-intersecting domains. Let the ring domain B go onto B' and B'', the set $(-B)_1$ onto $(-B)_1'$ and $(-B)_1''$. The complement of the set $(-B)_1' \cup (-B)_1''$ is a ring domain which we denote by A.

Since B' and B'' separate the sets $(-B)_1'$ and $(-B)_1''$ it follows from Lemma I.6.3 that $M(B') + M(B'') \leqq M(A)$. Here $M(B') = M(B'') = M(B)$, and therefore

$$M(B) \leqq \frac{1}{2} M(A) . \tag{1.7}$$

The ring domain A separates the points $\sqrt{z_1}$ and $\sqrt{z_2}$ from $-\sqrt{z_1}$ and $-\sqrt{z_2}$. Its module can therefore be estimated with the help of Teichmüller's module theorem. In order to obtain the normalization of Teichmüller's theorem we transform A by means of the linear mapping

$$w(\zeta) = \frac{\zeta - \sqrt{z_1}}{\zeta + \sqrt{z_1}} .$$

This maps A onto a ring domain A' which separates the points 0 and $w_1 = \left(\sqrt{z_2} - \sqrt{z_1}\right)/\left(\sqrt{z_2} + \sqrt{z_1}\right)$ from the points ∞ and $w_2 = 1/w_1$.

Thus, by Teichmüller's module theorem,

$$M(A) = M(A') \leqq 2\,\mu\left(\sqrt{\frac{|w_1|}{|w_1| + |w_2|}}\right) = 2\,\mu\left(\frac{|\sqrt{z_1} - \sqrt{z_2}|}{\sqrt{2\,(|z_1| + |z_2|)}}\right), \quad (1.8)$$

and our assertion (1.6) follows from this and (1.7).

1.5. Mori's module theorem. Under the additional hypothesis that z_1 and z_2 lie in the closed disc $|z| \leqq 1$, Theorem 1.2 yields a result due to Mori [1].

In this case we have

$$\frac{|\sqrt{z_1} - \sqrt{z_2}|}{\sqrt{2\,(|z_1| + |z_2|)}} \geqq \frac{|\sqrt{z_1} - \sqrt{z_2}|}{2}. \quad (1.9)$$

To estimate the right hand side we multiply both sides of the relation

$$\left|\sqrt{z_1} - \sqrt{z_2}\right|^2 + \left|\sqrt{z_1} + \sqrt{z_2}\right|^2 = 2\,(|z_1| + |z_2|) \leqq 4$$

by $\left|\sqrt{z_1} - \sqrt{z_2}\right|^2$ and obtain

$$\left|\sqrt{z_1} - \sqrt{z_2}\right|^4 - 4\left|\sqrt{z_1} - \sqrt{z_2}\right|^2 + |z_1 - z_2|^2 \leqq 0\,.$$

This yields

$$\left|\sqrt{z_1} - \sqrt{z_2}\right| \geqq \sqrt{2 - \sqrt{4 - |z_1 - z_2|^2}}\,. \quad (1.10)$$

Since μ is decreasing it follows from Theorem 1.2 and the inequalities (1.9) and (1.10) that

$$M(B) \leqq \mu\left(\frac{1}{2}\sqrt{2 - \sqrt{4 - |z_1 - z_2|^2}}\right). \quad (1.11)$$

We now investigate when equality holds in (1.11). First we see that equality is true in (1.9) if $|z_1| = |z_2| = 1$. If this is so and $(-B)_1$ is the shorter arc of the circle $|z| = 1$ joining z_1 and z_2, then equality holds in (1.10). In this case A' is Teichmüller's extremal domain or can be carried into this by a rotation, and so (1.8) holds as an equality. Finally equality holds in (1.7) if $(-B)_2$ is a half-line lying symmetrically with respect to $(-B)_1$.

The ring domain whose boundary consists of the half-line $x \leqq 0$ and the circular arc $\{z \mid |z| = 1, |z - 1| \leqq |z_1 - 1|\}$, $|z_1 - 1| \leqq \sqrt{2}$, is called *Mori's extremal domain* (Fig. 7). It satisfies the above conditions if $z_2 = \bar{z}_1$. The relation (1.11) therefore becomes an equality if B is Mori's extremal domain and z_1 and z_2 coincide with the endpoints of $(-B)_1$.

To sum up we obtain

Mori's module theorem. *If the ring domain B separates the points 0 and ∞ from the points z_1 and z_2 of the closed unit disc, then*

$$M(B) \leqq \mu\left(\frac{1}{2}\sqrt{2-\sqrt{4-|z_1-z_2|^2}}\right).$$

Equality holds if B is Mori's extremal domain.

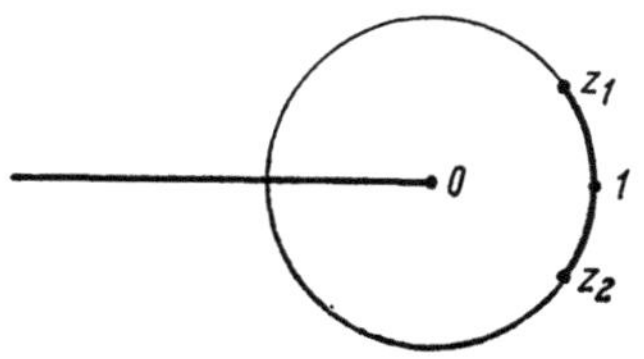

Fig. 7. Mori's extremal domain

§ 2. *Module of Grötzsch's Extremal Domain*

2.1. Representation by means of elliptic integrals. The modules of Teichmüller's and Mori's extremal domains can be simply expressed in terms of the module $\mu(r)$ of Grötzsch's extremal domain, i.e. the unit disc cut from 0 to r along the positive real axis. Since the function μ will also later play an important role, we shall study some of its properties more closely here[18].

For the sake of completeness we show first that $\mu(r)$ can be expressed in terms of elliptic integrals, although we shall not make use of this representation.

We remind the reader of the construction of the canonical mapping of $B(r)$. In 1.1 it was carried out by first mapping the upper half of the unit disc conformally onto the upper half of the annulus $1 < |w| < e^{\mu(r)}$. We now proceed in the same way but extend the mapping by reflection in $|z| = 1$ to a conformal mapping of the quadrilateral Q $(\infty, 0, r, 1/r)$ consisting of the upper half-plane. Since its image has module $2\,\mu(r)/\pi$, we obtain for $\mu(r)$ the expression

$$\mu(r) = \frac{\pi}{2} M(Q). \tag{2.1}$$

The function w,

$$w(z) = \frac{1}{2i\sqrt{r}} \int\limits_{\infty}^{z} \frac{dz}{\sqrt{z(z-r)(z-1/r)}}$$

[18] See Teichmüller [1], Hersch [1].

maps Q conformally onto the rectangle $R\left(0, K\left(\sqrt{1-r^2}\right), K\left(\sqrt{1-r^2}\right) + i\,K(r), i\,K(r)\right)$ where

$$K(r) = \int_0^1 \frac{dx}{\sqrt{(1-x^2)(1-r^2x^2)}}$$

is a Legendre normal integral. In view of (2.1) we thus obtain the representation

$$\mu(r) = \frac{\pi}{2} \frac{K\left(\sqrt{1-r^2}\right)}{K(r)}. \tag{2.2}$$

2.2. Functional equations. Proceeding from (2.2) we could derive various functional equations for $\mu(r)$. Another procedure will, however, enable us to obtain such results more directly.

The function w,

$$w(z) = \frac{-4z}{(1-z)^2},$$

maps Grötzsch's extremal domain $B(r)$ conformally onto the plane slit along the real axis from $-r_1 = -4r/(1-r)^2$ to 0 and from 1 to ∞. This domain is the Teichmüller extremal domain introduced in 1.2, and it has the module $2\mu\left(\sqrt{r_1/(1+r_1)}\right)$ by Theorem 1.1. On the other hand, the module is equal to $\mu(r)$, and we obtain the functional equation

$$\mu(r) = 2\mu\left(\frac{2\sqrt{r}}{1+r}\right). \tag{2.3}$$

This can be rewritten as

$$\mu(r) = \frac{1}{2}\mu\left(\left(\frac{1-\sqrt{1-r^2}}{r}\right)^2\right), \tag{2.4}$$

or, if we replace r by $\sqrt{1-r^2}$,

$$\mu\left(\sqrt{1-r^2}\right) = \frac{1}{2}\mu\left(\frac{1-r}{1+r}\right). \tag{2.5}$$

To obtain another important relation we consider the Teichmüller extremal domains B_1 and B_2 with boundary components $-r_1 \leqq x \leqq 0$, $x \geqq 1$ and $x \leqq -r_1$, $0 \leqq x \leqq 1$, respectively. By Theorem 1.1 we have

$$M(B_1) = 2\mu\left(\sqrt{\frac{r_1}{1+r_1}}\right), \qquad M(B_2) = 2\mu\left(\sqrt{\frac{1}{1+r_1}}\right).$$

The canonical mappings of both B_1 and B_2 can be chosen such that the upper half-plane goes into the upper half of the canonical annulus.

We thus conclude as in 2.1 that the quadrilaterals $Q_1 = Q(\infty, -r_1, 0, 1)$ and $Q_2 = Q(-r_1, 0, 1, \infty)$, formed from the upper half-plane, have modules $M(Q_1) = M(B_1)/\pi$ and $M(Q_2) = M(B_2)/\pi$. By the definition of the module, $M(Q_1)\, M(Q_2) = 1$, and so we obtain the equation

$$4\,\mu\left(\sqrt{\frac{r_1}{1+r_1}}\right)\mu\left(\sqrt{\frac{1}{1+r_1}}\right) = M(B_1)\,M(B_2) = \pi^2\,. \tag{2.6}$$

If we set $r = 1/\sqrt{1+r_1}$, then $\sqrt{r_1/(1+r_1)} = \sqrt{1-r^2}$, and we deduce from (2.6) that

$$\mu(r)\,\mu\left(\sqrt{1-r^2}\right) = \frac{\pi^2}{4}\,. \tag{2.7}$$

In particular, for $r = 1/\sqrt{2}$ this yields

$$\mu\left(\frac{1}{\sqrt{2}}\right) = \frac{\pi}{2}\,. \tag{2.8}$$

It follows from (2.5) and (2.7) that

$$\mu(r)\,\mu\left(\frac{1-r}{1+r}\right) = \frac{\pi^2}{2}\,. \tag{2.9}$$

Using equation (2.7) we can prove that μ is continuous in the interval $0 < r < 1$. By Lemma I.6.4, $\mu(t) \to \mu(r)$ when t decreases to r, i.e. μ is continuous on the right. Then $\tilde{\mu}$, $\tilde{\mu}(r) = \mu\left(\sqrt{1-r^2}\right)$, is continuous on the left, and so μ is continuous by (2.7).

2.3. Asymptotic behaviour. To investigate the behaviour of $\mu(r)$ for $r \to 0$ we map Grötzsch's extremal domain $B(r)$ conformally onto a ring domain B' of the $w = u + i\,v$-plane whose boundary consists of the segment $-1 \leqq u \leqq 1$ and the circle $|w| = R$. A simple calculation shows that $R = \left(1 + \sqrt{1-r^2}\right)/r$.

Now the function w, $w(\zeta) = (\zeta + 1/\zeta)/2$, maps the annulus $1 < |\zeta| < \varrho$ conformally onto an ellipse E_ϱ, with axes $\varrho \pm 1/\varrho$, slit along $-1 \leqq u \leqq 1$. If $\varrho + 1/\varrho \leqq 2R$, then E_ϱ lies in B'. It then follows from the monotonicity of the module (see I.6.6) that $\log \varrho = M(E_\varrho) < M(B') = \mu(r)$. For $\varrho - 1/\varrho \geqq 2R$ we have $B' \subset E_\varrho$ and $\log \varrho > \mu(r)$.

Since $1/(2R - r) < r$ the condition $\varrho + 1/\varrho \leqq 2R$ is satisfied for $\varrho = 2R - r = \left(1 + \sqrt{1-r^2}\right)^2/r$. The latter condition $\varrho - 1/\varrho \geqq 2R$ holds for $\varrho = 4/r$. Thus we obtain for $\mu(r)$ the estimation

$$\log\frac{(1+\sqrt{1-r^2})^2}{r} < \mu(r) < \log\frac{4}{r}\,. \tag{2.10}$$

It follows in particular that

$$\lim_{r \to 0} \left(\mu(r) - \log \frac{4}{r} \right) = 0 . \tag{2.11}$$

2.4. Successive approximations. By combining estimation (2.10) with the functional equation (2.4) the value of $\mu(r)$ can be determined in successive approximations, as follows:

We set

$$r_0 = r , \qquad r_{n+1} = \left(\frac{1 - \sqrt{1 - r_n^2}}{r_n} \right)^2 , \qquad n = 0, 1, 2, \ldots .$$

By (2.4) we then have $\mu(r_n) = 2^n \mu(r)$, and (2.10) yields

$$2^{-n} \log \frac{(1 + \sqrt{1 - r_n^2})^2}{r_n} < \mu(r) < 2^{-n} \log \frac{4}{r_n} \tag{2.12}$$

for every n. Now $r_n \to 0$ as $n \to \infty$. By (2.12) we obtain therefore the limiting relation

$$\mu(r) = \lim_{n \to \infty} 2^{-n} \log \frac{4}{r_n} .$$

Because of a later application we use (2.12) for $n = 1$. We first obtain

$$\log \frac{\left(1 + \sqrt[4]{1 - r^2}\right)^2}{r} < \mu(r) < \log \frac{2\,(1 + \sqrt{1 - r^2})}{r} .$$

From

$$2\left(1 + \sqrt{1 - r^2}\right) - \left(1 + \sqrt[4]{1 - r^2}\right)^2 = \left(1 - \sqrt[4]{1 - r^2}\right)^2 < r^4$$

we deduce that

$$\mu(r) = \log \left(\frac{2\,(1 + \sqrt{1 - r^2})}{r} - \delta(r) \right) , \tag{2.13}$$

where $0 < \delta(r) < r^3$ for $0 < r < 1$. Since

$$\frac{2\,(1 + \sqrt{1 - r^2})}{r} = \frac{4}{r} - r - \frac{r^3}{(1 + \sqrt{1 - r^2})^2} ,$$

we can also write (2.13) in the form

$$\mu(r) = \log \left(\frac{4}{r} - r - \delta_1(r) \right) , \tag{2.14}$$

where $r^3/4 < \delta_1(r) < 2\,r^3$.

Finally it should be mentioned that by combining (2.7) with (2.10—14) one finds estimations for $\mu(r)$ near $r = 1$. In particular, it follows that μ has the limit $\mu(1) = 0$. By successive applications of the functional equation (2.4) the estimations can be improved to an arbitrary degree.

§ 3. Distortion under a Bounded Quasiconformal Mapping of a Disc

3.1. Alteration of the distance from the origin. Using the module estimations of § 1 we can now take up the distortion problem mentioned in the introduction to Chapter II. In this section we restrict ourselves to quasiconformal mappings of the unit disc into itself. We begin with the following problem:

Let w be a K-quasiconformal mapping of the unit disc into itself, and let it be normalized by $w(0) = 0$. We want to find an upper bound for $|w(z)|$.

The problem can be transformed into Grötzsch's module theorem if $|z| < 1$ is slit radially from 0 to z. The module of the ring domain so formed is $\mu(|z|)$ while its image has a module $\leqq \mu(|w(z)|)$ by Grötzsch's module theorem. Consequently, by Theorem I.7.1, we obtain

$$\mu(|z|) \leqq K\,\mu(|w(z)|)\,.$$

If we write

$$\varphi_K(r) = \mu^{-1}(\mu(r)/K)\,, \tag{3.1}$$

then

$$|w(z)| \leqq \varphi_K(|z|)\,. \tag{3.2}$$

To construct a mapping for which equality holds in (3.2) at a point $z = r > 0$ we map Grötzsch's extremal domain $B(r)$ conformally onto the annulus $e^{-\mu(r)} < |\zeta| < 1$ by a function f_1 which is symmetric about the real axis and makes the boundary components $|z| = 1$ and $|\zeta| = 1$ correspond (cf. 1.1). The function f_2,

$$f_2(\zeta) = \zeta|\zeta|^{\frac{1}{K}-1}\,, \tag{3.3}$$

maps this annulus homeomorphically onto the annulus $e^{-\mu(r)/K} < |t| < 1$. This mapping is sense-preserving and, as can be easily calculated, has the constant dilatation quotient $D(\zeta) = K$. Thus by Theorem I.3.1, f_2 is K-quasiconformal. By the definition of φ_K, the annulus $e^{-\mu(r)/K} < |t| < 1$ can be conformally mapped onto the extremal domain $B(\varphi_K(r))$. We normalize this mapping f_3 so that it is symmetric with respect to the real axis and the unit circles correspond to each other. The composition $w = f_3 \circ f_2 \circ f_1$ can be extended to a homeomorphism of the disc $|z| < 1$ onto the disc $|w| < 1$, which leaves the origin invariant and carries the point r into $\varphi_K(r)$. By Theorem I.8.3 the mapping w is K-quasiconformal in the whole unit disc. Therefore w is the required extremal mapping for which (3.2) holds as an equality at the point $z = r$. Hence we have derived the following distortion theorem (Hersch-Pfluger [1]).

Theorem 3.1. *Let w be a K-quasiconformal mapping of the unit disc $|z| < 1$ onto a subdomain G' of $|w| < 1$, normalized by $w(0) = 0$. Then*

$$|w(z)| \leqq \varphi_K(|z|) .$$

The bound can be attained for every given z and K, $|z| < 1$, $K \geqq 1$, if G' is the whole disc $|w| < 1$.

If w is a K-quasiconformal mapping of the unit disc onto itself then the above theorem also holds for the inverse of w. Thus we have $|z| \leqq \varphi_K(|w(z)|)$. Since $\varphi_{1/K}$ is the inverse of φ_K, Theorem 3.1 can be complemented as follows:

Every K-quasiconformal mapping w of the unit disc onto itself which leaves the origin invariant satisfies the double inequality

$$\varphi_{1/K}(|z|) \leqq |w(z)| \leqq \varphi_K(|z|) . \tag{3.4}$$

Both bounds can be attained.

We remark further that Theorem 3.1 can obviously be generalized to the case when w, $w(0) = 0$, is a K-quasiconformal mapping of $|z| < r$ into $|w| < R$. Then we have

$$|w(z)| \leqq R\, \varphi_K(|z|/r) . \tag{3.5}$$

3.2. The distortion function φ_K. It follows immediately from the properties of the function μ that, for each fixed $K > 0$, φ_K is a strictly increasing continuous function of r in the interval $0 < r < 1$ and has the limits $\varphi_K(0) = 0$, $\varphi_K(1) = 1$. For $K \geqq 1$ we always have $\varphi_K(r) \geqq r$, while $\varphi_K(r) \leqq r$ for $K \leqq 1$. Further, by the definition of φ_K,

$$\varphi_{K_1 K_2} = \varphi_{K_1} \circ \varphi_{K_2} .$$

It follows from the functional equation (2.3) that

$$\varphi_2(r) = \frac{2\sqrt{r}}{1 + r} .$$

Iteration of this equality yields an expression for φ_{2^n}.

An upper estimate for $\varphi_K(r)$ is obtained as follows: For $0 < r < r' < 1$, consider the annulus $r/r' < |z| < 1$ and the ring domain bounded by the circle $|z| = r/r'$ and the segment $0 \leqq x \leqq r$. From the superadditivity of the module (Lemma I.6.3) it follows that

$$\log \frac{r'}{r} + \mu(r') \leqq \mu(r) .$$

Hence, $\log (1/r) - \mu(r)$ is increasing with r. Since $\log (4/r) - \mu(r)$ is positive by (2.10), we conclude that for every $K > 1$,

$$\frac{1}{K}\left(\log \frac{4}{r} - \mu(r)\right) < \log \frac{4}{r'} - \mu(r') .$$

Setting $r' = \varphi_K(r)$, we have $\mu(r)/K = \mu(r')$. Consequently,

$$\varphi_K(r) < 4^{1-1/K} r^{1/K} . \tag{3.6}$$

In order to study the behaviour of $\varphi_K(r)$ as $r \to 0$, we first deduce from $\mu(\varphi_K(r)) = \mu(r)/K$ and (2.14) that

$$\frac{\varphi_K(r)}{4 - \varphi_K^2(r) - \varphi_K(r)\,\delta_1(\varphi_K(r))} = \left(\frac{r}{4 - r^2 - r\,\delta_1(r)}\right)^{1/K} . \tag{3.7}$$

Because $\delta_1(\varphi_K(r)) < 2\,\varphi_K(r)$, we have

$$4 - \varphi_K^2(r) - \varphi_K(r)\,\delta_1(\varphi_K(r)) > 4 - 3\,\varphi_K^2(r) ,$$

and so by (3.7)

$$\varphi_K(r) > (r/4)^{1/K}\,(4 - 3\,\varphi_K^2(r)) .$$

Considering (3.6) we thus obtain the lower estimate

$$\varphi_K(r) > 4^{1-1/K} r^{1/K}\,(1 - 12\,r^{2/K}) . \tag{3.8}$$

In conjunction with (3.6) this shows that

$$\lim_{r \to 0} r^{-1/K} \varphi_K(r) = 4^{1-1/K} .$$

3.3. Distortion of hyperbolic distances in the disc. The above Theorem 3.1 deals with the situation $w(0) = 0$. We now consider a more general case in which w is an arbitrary K-quasiconformal mapping of the unit disc $|z| < 1$ onto a subdomain of the disc $|w| < 1$.

In order to estimate the distance between the images of two given points z_1, z_2 we first carry z_2 and $w(z_2)$ onto the origin by the linear mappings f_1 and f_2,

$$f_1(z) = \frac{z - z_2}{1 - \bar{z}_2 z} , \qquad f_2(w) = \frac{w - w(z_2)}{1 - \overline{w(z_2)}\, w} .$$

The composition $f = f_2 \circ w \circ f_1^{-1}$ then maps the unit disc onto its subdomain with the origin left invariant. Theorem 3.1 thus yields the following result:

Let w be a K-quasiconformal mapping of the unit disc $|z| < 1$ onto a subdomain of the disc $|w| < 1$. For every pair of points z_1, z_2 in $|z| < 1$ we then have

$$\left|\frac{w(z_1) - w(z_2)}{1 - \overline{w(z_2)}\, w(z_1)}\right| \leqq \varphi_K\left(\left|\frac{z_1 - z_2}{1 - \bar{z}_2 z_1}\right|\right) . \tag{3.9}$$

For a K-quasiconformal mapping w of $|z| < r$ into $|w| < R$ we have

$$R\left|\frac{w(z_1) - w(z_2)}{R^2 - \overline{w(z_2)}\, w(z_1)}\right| \leqq \varphi_K\left(r\left|\frac{z_1 - z_2}{r^2 - \bar{z}_2 z_1}\right|\right), \tag{3.10}$$

if $|z_1|, |z_2| < r$.

The inequalities (3.9) and (3.10) can be interpreted as statements about the distortion of the *hyperbolic distance.* The element of length of the hyperbolic metric in the disc $|z| < r$ is defined by

$$dh_r = \frac{r\,|dz|}{r^2 - |z|^2}\,. \tag{3.11}$$

The geodesics of this metric are circles orthogonal to $|z| = r$. Since the length element (3.11) is invariant under conformal mappings of the disc $|z| < r$ onto itself we may calculate the hyperbolic distance $h_r(z_1, z_2)$ of z_1 and z_2 by mapping one of these points into the origin. We then obtain

$$h_r(z_1, z_2) = \frac{1}{2}\log\frac{|r^2 - \bar{z}_2 z_1| + r\,|z_1 - z_2|}{|r^2 - \bar{z}_2 z_1| - r\,|z_1 - z_2|} = \operatorname{ar\,tanh}\left|\frac{r(z_1 - z_2)}{r^2 - \bar{z}_2 z_1}\right|. \tag{3.12}$$

The distortion formula (3.10) can thus be expressed in terms of the hyperbolic distance as

$$\tanh h_R(w(z_1), w(z_2)) \leqq \varphi_K(\tanh h_r(z_1, z_2))\,. \tag{3.13}$$

It follows from the discussion of 3.1 that this inequality is sharp.

3.4. Distortion of euclidean distance under normalized mappings. From the distortion formulae (3.9) and (3.10) we obtain an estimation for the euclidean distance $|w(z_1) - w(z_2)|$ in terms of $|z_1 - z_2|$ only in the case when z_1 and z_2 lie in a given compact subset of the domain in question. In the general case, when no special hypotheses about the quasiconformal mapping w are made, there are in fact no distortion results of the type $|w(z_1) - w(z_2)| < \varepsilon(|z_1 - z_2|)$, $\varepsilon(r) \to 0$ as $r \to 0$, which are valid in the neighbourhood of the boundary of the domain.

However, in special cases bounds for the distortion of the euclidean length can be found. This is possible for example in the case of normalized K-quasiconformal mappings of the unit disc onto a given bounded Jordan domain. We shall consider only self-mappings of the unit disc; for these we have the following distortion theorem, which in this formulation is due to Mori [1] (see also Lavrentieff [1], Ahlfors [1]).

Theorem 3.2. *Let w be a K-quasiconformal mapping of the unit disc onto itself, normalized by $w(0) = 0$. Then for every pair of points z_1, z_2 with $|z_1| \leqq 1$, $|z_2| \leqq 1$,*

$$|w(z_1) - w(z_2)| \leqq 16\,|z_1 - z_2|^{1/K}\,. \tag{3.14}$$

The number 16 cannot be replaced by any smaller bound if the inequality is to hold for all K.

Proof: Since $|w(z_1) - w(z_2)| \leqq 2$, (3.14) holds for $|z_1 - z_2| \geqq 1/8$. We may therefore suppose that $|z_1 - z_2| < 1/8$. The proof falls into two parts according as $|z_1 + z_2| \leqq 1$ or $|z_1 + z_2| > 1$.

In the former case $|1 - \bar{z}_2 z_1| > 1/2$. Since $|1 - \overline{w(z_2)}\, w(z_1)| \leqq 2$ we deduce from (3.9) that

$$|w(z_1) - w(z_2)| \leqq 2\,\varphi_K\,(2\,|z_1 - z_2|)\,.$$

The inequality (3.14) follows from this since $\varphi_K(r) \leqq 4\, r^{1/K}$ by (3.6).

In the case $|z_1 + z_2| > 1$ we first continue w to the whole plane by reflection. Consider the annulus $B = \{z|\; |z_1 - z_2|/2 < |z - (z_1 + z_2)/2| < 1/2\}$, which has the module

$$M(B) = \log \frac{1}{|z_1 - z_2|}\,. \tag{3.15}$$

The points 0 and ∞ belong to the same component of the complement of B and since w leaves both points invariant the image domain B' of B separates the points $w_1 = w(z_1)$ and $w_2 = w(z_2)$ from 0 and ∞. Hence by Theorem 1.2 we have

$$M(B') \leqq \mu\left(\frac{|\sqrt{w_1} - \sqrt{w_2}|}{\sqrt{2\,(|w_1| + |w_2|)}}\right) \leqq \mu\left(\frac{|w_1 - w_2|}{4}\right).$$

By (2.10), $\mu(r) < \log\,(4/r)$, and so

$$M(B') \leqq \log \frac{16}{|w_1 - w_2|}\,. \tag{3.16}$$

On the other hand, $M(B') \geqq M(B)/K$, and so (3.14) follows from (3.15) and (3.16).

We still have to prove that the number 16 in (3.14) cannot be replaced by any smaller number independent of K. For this purpose we consider the quadrilateral Q_α, consisting of the semidisc $\{z|\; |z| < 1,\ \operatorname{Im} z > 0\}$ and the vertices $0, 1, e^{i\alpha}, -1$. To calculate the module we map Q_α conformally by the function ζ, $\zeta(z) = (1 - z)^2/(1 + z)^2$, onto the quadrilateral Q_α' consisting of the lower half-plane and the vertices $1, 0, -\tan^2(\alpha/2), \infty$. This quadrilateral is the lower half of the Teichmüller extremal domain introduced in 1.2. Hence its module is

$$M(Q_\alpha) = M(Q_\alpha') = \frac{2}{\pi}\,\mu\left(\sqrt{\frac{\tan^2(\alpha/2)}{1 + \tan^2(\alpha/2)}}\right) = \frac{2}{\pi}\,\mu\left(\sin\frac{\alpha}{2}\right).$$

If the numbers α and β, $0 < \alpha < \pi$, $0 < \beta < \pi$, satisfy the equation

$$\mu\left(\sin\frac{\alpha}{2}\right) = K\,\mu\left(\sin\frac{\beta}{2}\right), \tag{3.17}$$

then by the remark at the end of I.3.4 we can find a K-quasiconformal mapping w of the upper half of the unit disc onto itself which has the boundary values $w(-1) = -1$, $w(0) = 0$, $w(1) = 1$, and $w(e^{i\alpha}) = e^{i\beta}$. By a reflection w can be extended to a self-mapping of the whole unit disc which preserves the origin and carries the points $z_1 = e^{i\alpha}$, $z_2 = e^{-i\alpha}$, into $w_1 = e^{i\beta}$, $w_2 = e^{-i\beta}$, respectively.

In order to compare the distances $|z_1 - z_2| = 2\sin\alpha$ and $|w_1 - w_2| = 2\sin\beta$ we write the equation (3.17) in the form $\sin(\beta/2) = \varphi_K(\sin(\alpha/2))$. It then follows from (3.6) and (3.8) that

$$|w_1 - w_2| = 16^{1-1/K}\,|z_1 - z_2|^{1/K}\left(1 + \varepsilon\,(|z_1 - z_2|)\right),$$

where $\varepsilon\,(|z_1 - z_2|) \to 0$ as $|z_1 - z_2| \to 0$. It follows that (3.14) cannot in general be true with a constant smaller than 16 and independent of K.

Remark. In the special case where z_1 and z_2 lie on the boundary of the unit disc the constant 16 in (3.14) can be replaced by $16^{1-1/K}$. In order to prove this we write $|z_1 - z_2| = r$, $|w_1 - w_2| = \varrho$, $\frac{1}{2}\sqrt{2 - \sqrt{4 - r^2}} = r'$, $\frac{1}{2}\sqrt{2 - \sqrt{4 - \varrho^2}} = \varrho'$. From Mori's module theorem and the estimation (3.6) it follows that

$$\varrho' \leqq \varphi_K(r') \leqq 4^{1-1/K}\,(r')^{1/K}.$$

We may assume that $\varrho \geqq r$, which implies that $\varrho' \geqq r'$. Hence,

$$\varrho = 4\,\varrho'\sqrt{1 - \varrho'^2} \leqq 4\,\varrho'\,(1 - r'^2)^{1/2K} \leqq 16^{1-1/K}\left[4\,r'\sqrt{1 - r'^2}\right]^{1/K},$$

which is the desired inequality

$$|w(z_1) - w(z_2)| \leqq 16^{1-1/K}\,|z_1 - z_2|^{1/K}. \tag{3.18}$$

It seems plausible that $|w(z_1) - w(z_2)|/|z_1 - z_2|^{1/K}$ will be maximized for our class of mappings when z_1 and z_2 lie on the boundary. Then (3.14) could be replaced by (3.18). This conjecture has in fact been made in the literature; we have, however, seen no proof for it.

§ 4. *Order of Continuity of Quasiconformal Mappings*

4.1. Equicontinuity. We now investigate the distortion of quasiconformal mappings of an arbitrary domain. Our aim is to derive local results, that is, to estimate the change of small distances.

Our first result will be of qualitative type; it concerns the *equicontinuity* of a family of quasiconformal mappings. This concept will be defined here with respect to the spherical metric k as follows:

A family $\mathcal{W}$ of mappings of a domain G into the plane is called *equicontinuous at the point* $z_0 \in G$ if to every $\varepsilon > 0$ there corresponds a neighbourhood U of z_0 such that

$$\sup_{w \in \mathcal{W},\, z \in U} k\big(w(z), w(z_0)\big) < \varepsilon .$$

The family $\mathcal{W}$ is *equicontinuous in a set* $E \subset G$ if it is equicontinuous at every point of E.

It is possible to give various criteria for the equicontinuity of a family of quasiconformal mappings. The following theorem is most suitable for our purposes.

Theorem 4.1. *Let $\mathcal{W}$ be a family of K-quasiconformal mappings of the domain G. If every mapping $w \in \mathcal{W}$ omits two values whose spherical distance is greater than a fixed positive number d, then $\mathcal{W}$ is equicontinuous in G.*

Proof: Given $z_0 \in G$ and ε, $0 < \varepsilon < d$, let a positive number r be chosen so that the disc $k(z, z_0) < r$ lies in G. Then choose $\delta < r$ such that the annulus $B_\delta = \{z | \delta < k(z, z_0) < r\}$ has the module

$$M(B_\delta) > K\pi^2/2\,\varepsilon^2 . \tag{4.1}$$

We denote the disc $k(z, z_0) < \delta$ by U and prove that $k\big(w(z_1), w(z_0)\big) < \varepsilon$ for every $z_1 \in U$, $w \in \mathcal{W}$.

To this end we derive an upper bound for the module of the ring domain $w(B_\delta) = B'_\delta$. By hypothesis, w omits two values a and b with $k(a, b) > d$. It follows that B'_δ separates a and b from $w(z_1)$ and $w(z_2)$. Application of Lemma I.6.1 then gives

$$M(B'_\delta) \leqq \pi^2/2\,\eta^2 , \tag{4.2}$$

where $\eta = \min\big(k(a, b), k(w(z_1), w(z_0))\big) \geqq \min\big(d, k(w(z_1), w(z_0))\big)$.

Since w is K-quasiconformal, $M(B'_\delta) \geqq M(B_\delta)/K$, and therefore by (4.1)

$$M(B'_\delta) > \pi^2/2\,\varepsilon^2 .$$

From (4.2) we see that $\eta < \varepsilon$. Since $\varepsilon < d$, we must have $\eta = k\big(w(z_1), w(z_0)\big)$, and the theorem is proved.

The following result can be immediately deduced from the above.

Theorem 4.2. *A family $\mathcal{W}$ of K-quasiconformal mappings of the domain G is equicontinuous in G if there exists a number $d > 0$ such that one of the following conditions is satisfied:*

1° *Every mapping $w \in \mathcal{W}$ omits one value a and for two fixed points z_1, $z_2 \in G$, the distances $k(w(z_i), a)$, $i = 1, 2$, are greater than d.*

2° *For three fixed points $z_1, z_2, z_3 \in G$ the distances $k(w(z_i), w(z_j))$, $i, j = 1, 2, 3$, $i \neq j$, are greater than d for every mapping $w \in \mathcal{W}$.*

Proof: By Theorem 4.1, $\mathcal{W}$ is equicontinuous in case 1° in the domains $G - \{z_i\}$, $i = 1, 2$; in case 2° in the domains $G - \{z_i, z_j\}$, $i, j = 1, 2, 3$, $i \neq j$. In both cases the domains cover G, and Theorem 4.2 follows.

If there exist two points z_1, $z_2 \in G$ such that $k(w(z_1), w(z_2)) > d > 0$ for every mapping $w \in \mathcal{W}$, then $\mathcal{W}$ is equicontinuous at every point of G other than z_1 and z_2. This need not be true at z_1 and z_2. The family of mappings w_n, $w_n(z) = 2^n z$, $n = 0, \pm 1, \ldots$, of the whole plane is an example, since it is not equicontinuous at $z = 0$ and $z = \infty$.

4.2. Hölder continuity. The family $\mathcal{W}$ of mappings of the domain G is called *Hölder continuous* with exponent $\alpha > 0$ at a point $z_0 \in G$ if the *Hölder condition*

$$k(w(z), w(z_0)) \leqq C(k(z, z_0))^\alpha \tag{4.3}$$

is satisfied with a fixed C for all $w \in \mathcal{W}$ and $z \in G$. If E is a subset of G and (4.3) holds with fixed α and C for all $w \in \mathcal{W}$, $z \in G$ and $z_0 \in E$, then $\mathcal{W}$ is called *uniformly*[19] *Hölder continuous* with exponent α in E.

If $\mathcal{W}$ is Hölder continuous at a point then it is also equicontinuous there. If $\mathcal{W}$ consists of K-quasiconformal mappings then the converse also holds: Hölder continuity follows from equicontinuity. In fact, the following sharper theorem is true:

Theorem 4.3. *A family $\mathcal{W}$ of K-quasiconformal mappings which is equicontinuous in a domain G is uniformly Hölder continuous with exponent $1/K$ in every compact subset of G.*

Proof: It is enough to prove that every point ζ of G possesses a neighbourhood $U(\zeta)$ in which $\mathcal{W}$ is uniformly Hölder continuous with exponent $1/K$. We may suppose without loss of generality that $\zeta = 0$, $w(\zeta) = 0$ for every mapping $w \in \mathcal{W}$ since this can always be attained by means of rotations of the sphere.

Since $\mathcal{W}$ is equicontinuous at $z = 0$ there is a disc $U_r = \{z \mid |z| < r < 1\} \subset G$ whose images $w(U_r)$, $w \in \mathcal{W}$, are contained in the unit disc. We

[19] This also defines the Hölder continuity and uniform Hölder continuity of a single mapping w if one considers w to be the family consisting of the single mapping w.

may therefore apply the estimations (3.10) and (3.6) obtained in § 3. If the point z_0 lies in the disc $U_{r/2} = \{z \mid |z| < r/2\}$, then for every $w \in \mathscr{W}$ and $z \in U_r$ we have

$$\frac{1}{2}|w(z) - w(z_0)| \leqq \left|\frac{w(z) - w(z_0)}{1 - \overline{w(z_0)}\,w(z)}\right| \leqq \varphi_K\left(\frac{r\,|z - z_0|}{|r^2 - \bar{z}_0\,z|}\right)$$

$$\leqq 4\left(\frac{r\,|z - z_0|}{|r^2 - \bar{z}_0 z|}\right)^{1/K} \leqq 4\left(\frac{2\,|z - z_0|}{r}\right)^{1/K} \leqq \frac{8}{r}\,|z - z_0|^{1/K}.$$

Now $|w(z) - w(z_0)| \geqq k(w(z), w(z_0))$ and $|z - z_0| \leqq 2\,k(z, z_0)$, since z and z_0 lie in the unit disc. Consequently, for $C \geqq 32/r$, we have

$$k(w(z), w(z_0)) \leqq C(k(z, z_0))^{1/K} \tag{4.4}$$

for all $w \in \mathscr{W}$, $z_0 \in U_{r/2}$ and $z \in U_r$.

If z lies outside U_r then $k(z, z_0) > \arctan r - \arctan(r/2)$ for $z_0 \in U_{r/2}$. Hence (4.4) also holds in this case if C is sufficiently large. The family $\mathscr{W}$ is therefore uniformly Hölder continuous in $U_{r/2}$ with exponent $1/K$.

The following result is contained in the above as a special case:

A K-quasiconformal mapping of a domain G is uniformly Hölder continuous with exponent $1/K$ in every compact subset of G.

The K-quasiconformal mapping w,

$$w(z) = z|z|^{\frac{1}{K} - 1},$$

shows that the exponent $1/K$ cannot in general be replaced by any greater number.

§ 5. *Convergence Theorems for Quasiconformal Mappings*

5.1. Normal families. From the equicontinuity of a family of mappings we can infer some important results on the convergence of sequences of its members. Before discussing quasiconformal mappings we present two general theorems of this type.

Lemma 5.1. *Let w_n, $n = 1, 2, \ldots$, be a sequence of mappings, equicontinuous in a domain G, and E an everywhere dense*[20] *subset of G. If the sequence w_n converges at every point of E then it is uniformly convergent in every compact subset of G.*

---

[20] A point set A is called everywhere dense in G if every point of G is a limit point of A. This topological concept of density should not be confused with the metrical one introduced in Chapter III.

Proof: Let $\varepsilon > 0$ and let z_0 be a point of G. Since the sequence w_n is equicontinuous at z_0, there is a neighbourhood $U \subset G$ of z_0 where $k(w_n(z), w_n(z_0)) < \varepsilon/5$ for every mapping w_n. By hypothesis there is a number $N(U)$ such that $k(w_n(a), w_m(a)) < \varepsilon/5$ at a point $a \in E \cap U$ whenever $m, n \geqq N(U)$. Consequently, we have

$$k(w_n(z), w_m(z)) \leqq k(w_n(z), w_n(z_0)) + k(w_n(z_0), w_n(a)) + k(w_n(a), w_m(a)) + k(w_m(a), w_m(z_0)) + k(w_m(z_0), w_m(z)) < \varepsilon$$

for $z \in U$ and $m, n \geqq N(U)$.

Every compact set $F \subset G$ can be covered by finitely many neighbourhoods U_i of the above type. If N denotes the greatest of the numbers $N(U_i)$, then $k(w_n(z), w_m(z)) < \varepsilon$ for $z \in F$ and $m, n \geqq N$. The sequence w_n is therefore uniformly convergent in F.

Let $\mathscr{W}$ be a family of continuous mappings of the domain G. We say that $\mathscr{W}$ is *normal* if every sequence of elements of $\mathscr{W}$ contains a subsequence which converges uniformly in every compact subset of G. A normal family which contains all its limit functions is called *closed*.

It is easy to see that a *normal family is equicontinuous*. From the above Lemma 5.1 it follows that the converse also holds:

Lemma 5.2. *A family of mappings which is equicontinuous in a domain G is normal.*

Proof: In view of Lemma 5.1, it is sufficient to show that every sequence w_n of mappings of G contains a subsequence which converges in a point set E everywhere dense in G. For E we can choose a countable set $\{a_n\}$, $n = 1, 2, \ldots$, for example, the points of G with rational coordinates.

Since the plane is compact, the sequence $w_n(a_1)$ possesses an accumulation point. Hence, w_n contains a subsequence w_{1n} which converges at a_1. From this sequence we can take a subsequence which converges at a_2, and repeating this process k times we obtain a sequence w_{kn} which converges at the points $a_1, \ldots, a_k$. The diagonal sequence w_{nn} then converges at all points of E, and the lemma is proved.

5.2. Normality criteria for families of quasiconformal mappings. In combination with Theorem 4.3, the above general results yield the following theorem on quasiconformal mappings:

The equicontinuity, Hölder continuity and normality of a family of K-quasiconformal mappings are equivalent concepts.

Thus, for example, a family of K-quasiconformal mappings is normal if the hypotheses either of Theorem 4.1 or of Theorem 4.2 are fulfilled.

Theorem 5.1. *A family $\mathcal{W}$ of K-quasiconformal mappings of the domain G is normal if there is a number $d > 0$ such that one of the following conditions is satisfied:*

1. *Every mapping $w \in \mathcal{W}$ omits two values whose spherical distance is greater than d.*
2. *Every mapping $w \in \mathcal{W}$ omits one value a and at two fixed points z_1, $z_2 \in G$ takes values such that the distances $k(w(z_i), a)$, $i = 1, 2$, are greater than d.*
3. *Every mapping $w \in \mathcal{W}$ takes values at three fixed points z_1, z_2, $z_3 \in G$ such that the distances $k(w(z_i), w(z_j))$, $i, j = 1, 2, 3$, $i \neq j$, are greater than d.*

In particular $\mathcal{W}$ is normal if all mappings $w \in \mathcal{W}$ omit two *fixed* values.

5.3. Classification of the limit functions. We proved in I.5.2 that the limit mapping of a uniformly convergent sequence of K-quasiconformal mappings is either K-quasiconformal or non-topological. This result can now be given in a more complete form. We begin with a somewhat more general discussion and study the limit functions of a not necessarily uniformly convergent sequence w_n of K-quasiconformal mappings of a domain G.

We suppose first that w_n tends to a limit function in an everywhere dense subset E of G. Then the following three cases can be distinguished.

Case A. The limit function w takes at least three values in E. Then there are three points $a_1, a_2, a_3 \in E$ at which the values of w are different. Thus all distances $k(w_n(a_i), w_n(a_j))$, $i \neq j$, are greater than a fixed positive bound, and by Theorem 4.2 the sequence w_n is equicontinuous in G. By Lemma 5.1 it therefore converges uniformly in every compact subset of G and the limit function w is therefore continuous in G. We shall show in 5.4 (Theorem 5.3) *that in this case w is a K-quasiconformal mapping.*

Case B. It is possible that the limit function w assumes exactly two values $w(a_1) = c_1$ and $w(a_2) = c_2$ in E. Then by Theorem 4.1 (see also the remark at the end of 4.1) the sequence is equicontinuous in G with the possible exception of the points a_1 and a_2. If $w(z) = c_1$ at a point $z \in E$, $z \neq a_1$, then the sequence w_n is also equicontinuous in $G - \{z\} - \{a_2\}$. Hence w_n is equicontinuous in $G - \{a_2\}$. By Lemma 5.1, w_n is then convergent in G and uniformly convergent in every compact subset of $G - \{a_2\}$. The limit function is therefore continuous in G

except at a_2. Since it assumes only two values in E it follows that in G, $w(z) = c_1$ for $z \neq a_2$ and $w(a_2) = c_2$, that is, *the limit function takes one of its two values at only one point of G.*

As an example we quote the sequence w_n, $w_n(z) = n\,z$, of mappings of the plane. The limit function w of this sequence has the values $w(0) = 0$, $w(z) = \infty$ for $z \neq 0$.

Case C. In this case $w(z) = c$ at all points of E. *However, the sequence need not converge in G.* For example, the sequence $z, z-1, 2\,z, 2\,(z-1), 2\,(z-1/2), \ldots, n\,z, n\,(z-1), \ldots, n\,(z-1/n), \ldots$ is not convergent at any of the points $0, 1/n, n = 1, 2, \ldots$, though it tends to ∞ at the other points of the plane.

It is also possible that w_n tends to a constant at every point of G but the convergence in a compact subset F of G is not uniform. This can be seen from the example $w_n(z) = z - n$, if one chooses F to be the whole plane.

In the case $E = G$ we can sum up the above discussion as follows:

Theorem 5.2. *If w_n is a sequence of K-quasiconformal mappings of the domain G and w_n converges everywhere in G to a limit function w, then w is either a continuous non-constant function, or a mapping of G onto two points, or a constant. In the first case the sequence w_n converges uniformly in every compact subset of G, in the second the limit function w takes one of its values at only one point a of G and the convergence is uniform in every compact subset of $G - \{a\}$.*

5.4. Quasiconformal limit functions. To complete the results of Theorem I.5.2 and the above Theorem 5.2, we prove that the limit function in the first case mentioned in Theorem 5.2 is K-quasiconformal.

Theorem 5.3. *The limit function w of a sequence w_n of K-quasiconformal mappings convergent in G is either a constant, a mapping of G onto two points, or a K-quasiconformal mapping of G.*

Proof: It is sufficient to consider the above case A. Then the sequence w_n is equicontinuous in G and therefore converges uniformly in every compact subset of G. Further, it is enough to show that w is one-to-one in G. For by Lemma I.1.1 it is then a homeomorphism which is K-quasiconformal by Theorem I.5.2.

Let a and b be two points of G such that $w(a) = w(b)$. We derive a contradiction by showing that in this case w has the value $w(a)$ at all points of G. The proof is divided into three parts.

We show first that *every neighbourhood of a contains points $z \neq a$ at which $w(z) = w(a)$.* To prove this we choose $\varepsilon > 0$ so small that

$k(a, b) > \varepsilon$ and the circle $C_\varepsilon = \{z | k(z, a) = \varepsilon\}$ together with its interior lies in G. The curve $w_n(C_\varepsilon)$ then separates $w_n(a)$ and $w_n(b)$ for every n. The minimum of $k(w_n(z), w_n(a))$ on C_ε is equal to $k(w(a), w(b)) = 0$, and the first assertion is proved.

Secondly, we show that *every point* $z_0 \in G$ *possesses a neighbourhood* $U(z_0) \subset G$ *in which* w *is either one-to-one or constant.* In particular, we then have $w(z) = w(a)$ in a neighbourhood $U(a)$ of a, since w is not one-to-one in any neighbourhood of a by the first part of the proof.

We choose a neighbourhood $U(z_0) = \{z | k(z, z_0) < r\} \subset G$ such that $k(w_n(z), w_n(z_0)) < \pi/4$ for every n and $z \in U(z_0)$; this is possible by the equicontinuity of $\{w_n\}$. If $U(z_0)$ does not possess the required property we can choose three points $z_1, z_2, z_3 \in U(z_0)$ such that $w(z_1) \neq w(z_2) = w(z_3)$. To show that this leads to a contradiction we use a simple module estimation.

We join the points z_1 and z_2 to one another and the point z_3 to the boundary of $U(z_0)$ by Jordan arcs C_1 and C_2, respectively, so that C_1 and C_2 have no common points and lie in $U(z_0)$ apart from one end-point of C_2 (Fig. 8). The point set $U(z_0) - (C_1 \cup C_2) = B$ is a ring domain. The image domain $w_n(B)$ of B lies in the disc $k(w, w_n(z_0)) < \pi/4$ and separates $w_n(z_1)$ and $w_n(z_2)$ from $w_n(z_3)$ and the circle $k(w, w_n(z_0)) = \pi/4$. For $n \to \infty$ we have $\lim k(w_n(z_2), w_n(z_3)) = 0$, $\lim k(w_n(z_1), w_n(z_2)) > 0$, and the diameters of the boundary components of $w_n(B)$ are not less than $k(w_n(z_1), w_n(z_2))$ for any n. It follows from Lemma I.6.2 that $\lim M(w_n(B)) = 0$. On the other hand, $M(w_n(B)) \geqq M(B)/K$ by the K-quasiconformality of w_n. This is a contradiction, and the second part of the proof is complete.

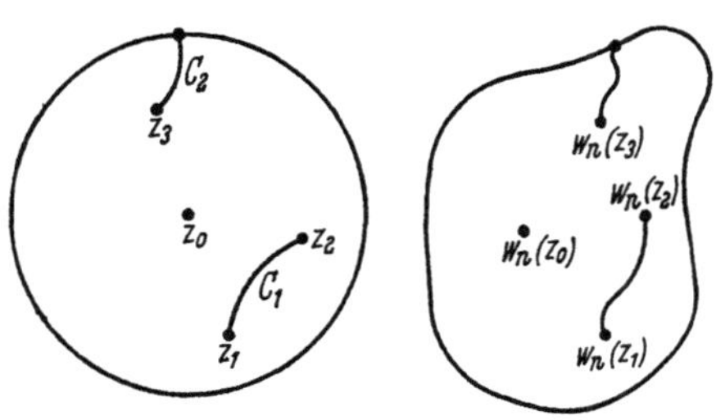

Fig. 8

Finally we prove: *If* w *has the constant value* $w(a)$ *in* $U(a)$ *then* $w(z) = w(a)$ *throughout* G. To this end we divide G into two parts E_1 and E_2 in the following way: If a point $z \in G$ has a neighbourhood $U(z)$ in which w has the constant value $w(a)$, then we take z to be a point of

E_1. On the other hand, if there exists a neighbourhood $U(z)$ of z in which w is either one-to-one or has a constant value different from $w(a)$, then z is to be a point of E_2. It follows from the second part of the proof that $E_1 \cup E_2 = G$. Since both sets are open and G is connected, E_2 must be empty. This gives the desired contradiction and Theorem 5.3 is proved.

5.5. Kernel of a sequence of sets. Let $w_n : G \to G'_n$ be a sequence of K-quasiconformal mappings converging to a limit mapping w. We want to investigate the dependence of the image $w(G) = G'$ on the sequence of domains G'_n.

To do this we introduce a topological concept:

The *kernel* $N\{E_n\}$ of a sequence of point sets E_n, $n = 1, 2, \ldots$, is the set

$$\bigcup_{m=1}^{\infty} \left(\bigcap_{n=m}^{\infty} E_n \right)^{\circ},$$

where $(\bigcap E_n)^{\circ}$ is the interior of $\bigcap E_n$. Thus a point z belongs to $N\{E_n\}$ if and only if there exists a neighbourhood U of z and a number m such that $U \subset E_n$ for $n \geqq m$.

The kernel, being the union of open sets, is open. However $N\{E_n\}$ need not be connected, even if the sets E_n are domains.

5.6. Mappings of a domain with at least two boundary points. In order to examine the dependence of $w(G) = G'$ on the sequence $w_n(G) = G'_n$ we suppose first that G has at least two boundary points. We again distinguish the three possible cases A, B and C studied in 5.3.

Case A. We prove the following theorem:

Theorem 5.4. *Let $w_n : G \to G'_n$ be K-quasiconformal mappings of a domain G with at least two boundary points. If the sequence w_n converges in G to a quasiconformal mapping w, then $w(G) = G'$ coincides with a component of $N\{G'_n\}$.*

Proof: We show first that G' is a subdomain of $N\{G'_n\}$. Let z_0 be an arbitrary point of G and U, $\overline{U} \subset G$, a neighbourhood of z_0. By Theorem 5.2 the sequence w_n converges uniformly to w in U. Since the boundary of $w(U)$ lies at a positive distance r from the point $w(z_0)$, there exists a number n_0 such that the disc $V' = \{w \mid k(w, w(z_0)) < r/2\}$ contains no boundary point of $w_n(U)$ for $n \geqq n_0$, while the interior point $w_n(z_0)$ of $w_n(U)$ lies in V'. Then V' must be contained in $w_n(U) \subset G'_n$ for $n \geqq n_0$. Therefore $w(z_0) \in V' \subset N\{G'_n\}$, and so $G' \subset N\{G'_n\}$.

As a connected subset of $N\{G_n'\}$, G' must be contained in one component of $N\{G_n'\}$. If G' is a proper subset of this component then $N\{G_n'\}$ contains a boundary point a' of G'. We show that this assumption leads to a contradiction.

By the definition of the kernel there exists a number n_0 and a connected neighbourhood U' of a' which lies in G_n' for $n \geqq n_0$. Let $a_k' = w(a_k)$ be a sequence of points lying in $U' \cap G'$ and converging to a'.

The inverses w_n^{-1} of the functions w_n, $n \geqq n_0$, omit two fixed values (boundary points of G) in U'. The family $\{w_n^{-1} | n \geqq n_0\}$ is therefore equicontinuous by Theorem 4.1 and normal by Theorem 5.1. Consequently, we can pick out a subsequence $w_{n_i}^{-1}$ which converges uniformly in every compact subset of U' to a continuous limit function φ. By Theorem 5.3, φ is either a K-quasiconformal mapping or a constant. Since $\lim w_n(a_k) = a_k'$, it follows from the equicontinuity of the family $\{w_n^{-1} | n \geqq n_0\}$ that $\varphi(a_k') = \lim w_{n_i}^{-1}(w_{n_i}(a_k)) = a_k$, $k = 1, 2, \ldots$ The function φ must therefore be a K-quasiconformal mapping of U'. We proved above that in this case the image $\varphi(U')$ is a subdomain of $N\{w_n^{-1}(U')\} \subset G$. Hence the point $\varphi(a') = a$ also belongs to G and it follows from $w(a_k) = a_k'$, $k = 1, 2, \ldots$, that $w(a) = a'$. This contradicts the assumption that a' is a boundary point of G'. Hence G' must coincide with a component of $N\{G_n'\}$.

To investigate the relationships between G' and $N\{G_n'\}$ in cases B and C where the limit function w is not one-to-one we first prove the following preparatory result.

Lemma 5.3. *Let G be a domain possessing at least two boundary points and $w_n : G \to G_n'$ a sequence of K-quasiconformal mappings which converges in G. Then the limit function $w = \lim w_n$ takes no value lying in $N\{G_n'\}$ at two different points of G.*

Proof: Let $w_0 \in N\{G_n'\}$ and $w(z_0) = w_0$. Let U' be a connected neighbourhood of w_0 which lies in G_n' for $n \geqq n_0$. The inverses w_n^{-1} of the mappings w_n, $n \geqq n_0$, then omit two fixed values, and the family $\{w_n^{-1} | n \geqq n_0\}$ is therefore equicontinuous in U' by Theorem 4.1. Since $\lim w_n(z_0) = w_0$ it follows that $w_n^{-1}(w_0)$ tends to $w_n^{-1}(w_n(z_0)) = z_0$ as $n \to \infty$.

Suppose that z_0 and z_1 are two points of G such that $w(z_0) = w(z_1) = w_0$. From the above it then follows that $\lim w_n^{-1}(w_0) = z_0 = z_1$, and the lemma is proved.

Case B. The limit function w is now a mapping of G onto two points c_1 and c_2. By 5.3, w assumes one of the values c_i, say c_1, at a single point $z_1 \in G$ while $w(z) = c_2$ at all other points $z \in G$.

By Lemma 5.3 the point c_2 *cannot* lie in $N\{G'_n\}$. On the other hand c_1 always lies in $N\{G'_n\}$. In fact the following is true: *The set $N\{G'_n\}$ is the whole plane with c_2 removed.*

To see this we remark first that no infinite subset of the family $\{w_n\}$ can be equicontinuous in G since otherwise the limit function w would be continuous by Lemma 5.1. By Theorem 4.1 the diameter of the complement $-G'_n$ of G'_n must tend to zero as $n \to \infty$. If we choose a subsequence $-G'_{n_i}$ of the sequence $-G'_n$ which converges to a point a', then a' is the only boundary point of $N\{G'_{n_i}\}$. By the above, $a' = c_2$. Since the limit point does not depend on the choice of the subsequence, $-G'_n$ converges to c_2 for $n \to \infty$. Consequently, $N\{G'_n\}$ is the plane punctured at c_2, as we asserted.

Case C. The limit function w is now a mapping of G onto a single point c. By Lemma 5.3, c must be either a boundary point or an exterior point of $N\{G'_n\}$. On the other hand, every neighbourhood of c has points in common with infinitely many of the domains G'_n. Thus the point c belongs neither to $N\{G'_n\}$ nor to $N\{-G'_n\}$.

The case when G has at least two boundary points is thus completely solved. If all the domains G'_n coincide with a fixed domain G', then $N\{G'_n\} = G'$ and $N\{-G'_n\} = -\overline{G}'$. The case B cannot occur, and we obtain the following result as a special case:

Theorem 5.5. *Let G be a domain possessing at least two boundary points and w_n a sequence of K-quasiconformal mappings of G onto a fixed domain G'. If w_n converges in G, the limit function $w = \lim w_n$ is either a K-quasiconformal mapping of G onto G' or a mapping of G onto a boundary point of G'.*

5.7. Mappings of a domain with at most one boundary point. If the domain G has at most one boundary point, the following is true:

If G is the whole plane, then all G'_n and also their kernel $N\{G'_n\}$ must coincide with the whole plane. Hence the set $w(G)$ lies in $N\{G'_n\}$ independently of whether w is a K-quasiconformal mapping or $w(G)$ contains only one or two points.

If G has a single boundary point a, then every mapping w_n can be extended to a K-quasiconformal mapping of the plane (Theorem I.8.1). If the sequence w_n converges in G to a K-quasiconformal mapping or a mapping of G onto two points, then it also converges in the whole plane. This follows from the discussion of the cases A and B in 5.3, since G is everywhere dense in the plane. Hence in both these cases $N\{G'_n\}$ is the whole plane punctured at $w(a)$. The limit function w is either a K-quasiconformal mapping of G onto $N\{G'_n\}$ or a mapping of

G onto two points, of which one coincides with the single boundary point of $N\{G'_n\}$ and the other is assumed by w at a single point.

The results derived above for domains G with at least two boundary points thus remain true in the cases A and B if G has only one boundary point. If w_n converges to a constant the situation is different. In this case both the limit $c = \lim w_n$ and the open set $N\{G'_n\}$ can be arbitrarily prescribed. For example let G be the plane punctured at the origin, $c = \infty$, and E an arbitrary open set of the plane. We choose a sequence of points ζ_n, $n = 1, 2, \ldots$, such that the set of their limit points coincides with $-E$. The function w_n,

$$w_n(z) = n\,(1 + |\zeta_n|)\cdot z + \zeta_n\,,$$

maps G conformally onto $G'_n = \{w | w \neq \zeta_n\}$. Hence $N\{G'_n\} = E$ and $\lim w_n(z) = \infty$ for every $z \in G$.

§ 6. Boundary Values of a Quasiconformal Mapping

6.1. Statement of the boundary value problem. The results given in I.8 on the boundary behaviour of a quasiconformal mapping concern the existence and continuity of the boundary values. In this section we consider boundary correspondence from a different viewpoint. We investigate the solvability of the *boundary value problem*, that is, the possibility of constructing a quasiconformal mapping with given boundary values.[21]

This problem plays an important role in our considerations. Its solution will be used in the applications in §§ 7 and 8 of this chapter as well as in Chapter V, when we prove the existence of quasiconformal mappings with prescribed complex dilatation.

With the exception of the discussion in 6.6 we restrict ourselves to mappings between Jordan domains. It follows from Theorem I.8.2 that every quasiconformal mapping can then be extended to a homeomorphism between the closures of these domains. By the orientation theorem (cf. I.1.5) the extension is sense-preserving, and the boundary value problem therefore reads as follows:

Let G and G' be two Jordan domains with boundaries C, C', respectively, and φ a homeomorphism of C onto C' which preserves the positive orientation (cf. I.1.4). It is required to find necessary and sufficient

---

[21] The contents of this section are largely based on the work [1] of Beurling and Ahlfors.

conditions on φ such that there exists a quasiconformal mapping $w: G \to G'$ with the boundary values $w(z) = \varphi(z)$.

6.2. A general boundary condition. In order to find a *necessary* condition, we suppose that the boundary value problem has a K-quasiconformal solution. If $Q = G(z_1, z_2, z_3, z_4)$ denotes a quadrilateral which consists of G and the vertices z_i and $Q' = G'(w_1, w_2, w_3, w_4)$ is the image of Q under w, then by Lemma I.5.1 we have

$$M(Q)/K \leqq M(Q') \leqq K\, M(Q)\,. \tag{6.1}$$

For every choice of the vertices $z_i \in C$ the image points $w_i = \varphi(z_i)$ must therefore lie on C' so that the double inequality (6.1) holds.

The condition (6.1) assumes a weaker form, better suited to our applications, if we fix the vertex z_4 and choose the other three vertices such that Q is conformally equivalent to a square. Then $M(Q) = 1$, and so Q' must satisfy the condition

$$1/K \leqq M(Q') \leqq K\,. \tag{6.2}$$

On the other hand, we shall show in 6.5 that this necessary condition is also sufficient in the following sense: it guarantees the existence of a quasiconformal, though not necessarily K-quasiconformal solution. Thus the following basic result is true:

Theorem 6.1. *Let φ be a homeomorphism between the boundaries of the Jordan domains G and G' which preserves the positive orientation, and $Q = G(z_1, z_2, z_3, z_4)$ a quadrilateral. If there exists a K-quasiconformal mapping $w: G \to G'$ with the boundary values φ, then* (6.1) *holds for the quadrilateral $Q' = G'(\varphi(z_1), \varphi(z_2), \varphi(z_3), \varphi(z_4))$.*

Conversely, if (6.2) *holds for the family of quadrilaterals Q with unit module and fixed z_4, then there exists a quasiconformal mapping $w: G \to G'$ with the boundary values φ and with a maximal dilatation bounded above by a number depending only on K.*

In the proof of this theorem, which will be completed in 6.5, there is no loss of generality in supposing that G and G' are half-planes; this follows from the conformal invariance of the module and the boundary correspondence theorem quoted in I.2.2. In this special case condition (6.2) can be given in an explicit form as we shall now show.

6.3. Necessary condition for the half-plane. Let G and G' be upper half-planes. To transform condition (6.2) to a simple form we set $z_4 = \infty$ and normalize all our mappings so that they preserve the point at infinity.

The quadrilateral $Q = G(z_1, z_2, z_3, \infty)$ is conformally equivalent to a square if $z_3 - z_2 = z_2 - z_1$. We then write $z_2 = x$, $z_1 = x - t$, $z_3 = x + t$, where $t > 0$ since the orientation z_1, z_2, z_3 is positive with respect to G.

For the module of the quadrilateral $Q' = G'(w_1, w_2, w_3, \infty)$ we have by 1.2,

$$M(Q') = \frac{2}{\pi} \mu \left(\sqrt{\frac{w_3 - w_2}{w_3 - w_1}} \right).$$

If we set this value into (6.2) we obtain for the boundary values of a K-quasiconformal mapping $w : G \to G'$, $w(\infty) = \infty$, the double inequality

$$\mu^{-1}\left(\frac{\pi K}{2}\right) \leqq \sqrt{\frac{w(x+t) - w(x)}{w(x+t) - w(x-t)}} \leqq \mu^{-1}\left(\frac{\pi}{2K}\right). \tag{6.3}$$

To reduce this condition to a symmetric form we write

$$\lambda(K) = \frac{1}{(\mu^{-1}(\pi K/2))^2} - 1 . \tag{6.4}$$

Then $\mu\left(1/\sqrt{1 + \lambda(K)}\right) = \pi K/2$, and so

$$\mu\left(1/\sqrt{1 + \lambda(K)}\right) \mu\left(1/\sqrt{1 + \lambda(1/K)}\right) = \pi^2/4 .$$

By (2.7) this equation remains true if $\lambda(1/K)$ is replaced by $1/\lambda(K)$. Thus

$$\lambda(1/K) = 1/\lambda(K) . \tag{6.5}$$

Expressing $\mu^{-1}(\pi K/2)$ in (6.3) in terms of $\lambda(K)$ and applying (6.5) we obtain the following form of the first part of Theorem 6.1, in the special case of half-planes:

Theorem 6.2. *The boundary values of a K-quasiconformal self-mapping w of the upper half-plane preserving the point ∞ satisfy the double inequality*

$$\frac{1}{\lambda(K)} \leqq \frac{w(x+t) - w(x)}{w(x) - w(x-t)} \leqq \lambda(K) \tag{6.6}$$

for all real x and t, $t \neq 0$.

The inequality (6.6) *is sharp.* This follows immediately from the fact that a square can be mapped K-quasiconformally onto any quadrilateral whose module is either K or $1/K$ (cf. the remark at the end of I.3.4).

6.4. The distortion function λ. Before dealing with the converse of Theorem 6.2 we give some estimations concerning the distortion function λ, which will also occur in another connection in § 9.

From (2.4) we obtain for the inverse of μ the functional equation

$$\mu^{-1}(2x) = \left(\frac{1-\sqrt{1-(\mu^{-1}(x))^2}}{\mu^{-1}(x)}\right)^2.$$

Because $1/\sqrt{1+\lambda(K)} = \mu^{-1}(\pi K/2)$, λ satisfies the equation

$$1+\lambda(2K) = \left(\sqrt{1+\lambda(K)}+\sqrt{\lambda(K)}\right)^4 \tag{6.7}$$

for every K.

Since $\lambda(1) = 1$ we obtain from (6.7)

$$\lambda(2) = 16 + 12\sqrt{2}.$$

Repeated application of (6.7) yields an expression for $\lambda(2^n)$ for every positive integer n.

Using the estimations of $\mu(r)$ set out in 2.4 we can derive an approximation formula for the asymptotic behaviour of $\lambda(K)$ as $K \to \infty$. To this end insert $r = 1/\sqrt{1+\lambda(K)}$ in (2.13). Then $\mu(r) = \pi K/2$ and so

$$2\left(\sqrt{1+\lambda(K)}+\sqrt{\lambda(K)}\right) = e^{\pi K/2} + \delta, \tag{6.8}$$

where $0 < \delta < (1+\lambda(K))^{-3/2}$.

We write

$$\sqrt{1+\lambda(K)} = \frac{1}{4} e^{\pi K/2} + e^{-\pi K/2} + t\, e^{-3\pi K/2}, \tag{6.9}$$

where t is to be estimated for $K \geqq 1$. Were $t \leqq 0$ we could infer that

$$2\left(\sqrt{1+\lambda(K)}+\sqrt{\lambda(K)}\right) \leqq e^{\pi K/2}.$$

This contradicts (6.8), and hence $t > 0$. It follows that

$$2\left(\sqrt{1+\lambda(K)}+\sqrt{\lambda(K)}\right) > e^{\pi K/2} + 4t\, e^{-3\pi K/2}, \qquad \delta < 64\, e^{-3\pi K/2}.$$

A comparison with (6.8) shows that $t < 16$.

Since $0 < t < 16$ we obtain from (6.9) the asymptotic expansion

$$\lambda(K) = \frac{1}{16} e^{\pi K} - \frac{1}{2} + O(e^{-\pi K}), \tag{6.10}$$

where the remainder is positive.

6.5. Solution of the boundary value problem. Following Beurling and Ahlfors [1] we now construct a quasiconformal self-mapping of the upper half-plane with given boundary values. By Theorem 6.2 the construction is only possible if the boundary values satisfy a condition of form (6.6). The following theorem shows that this condition is also sufficient, although the maximal dilatation of the constructed mapping turns out to be larger than the value arising from (6.6).

The result can be immediately generalized to the case dealt with in Theorem 6.1. We thus obtain the required proof for the sufficiency of condition (6.2).

Theorem 6.3. *Let φ be a function which is continuous and strictly increasing on the real axis and which satisfies the condition*

$$\frac{1}{\varrho} \leqq \frac{\varphi(x+t)-\varphi(x)}{\varphi(x)-\varphi(x-t)} \leqq \varrho \tag{6.11}$$

for every real x and t, $t > 0$. Then there exists a quasiconformal mapping w of the upper half-plane onto itself, which has the boundary values $w(x) = \varphi(x)$ and a maximal dilatation less than a bound depending on ϱ.

Proof: The function

$$w = u + i\,v = \frac{1}{2}(\alpha + \beta) + \frac{1}{2} i\,(\alpha - \beta) = \frac{1}{2}(1 + i)\,(\alpha - i\,\beta)\,,$$

where

$$\alpha(x, y) = \int_0^1 \varphi(x + t\,y)\,dt\,, \qquad \beta(x, y) = \int_0^1 \varphi(x - t\,y)\,dt\,, \tag{6.12}$$

is defined and continuous in the whole finite $z = x + i\,y$-plane. Since φ is strictly increasing, w carries every point of the upper half-plane $y > 0$ into the upper half-plane $v > 0$. On the real axis, $w(x) = \varphi(x)$, and at conjugate points $z = x + i\,y$, $\bar{z} = x - i\,y$ we have $w(\bar{z}) = \overline{w}(z)$.

We have to prove that w is a quasiconformal mapping of the finite plane. Then the image domain will also consist of the whole finite plane (cf. the remark at the end of I.8.1), and it follows from the above that w maps the upper half-plane onto itself and has the right boundary values $\varphi(x)$.

We show first that w is a homeomorphism of the finite plane. It is enough to prove that w assumes every value at no more than one point. By Lemma I.1.1 the continuity of the inverse w^{-1} of w then follows from the one-to-one character and continuity of w.

Let $z_1 = x_1 + i\,y_1$, $z_2 = x_2 + i\,y_2$ be two points such that $w(z_1) = w(z_2)$. It follows from the above that z_1 and z_2 lie in the same half-plane bounded by $y = 0$. Since $w(\bar{z}) = \overline{w}(z)$ we may suppose that y_1 and y_2 are positive.

We write (6.12) in the form

$$\alpha(x, y) = \frac{1}{y} \int_x^{x+y} \varphi(\xi)\, d\xi\,, \qquad \beta(x, y) = \frac{1}{y} \int_{x-y}^{x} \varphi(\xi)\, d\xi\,. \tag{6.13}$$

We see that the mean values of φ are equal on the intervals $(x_1, x_1 + y_1)$ and $(x_2, x_2 + y_2)$ as well as on $(x_1 - y_1, x_1)$ and $(x_2 - y_2, x_2)$. Since φ is strictly increasing, one of the first mentioned intervals must be contained in the other, and the same holds for the last two. If for example $x_2 \geqq x_1$ it follows that $x_2 + y_2 \leqq x_1 + y_1$ and $x_2 - y_2 \leqq x_1 - y_1$. This implies that $z_1 = z_2$, and so the mapping w is a homeomorphism.

Finally we show that w is quasiconformal in the upper half-plane. From (6.13) it follows that α and β are continuously differentiable and have the partial derivatives

$$\alpha_x(x, y) = \frac{1}{y}\left(\varphi(x + y) - \varphi(x)\right), \qquad \beta_x(x, y) = \frac{1}{y}\left(\varphi(x) - \varphi(x - y)\right),$$

$$\alpha_y(x, y) = \frac{1}{y}\left(\varphi(x + y) - \alpha(x, y)\right), \quad \beta_y(x, y) = \frac{1}{y}\left(\varphi(x - y) - \beta(x, y)\right).$$

Since φ is strictly increasing we have

$$\alpha_x > 0\,, \quad \beta_x > 0\,, \quad \alpha_y > 0\,, \quad \beta_y < 0\,. \tag{6.14}$$

The Jacobian $J = (\alpha_y \beta_x - \alpha_x \beta_y)/2$ of w is therefore positive. Hence all points of the upper half-plane are regular for w.

To obtain an upper bound for the dilatation quotient $D = (|w_z| + |w_{\bar{z}}|)/(|w_z| - |w_{\bar{z}}|)$ of w (cf. I.9.4) we write

$$D \leqq 2\,\frac{|w_z|^2 + |w_{\bar{z}}|^2}{|w_z|^2 - |w_{\bar{z}}|^2} = \frac{\alpha_x^2 + \alpha_y^2 + \beta_x^2 + \beta_y^2}{\alpha_y \beta_x - \alpha_x \beta_y}$$

$$= \frac{\alpha_x \beta_x}{-\alpha_y \beta_y} \cdot \frac{\alpha_x/\beta_x + \alpha_y^2/(\alpha_x \beta_x) + \beta_x/\alpha_x + \beta_y^2/(\alpha_x \beta_x)}{-\beta_x/\beta_y + \alpha_x/\alpha_y}\,. \tag{6.15}$$

To estimate the partial derivatives of α and β we make use of the assumption (6.11). From

$$\varphi\left(x + \frac{1}{2} y\right) - \varphi(x) \leqq \varrho \left(\varphi(x + y) - \varphi\left(x + \frac{1}{2} y\right)\right)$$

it follows that

$$\varphi(x + y) - \varphi(x) \leqq (1 + \varrho)\left(\varphi(x + y) - \varphi\left(x + \frac{1}{2} y\right)\right).$$

Between α_x and α_y we have therefore the double inequality

$$\alpha_x(x, y) > \alpha_y(x, y) = \frac{1}{y} \int_0^1 (\varphi(x + y) - \varphi(x + t\,y))\, dt$$

$$\geqq \frac{1}{2y}\left(\varphi(x + y) - \varphi\left(x + \frac{1}{2} y\right)\right) \geqq \frac{\alpha_x(x, y)}{2\,(1 + \varrho)} \,. \qquad (6.16)$$

In the same way we obtain

$$\beta_x(x, y) > -\beta_y(x, y) \geqq \frac{1}{2y}\left(\varphi\left(x - \frac{1}{2} y\right) - \varphi\,(x - y)\right) \geqq \frac{\beta_x(x, y)}{2\,(1 + \varrho)} \,. \qquad (6.17)$$

Finally it follows from (6.11) that

$$\frac{\beta_x}{\varrho} \leqq \alpha_x \leqq \varrho\, \beta_x \,. \qquad (6.18)$$

Considering the estimations (6.16—18) we deduce from (6.15) that

$$D(z) \leqq 8\,\varrho\,(1 + \varrho)^2 \,. \qquad (6.19)$$

By Theorem I.3.1, w is quasiconformal in the upper half-plane.

To complete the proof, we conclude from symmetry property $w(\bar{z}) = \overline{w}(z)$ that (6.19) holds in the lower half-plane also. By Theorem I.8.3 w is therefore a quasiconformal mapping of the whole plane, with a maximal dilatation not exceeding $8\,\varrho\,(1 + \varrho)^2$.

Remark. Beurling and Ahlfors [1] showed that under condition (6.11) the boundary value problem has a solution with maximal dilatation $\leqq \varrho^2$. Reed [1] has proved that the above mapping w is $8\,\varrho$-quasiconformal.

6.6. Boundary value problem for a boundary component of a multiply connected domain. In the case of multiply connected domains we restrict ourselves to the problem of finding a mapping whose boundary values are given on a single boundary component.

Let D be a domain having the Jordan curve C as a free boundary curve; C is then a boundary component of D (cf. I.1.9). Let G' be a Jordan domain and f a homeomorphism of C onto the boundary curve C' of G'. We shall investigate the conditions under which there exists a quasiconformal mapping w of D into G' (i.e. onto a subdomain of G') which has the boundary values $w = f$ on C.

The problem goes back to the above boundary value problem for Jordan domains. In one direction this is clear: If G denotes the domain bounded by C and containing D, and if w is a quasiconformal mapping of G onto G' with the boundary values $w = f$ on C, then the restriction of w to D is the required solution. What is interesting is that the following converse is also true (see Väisälä [2]):

Theorem 6.4. *Let D and D' be domains with the free boundary curves C and C', respectively, and $w: D \to D'$ a K-quasiconformal mapping which can be extended to a homeomorphism of $D \cup C$ onto $D' \cup C'$. Let G and G' be the components of the complement of C and C' which contain D and D', respectively. Then there exists a quasiconformal mapping of G onto G' which has the same boundary values as w on C and whose maximal dilatation is less than a bound depending only on K and the domain D.*

We remark that this theorem is closely connected with the problem of quasiconformal continuation to be dealt with in § 8. It states that the boundary problem is solvable for the whole domain G if there exists a solution in a neighbourhood of the boundary of G.

Proof of Theorem 6.4: By means of a conformal mapping the theorem can be reduced to the case when G is the unit disc and C the unit circle. Since C is a free boundary curve of D there exists a number $r < 1$ such that the set $\{z | r \leqq |z| < 1\}$ is contained in D. Let z_1, z_2, z_3 and z_4 be four points on C which lie in such a manner that the module of the quadrilateral $Q_1 = G(z_1, z_2, z_3, z_4)$ is equal to one. By Theorem 6.1 it remains to find an upper bound, depending only on r and K, for the module of the quadrilateral $Q_1' = G'(w_1, w_2, w_3, w_4)$, where w_i denotes the boundary value of w at z_i.

To do this we join the endpoints of the longer a-side of Q_1, say z_1 and z_2, to the circle $|z| = r$ by means of radial segments. They divide the annulus $r < |z| < 1$ into two Jordan domains, of which one has z_3 and z_4 as boundary points. Denote by Q_2 the quadrilateral consisting of this domain and the vertices z_1, z_2, z_3, z_4 (Fig. 9).

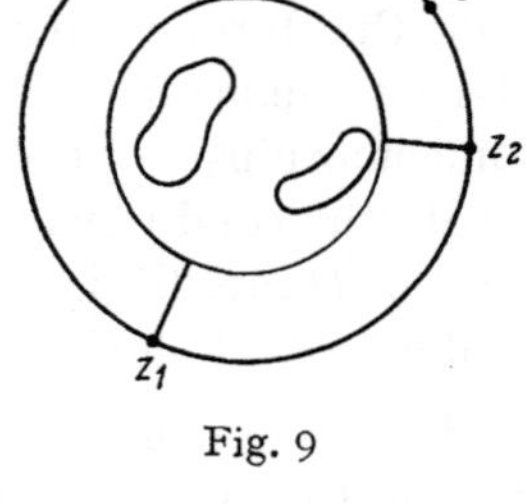

Fig. 9

The image $w(Q_2)$ of Q_2 has the same b-sides as Q_1', and by the monotonicity theorem in I.4.6 its module is greater than $M(Q_1')$. Since w is

K-quasiconformal it follows that

$$M(Q_1') < K\, M(Q_2)\,. \tag{6.20}$$

To find an upper bound for $M(Q_2)$ we make use of Lemma I.4.1. We denote by $s_a(Q_i)$ and $s_b(Q_i)$ the distances between the a- and b-sides of Q_i both measured in Q_i, $i = 1, 2$. Since $\widehat{z_1 z_2}$ is the longer a-side of Q_1, it follows that $s_b(Q_1) = |z_3 - z_4|$ and

$$s_b(Q_2) < \frac{\pi}{2}\, s_b(Q_1)\,.$$

On the other hand,

$$s_a(Q_2) \geqq \min\left(1 - r, \frac{1}{2}\, s_a(Q_1)\right) > \frac{1-r}{2}\, s_a(Q_1)\,.$$

Hence, we obtain

$$\frac{s_a(Q_2)}{s_b(Q_2)} > \frac{1-r}{\pi} \cdot \frac{s_a(Q_1)}{s_b(Q_1)}\,. \tag{6.21}$$

Since the majorant (4.7) for $M(Q)$ introduced in Lemma I.4.1 is a decreasing function of s_a/s_b, it follows from (6.21) that

$$M(Q_2) < \pi\, \frac{1 + 2\log\left(1 + 2\,\dfrac{(1-r)\, s_a(Q_1)}{\pi\, s_b(Q_1)}\right)}{\left(\log\left(1 + 2\,\dfrac{(1-r)\, s_a(Q_1)}{\pi\, s_b(Q_1)}\right)\right)^2}\,. \tag{6.22}$$

On the other hand, interchanging the roles of the a- and b-sides of Q_1 we deduce from Lemma I.4.1 that

$$1 = \frac{1}{M(Q_1)} \leqq \pi\, \frac{1 + 2\log\left(1 + 2\,\dfrac{s_b(Q_1)}{s_a(Q_1)}\right)}{\left(\log\left(1 + 2\,\dfrac{s_b(Q_1)}{s_a(Q_1)}\right)\right)^2}$$

or

$$\frac{s_a(Q_1)}{s_b(Q_1)} \geqq \frac{2}{e^{\pi + \sqrt{\pi^2 + \pi}} - 1} > 10^{-3}\,. \tag{6.23}$$

Setting this estimate[22] in (6.22) we obtain an upper bound for $M(Q_2)$ which depends only on r. By (6.20) it follows that $M(Q_1')$ is bounded above by a number depending only on r and K. From Theorem 6.1 we infer the existence of the desired quasiconformal mapping of G having the same boundary values as w on C.

---

[22] In view of later applications we remark that (6.23) holds for every quadrilateral with module one.

§ 7. *Quasisymmetric Functions*

7.1. Definition of a quasisymmetric function. Let φ be a continuous, strictly increasing function defined on the bounded or unbounded interval I of the (finite) x-axis. We call φ *k-quasisymmetric on I* if there exists a positive constant k such that

$$\frac{1}{k} \leqq \frac{\varphi(x+t)-\varphi(x)}{\varphi(x)-\varphi(x-t)} \leqq k \tag{7.1}$$

for $x, x-t, x+t \in I$ and $t > 0$. A function φ is called *quasisymmetric on I* if φ is k-quasisymmetric on I for some k.[23]

By Theorems 6.2 and 6.3 a finite real function φ is quasisymmetric on the real axis *if and only if there exists a quasiconformal mapping of the upper half-plane onto itself with the boundary values* φ. It follows immediately that the inverses and compositions of functions quasisymmetric on the real axis are also quasisymmetric.

A linear polynomial $\varphi(x) = a\,x + b$, $a > 0$, is 1-quasisymmetric on every interval. Conversely, it is easy to see that every function which is 1-quasisymmetric on an interval is linear.

The k-quasisymmetry of a function is obviously invariant with respect to 1-quasisymmetric mappings. In other words, if φ is k-quasisymmetric on I and φ_1, φ_2 are 1-quasisymmetric (thus linear) then the composition $\varphi_2 \circ \varphi \circ \varphi_1$ is k-quasisymmetric on $\varphi_1^{-1}(I)$.

The quasisymmetry of a function φ on the whole x-axis cannot be characterized by the local properties of φ, i.e. it is not enough to assume (7.1) only for small values of t. An example is provided by the exponential function, which is e-quasisymmetric on every segment of length 2 but not quasisymmetric on the whole x-axis.

On the other hand, on bounded intervals the quasisymmetry of a function can be deduced from its local properties. In fact if (7.1) holds for $0 < t \leqq \delta$, then we have for $0 < t \leqq 2\delta$,

$$\varphi(x+t) - \varphi\left(x+\frac{t}{2}\right) \leqq k\left(\varphi\left(x+\frac{t}{2}\right) - \varphi(x)\right) \leqq k^2\left(\varphi(x) - \varphi\left(x-\frac{t}{2}\right)\right),$$

$$\varphi\left(x+\frac{t}{2}\right) - \varphi(x) \leqq k\left(\varphi(x) - \varphi\left(x-\frac{t}{2}\right)\right) \leqq k^2\left(\varphi\left(x-\frac{t}{2}\right) - \varphi(x-t)\right).$$

By addition we obtain

$$\frac{\varphi(x+t)-\varphi(x)}{\varphi(x)-\varphi(x-t)} \leqq k^2\,,$$

[23] The term quasisymmetric is due to Kelingos [1].

and a similar reasoning yields the lower bound $1/k^2$. By repeating this process n times we arrive at the following result: If φ is continuous and strictly increasing on the x-axis and if (7.1) holds for $0 < t \leqq \delta$, then φ is k^{2n}-quasisymmetric on every interval of length $2^{n+1}\,\delta$.

Remark. The results can also be expressed in the following form: If φ is continuous and strictly increasing on the x-axis and (7.1) holds for $0 < t \leqq \delta$ and $t > 2^n\delta$, where n is a positive integer, then φ is k^{2n}-quasisymmetric on the whole x-axis.

7.2. Extension of a quasisymmetric function. Apart from the boundary problem solved in 6.5, we are interested in the case when the boundary values are given only on a segment I of the real axis. By Theorem 6.3 a boundary problem of this type is solvable if the given boundary values can be extended to a function which is quasisymmetric on the whole x-axis. The following lemma states that this is always so if the boundary function is quasisymmetric on I.

Lemma 7.1. *Every function φ which is k-quasisymmetric on an interval $I = \{x|a \leqq x \leqq b\}$ can be extended to a K-quasisymmetric function on the whole x-axis. The constant K is less than a number depending only on k.*

Proof: Since k-quasisymmetry is invariant with respect to linear (polynomial) transformations there is no loss of generality in assuming that $a = \varphi(a) = 0$ and $b = \varphi(b) = 1$. We can then extend φ by means of the formulae

$$\varphi(-x) = -\varphi(x)\,, \qquad \varphi(x+2) = \varphi(x) + 2 \tag{7.2}$$

to a continuous strictly increasing function on the whole x-axis. We prove that this extension satisfies an inequality of form (7.1) for all x and t, $t > 0$, with a constant K in place of k.

First, let $t \leqq 1/2$. In view of (7.2) we may suppose that $x - t < 0 \leqq x < x + t$. Then $x + t \geqq \max(x, t - x) \geqq (x+t)/3$ and

$$2\,\varphi(x+t) \geqq \varphi(x) - \varphi(x-t) = \varphi(x) + \varphi(t-x) \geqq \varphi\left(\frac{1}{3}(x+t)\right). \tag{7.3}$$

By hypothesis φ is k-quasisymmetric on the segment $\{x|0 \leqq x \leqq 1\}$. Since $\varphi(0) = 0$ we therefore have

$$\begin{aligned}&\frac{1}{k^2}\,\varphi\left(\frac{1}{3}(x+t)\right)\\ &\leqq \frac{1}{k}\left(\varphi\left(\frac{2}{3}(x+t)\right) - \varphi\left(\frac{1}{3}(x+t)\right)\right) \leqq \varphi(x+t) - \varphi\left(\frac{2}{3}(x+t)\right)\\ &\leqq k\left(\varphi\left(\frac{2}{3}(x+t)\right) - \varphi\left(\frac{1}{3}(x+t)\right)\right) \leqq k^2\,\varphi\left(\frac{1}{3}(x+t)\right).\end{aligned} \tag{7.4}$$

It follows that

$$\varphi(x+t) \leqq (1+k+k^2)\,\varphi\left(\frac{1}{3}(x+t)\right). \tag{7.5}$$

Since

$$\varphi(x+t) \geqq \varphi(x+t) - \varphi(x) > \varphi(x+t) - \varphi\left(\frac{2}{3}(x+t)\right)$$

we deduce from (7.3—5) that

$$\frac{1}{2k^2(1+k+k^2)} \leqq \frac{\varphi(x+t)-\varphi(x)}{\varphi(x)-\varphi(x-t)} \leqq 1+k+k^2. \tag{7.6}$$

This double inequality was proved under the hypothesis that $t \leqq 1/2$. In the case $t \geqq 2$ it follows from (7.2) that both $\varphi(x+t) - \varphi(x)$ and $\varphi(x) - \varphi(x-t)$ lie between $t/3$ and $3\,t$. Then we have

$$\frac{1}{9} \leqq \frac{\varphi(x+t)-\varphi(x)}{\varphi(x)-\varphi(x-t)} \leqq 9. \tag{7.7}$$

From the inequalities (7.6) and (7.7) and from the remark at the end of 7.1 we conclude that φ is K-quasisymmetric on the x-axis with $K = \max\left(9^4, 2^4(k^2+k^3+k^4)^4\right)$.

7.3. Hölder continuity of a quasisymmetric function. Let φ be a k-quasisymmetric function on the x-axis. By Theorem 6.3 there exists a quasiconformal mapping w of the upper half-plane onto itself with the boundary values $\varphi(x)$. The mapping can be extended to the whole plane by reflection, and it follows from Theorem 4.3 (cf. the remark at the end of 4.2) that w, and consequently φ, are uniformly Hölder continuous with respect to the spherical metric.

In every compact subset of the finite plane the ratio of the euclidean and spherical distances is bounded. Since $\varphi(\infty) = \infty$, φ therefore satisfies a condition

$$|\varphi(x) - \varphi(y)| < C_1 |x-y|^{q_1}, \tag{7.8}$$

in every bounded interval I, where q_1, $0 < q_1 \leqq 1$, depends only on k, while the finite constant C_1 also depends on I and the function φ.

To estimate $|\varphi(x)|$ for large values of $|x|$ we note that the inverse of φ is also quasisymmetric and therefore Hölder continuous with respect to the spherical metric. Since the spherical distance between x and ∞ is of the same order as $1/|x|$ for large $|x|$, there corresponds to every $r > 0$ a finite C_2 such that

$$\frac{1}{|x|} < \frac{C_2}{|\varphi(x)^{q_2}|} \tag{7.9}$$

holds for $|x| > r$, where $0 < q_2 \leqq 1$. It is easy to see that q_2 too depends only on k.

If finally we take two real numbers x, y such that y belongs to a bounded interval I and x is arbitrary, then it follows from (7.8) and (7.9) that

$$|\varphi(x) - \varphi(y)| < C\,(|x-y|^{q_1} + |x-y|^{1/q_2})\,, \tag{7.10}$$

where C depends on I and φ.[24]

7.4. Approximation of a quasisymmetric function. As above let φ be a k-quasisymmetric function on the x-axis. For every positive integer n we introduce the integral

$$\varphi_n(z) = \int_{-\infty}^{\infty} c_n\, \varphi(r)\, e^{-n(r-z)^2}\, dr\,,$$

where

$$c_n = \left(\int_{-\infty}^{\infty} e^{-nr^2}\, dr\right)^{-1} = \sqrt{\frac{n}{\pi}}\,.$$

It follows from inequality (7.10) that this integral converges uniformly in every compact set of the finite plane and defines an analytic function φ_n of the complex variable $z = x + iy$.

The functions φ_n have the following properties:

1°. *The restriction of φ_n to the x-axis is k-quasisymmetric.* For we have

$$\varphi_n(x+t) - \varphi_n(x) = c_n \int_{-\infty}^{\infty} (\varphi(x+t+r) - \varphi(x+r))\, e^{-nr^2}\, dr$$

for all real x and t.

2°. *The derivative of φ_n is positive at every point of the x-axis.* Indeed for real x we have

$$\varphi_n'(x) = 2nc_n \int_{-\infty}^{\infty} \varphi(r)\,(r-x)\, e^{-n(r-x)^2}\, dr$$

$$= 2nc_n \int_{-\infty}^{\infty} (\varphi(r) - \varphi(x))\,(r-x)\, e^{-n(r-x)^2}\, dr\,.$$

The last expression here is positive since $\varphi(x)$ is increasing.

3°. *As $n \to \infty$, $\varphi_n(x)$ converges uniformly to $\varphi(x)$ on every bounded segment of the real axis.*

This assertion can be proved as follows: If x lies in a bounded interval I, then by (7.10)

$$|\varphi_n(x) - \varphi(x)| = c_n \left| \int_{-\infty}^{\infty} (\varphi(x+r) - \varphi(x))\, e^{-nr^2}\, dr \right|$$

$$\le c_n \int_{-n^{-1/3}}^{n^{-1/3}} |\varphi(x+r) - \varphi(x)|\, e^{-nr^2}\, dr + 2c_n C \int_{n^{-1/3}}^{\infty} (r^{q_1} + r^{1/q_2})\, e^{-nr^2}\, dr\,.$$

---

[24] Despite the Hölder continuity a quasisymmetric function need not be absolutely continuous (cf. the remark in IV.1.4).

Since φ is uniformly continuous on every bounded interval the first integral on the right tends uniformly to zero as $n \to \infty$. This is also true for the second integral as we see from the following inequality, valid for $q > 0$ and $n > (q/2)^3$,

$$\int_{n^{-1/3}}^{\infty} r^q e^{-nr^2} dr = \int_{n^{-1/3}}^{\infty} e^{-nr^2 + q \log r} dr$$

$$\leq \int_{n^{-1/3}}^{\infty} \frac{2nr - \frac{q}{r}}{2n^{2/3} - q n^{1/3}} e^{-nr^2 + q \log r} dr = n^{-q/3} \cdot (2 n^{2/3} - 2 n^{1/3})^{-1} e^{-n^{1/3}}.$$

Thus the functions φ_n have all the properties 1°—3°. This approximation to a quasisymmetric function by a sequence of analytic quasisymmetric functions will be applied in order to solve a problem on conformal mappings.

7.5. A sewing theorem for conformal mappings. We now look at a purely function-theoretical problem. It concerns mapping two Jordan domains with a common boundary curve conformally onto the upper and lower half-planes in such a way that a given relation holds between the boundary values of the mappings. The following theorem states that quasisymmetry of the relation is a sufficient condition; for a partial converse see 8.6.

Sewing theorem.[25] *Let φ be a quasisymmetric function on the real axis. Then the upper and lower half-planes can be mapped conformally onto disjoint Jordan domains such that the boundary values f_1 and f_2 satisfy the relation $f_1(x) = f_2(\varphi(x))$.*

Proof: By applying the approximation process of 7.4 we can carry out the proof in two steps. We first suppose that φ can be extended to an analytic function in the finite plane which has the properties 1° and 2° of 7.4. We then construct two conformal mappings f_1 and f_2 of the upper and lower half-planes, respectively, which satisfy the condition $f_1(x) = f_2(\varphi(x))$ on an arbitrarily given interval I. In the second step an arbitrary quasisymmetric function is approximated by analytic functions of the above type and the interval I tends to the whole x-axis.

Thus let φ be an analytic function in the whole plane, having the properties 1° and 2°. There is no loss of generality in assuming that $\varphi(I) = I$. In fact, if we can find the required mapping in the case $\varphi(I) \neq I$, we

[25] Lehto-Virtanen [1], Pfluger [3]. The sewing theorem will be used in the proof of the so-called existence theorem in V.1, whereas Pfluger's proof depends on an application of the existence theorem.

need only write $\tilde{\varphi}(z) = a\,\varphi(z) + b$ with $a > 0$ and b chosen so that $\tilde{\varphi}(I) = I$. If f_1 and f_2 satisfy the relation $f_1(x) = f_2(\tilde{\varphi}(x))$ then $f_1(x) = \tilde{f}_2(\varphi(x))$ where $\tilde{f}_2(z) = f_2(a\,z + b)$.

For every $\nu = 0, 1, 2, \ldots$, let $S_{1,\nu}$ be the subdomain of the upper half-plane whose boundary consists of $\bar{I}$ and a circular arc which meets $\bar{I}$ at its end-points in angles $(2/3)^\nu \pi$. Let $S_{2,\nu}$ denote the reflection of $S_{1,\nu}$ in I. From some $\nu = n$ onwards the domains $S_{1,\nu}$ will be mapped by φ conformally on a Jordan domain lying in the upper half-plane. If this were not so, there would exist in every $S_{1,\nu}$ either two points z_1, z_2 such that $\varphi(z_1) = \varphi(z_2)$ or a point z where the imaginary part of φ is non-positive. These points would have limit points on $\bar{I}$. This leads to a contradiction since φ maps $\bar{I}$ topologically and since $\varphi'(x) > 0$ implies that every point $x \in \bar{I}$ has a disc neighbourhood whose upper half φ maps conformally into the upper half-plane.

For $\nu = n$ we now define two conformal mappings $f_{1,n}$, $f_{2,n}$ of $S_{1,n}$ and $S_{2,n}$, respectively, by setting

$$f_{1,n}(z) = \varphi(z)\,, \qquad f_{2,n}(z) = z\,.$$

Then $f_{1,n}$ maps $S_{1,n}$ onto a domain lying in the upper half-plane while $f_{2,n}$ maps the sub-domain $S_{2,n}$ of the lower half-plane onto itself. Thus the image domains are non-intersecting Jordan domains. It follows from the definition of $f_{1,n}$ and $f_{2,n}$ that the boundary condition $f_{1,n}(x) = f_{2,n}(\varphi(x))$ is satisfied on $\bar{I}$.

If the above holds for $n = 0$, then $S_{1,n}$ coincides with the upper half-plane. In this special case the pair of functions $f_{1,n}$, $f_{2,n}$ is thus a solution of our problem. The general case can be reduced to this in the following way:

We start from some value of ν, $\nu \geqq 1$, for which there exist two conformal mappings $f_{1,\nu} : S_{1,\nu} \to S'_{1,\nu}$, $f_{2,\nu} : S_{2,\nu} \to S'_{2,\nu}$ satisfying the above conditions, i.e. $S'_{1,\nu}$ and $S'_{2,\nu}$ are disjoint Jordan domains and the boundary values of the mappings satisfy $f_{1,\nu}(x) = f_{2,\nu}(\varphi(x))$ on $\bar{I}$. If we can construct corresponding mappings of $S_{1,\nu-1}$ and $S_{2,\nu-1}$, repeating the process ν times leads to the case where the domains considered are the upper and lower half-planes, respectively.

Let $T_{1,\nu}$ be the sub-domain of $S_{1,\nu}$ whose boundary consists of $\bar{I}$ and the circular arc $C_{1,\nu}$ which meets $\bar{I}$ at its endpoints in an angle $(3/4)\,(2/3)^\nu \pi$. Denote the reflections of $T_{1,\nu}$ and $C_{1,\nu}$ in I by $T_{2,\nu}$ and $C_{2,\nu}$, respectively.

We consider the sets $D_\nu = T_{1,\nu} \cup I \cup T_{2,\nu}$ and $D'_\nu = T'_{1,\nu} \cup I' \cup T'_{2,\nu}$, where $T'_{i,\nu} = f_{i,\nu}(T_{i,\nu})$ and I' denotes the Jordan arc $f_{1,\nu}(I) = f_{2,\nu}(I)$, (Fig. 10). Then D'_ν is a simply connected domain whose boundary consists of the Jordan arcs $f_{1,\nu}(C_{1,\nu}) = C'_{1,\nu}$, $f_{2,\nu}(C_{2,\nu}) = C'_{2,\nu}$ and their common endpoints. The Jordan arc $\bar{I}'$ joins these endpoints.[26]

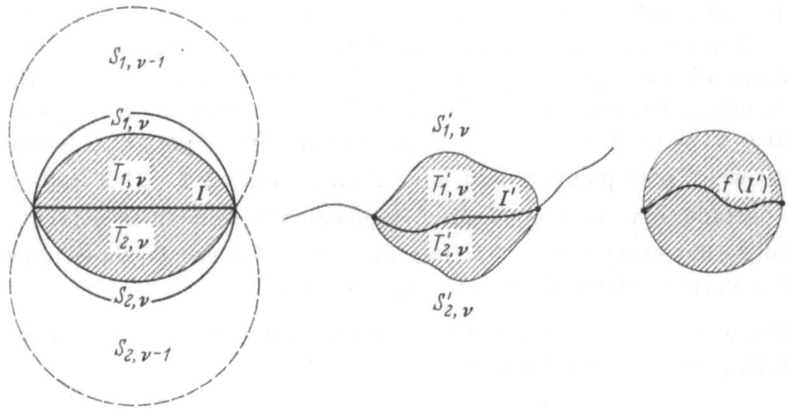

Fig. 10

Let f be a conformal mapping of D'_ν onto the unit disc. The composite functions $f_{1,\nu-1} = f \circ f_{1,\nu}$ and $f_{2,\nu-1} = f \circ f_{2,\nu}$ then map $T_{1,\nu}$ and $T_{2,\nu}$ conformally onto disjoint Jordan domains and satisfy the boundary condition $f_{1,\nu-1}(x) = f_{2,\nu-1}((\varphi(x))$ on $\bar{I}$. The circular arcs $C_{1,\nu}$ and $C_{2,\nu}$ are mapped onto the arcs $f(C'_{1,\nu})$ and $f(C'_{2,\nu})$ respectively, lying on the circumference of the unit disc. Hence $f_{1,\nu-1}$ and $f_{2,\nu-1}$ can be continued by reflection to conformal mappings of $S_{1,\nu-1}$ and $S_{2,\nu-1}$. The image domains $f_{1,\nu-1}(S_{1,\nu-1})$ and $f_{2,\nu-1}(S_{2,\nu-1})$ are disjoint and their common boundary consists of $f(\bar{I}')$ and its reflection in the unit circle. The mappings $f_{1,\nu-1}$ and $f_{2,\nu-1}$ thus possess the required properties, and we have proved the theorem in the case when φ is an analytic function with the properties 1° and 2° listed in 7.4.

Let us now consider the general case when φ is an arbitrary k-quasisymmetric function on the real axis. By 7.4 we can approximate φ by a sequence of analytic functions φ_n, $n = 1, 2, \ldots$, having the properties 1°–3°.

By the first part of the proof, there exist for each $n = 1, 2, \ldots$ two conformal mappings $f_{n,1}$ and $f_{n,2}$ which map the upper half-plane H_1

[26] The change from the domains $S_{i,\nu}$ to the $T_{i,\nu}$ is necessary, since the set $S'_{1,\nu} \cup I' \cup S'_{2,\nu}$ need not be a Jordan domain.

and the lower half-plane H_2 onto disjoint Jordan domains $G_{n,1}$ and $G_{n,2}$, respectively, and satisfy $f_{n,1}(x) = f_{n,2}(\varphi(x))$ on the interval $I_n = \{x \mid -n \leqq x \leqq n\}$. Using a linear mapping we can ensure that $f_{n,1}(0) = 0$, $f_{n,1}(1) = 1$ and $f_{n,1}(-1) = -1$. We show that $f_{n,1}$ and $f_{n,2}$ then converge to conformal mappings as $n \to \infty$.

Since all functions φ_n are k-quasisymmetric on the real axis there exist by Theorem 6.3 K-quasiconformal self-mappings w_n of H_1 with the boundary values $\varphi_n(x)$, where K is a number depending on k. By reflection each w_n can be extended to a K-quasiconformal mapping of the plane. This will still be denoted by w_n.

The mapping W_n, $n = 1, 2, \ldots$, defined by the formula

$$W_n(z) = \begin{cases} f_{n,1}(z) & \text{for } z \in H_1 \cup I_n, \\ f_{n,2}(w_n(z)) & \text{for } z \in H_2 \cup I_n, \end{cases} \tag{7.11}$$

is a homeomorphism of the slit-domain $\tilde{H}_n = H_1 \cup I_n \cup H_2$. By Theorem I.8.3 on the removability of analytic arcs it follows that W_n is K-quasiconformal.

We now make use of the convergence theorems proved in § 5. The mappings W_n are defined in every bounded domain G from some $n = n_0$ onwards and satisfy the normalization conditions $W_n(0) = 0$, $W_n(1) = 1$ and $W_n(-1) = -1$. It follows from Theorem 5.1 that $\{W_n \mid n \geqq n_0\}$ is a normal family in G. Hence, it contains a subsequence which converges uniformly to a limit function W in every bounded domain. By Theorem 5.3, W is K-quasiconformal. Since the point at infinity is removable (Theorem I.8.1), W can be extended to a K-quasiconformal mapping of the whole plane.

As regards the mappings w_n and their inverses w_n^{-1}, these converge on the real axis to φ and φ^{-1}, respectively. From Theorem 5.1 we deduce therefore that the families $\{w_n\}$ and $\{w_n^{-1}\}$ are normal. By Theorem 5.3 we can choose a subsequence such that as $n \to \infty$ all three sequences W_n, w_n and w_n^{-1} tend to K-quasiconformal mappings W, w and w^{-1} of the plane, the first converging uniformly in every bounded domain and the last two uniformly in the whole plane.

If we write $W = f_1$, $W \circ w^{-1} = f_2$, then it follows from (7.11) that $f_{n,i}$, $i = 1, 2$ converge uniformly to f_i as $n \to \infty$ in every bounded subset of $\overline{H}_i$. Thus the restictions of the functions f_1 and f_2 to H_1 and H_2 are conformal mappings which map H_1 and H_2 onto the disjoint Jordan domains $W(H_1)$ and $W(H_2)$. From $\lim \varphi_n(x) = \varphi(x)$ it follows that

$$f_1(x) = \lim f_{n,1}(x) = \lim f_{n,2}(\varphi_n(x)) = f_2(\varphi(x))$$

for every real x, and the sewing theorem is proved.

§ 8. *Quasiconformal Continuation*

8.1. Continuation to the exterior of a compact set. In this section we are concerned with extending a quasiconformal mapping $w : G \to G'$ to a quasiconformal mapping of a domain G_1 containing G as a proper subset. Such an extension is called a *quasiconformal continuation.*

The reflection principle discussed in I.8.4 gives us an example of quasiconformal continuation. This method has the advantage that the maximal dilatation remains unchanged. Here we shall prove the existence of a quasiconformal continuation under much more general hypotheses but cannot determine the best possible bound for the maximal dilatation of the extended mapping.

We begin with the following extension theorem which, apart from its intrinsic interest, is important in later applications.

Theorem 8.1. *Let $w_0 : G \to G'$ be a K-quasiconformal mapping and F a compact subset of the domain G. Then there exists a quasiconformal mapping of the whole plane which coincides with w_0 in F and whose maximal dilatation is bounded by a number depending only on K, G and F.*

Proof. We cover the given set F by finitely many closed discs lying in G. Let these discs be joined in G by Jordan arcs such that a connected compact set $F_1 \subset G$ is formed. The components of the complement of F_1 are simply connected domains (cf. I.1.9). We choose a finite number $U_1, U_2, \ldots, U_n$ of these domains such that their union covers the compact set $-G$. Finally, by Lemma I.1.4, we can separate the closed set $-G \cap U_i$ from $-U_i$ by a closed polygon C_i for each i. Denote by D_i the Jordan domain lying in U_i and bounded by C_i. The complement of $\bigcup D_i$ is a compact set $\tilde{F}$ lying in G and containing F.

If the theorem is proved for $\tilde{F}$ instead of F, then it is automatically true for F, since all ways of choosing $\tilde{F}$ depend on F and G only. Hence, there is no loss in generality in assuming from now on that the complement of F consists of n disjoint Jordan domains bounded by closed polygons. The required mapping is then obtained by applying Theorem 6.4 n times in the following way.

Since the boundary curves C_i of D_i lie in G, each of the sets $G \cup D_i$ is a domain. As a first step in the proof we show that there exists a quasiconformal mapping of $G \cup D_1$ which coincides with w_0 in F. To this end we note that the sets $G \cap D_1$ and $G \cap (-\overline{D}_1)$ are domains. This can be inferred from Lemma I.1.3 in the same way as in I.1.8.

The polygon C_1, which is a free boundary curve of $G \cap D_1$, is mapped by w_0 onto a Jordan curve C_1'. Let D_1' denote the domain bounded by C_1' which contains the set $w_0(G \cap D_1)$. By Theorem 6.4 there exists a quasiconformal mapping $w_1 : D_1 \to D_1'$ having the boundary values $w_1 = w_0$ on C_1 and a maximal dilatation K_1 which is less than a bound depending only on K and $G \cap D_1$.

The mapping

$$w(z) = \begin{cases} w_1(z) & \text{for} \quad z \in D_1 \\ w_0(z) & \text{for} \quad z \in (-D_1) \cap G \end{cases}$$

is a homeomorphism of the domain $G \cup D_1$ and has a maximal dilatation at most $\max(K, K_1)$ in $G \cup D_1 - C_1$. From Theorem I.8.3 on the removability of analytic arcs it now follows that the maximal dilatation of w in $G \cup D_1$ is also less than the same bound. Since w coincides with w_0 in F, the first part of the proof is complete.

Since the boundary of D_1 lies in F, $F_1 = F \cup D_1 = F \cup \overline{D}_1$ is a compact subset of the domain $G \cup D_1$ and its complement consists of the domains $D_2, \ldots, D_n$. In the same way as above we can construct a quasiconformal mapping of $G \cup D_1 \cup D_2$ which coincides with w in F_1 and therefore with w_0 in F. By repeating this method n times we obtain the required mapping of the whole plane.

8.2. Quasiconformal arcs and curves. By analogy with the concept of analytic arc we can define quasiconformal arcs and curves:

A Jordan arc or a Jordan curve C is called *quasiconformal* if there exists a quasiconformal mapping of a domain $G \supset C$ which carries C into a line segment or a circle.

The definition can be stated in a simpler form if C is compact. In fact, by Theorem 8.1 we have:

Every closed quasiconformal arc and every quasiconformal curve is the image of a line segment or a circle under a quasiconformal mapping of the whole plane.

A subarc of a quasiconformal curve is obviously quasiconformal. Conversely let C be a closed quasiconformal arc and w a quasiconformal mapping of the plane which carries a line segment I onto C. If I lies on the line L, then C lies on the quasiconformal curve $w(L)$, and we conclude: *A closed Jordan arc is quasiconformal if and only if it is a subarc of a quasiconformal curve.*

All analytic arcs and curves are quasiconformal by definition. We give some further examples in 8.10.

8.3. Continuation over a boundary arc. Quasiconformal arcs and curves play an important role in the study of the conditions under which a quasiconformal mapping can be continued across the boundary of a given domain.

Theorem 8.2. *Let G and G' be two domains having C and C' as either free boundary arcs or free boundary curves. Further let $w : G \to G'$ be a quasiconformal mapping under which C and C' correspond. If C and C' are quasiconformal, then w can be continued to a quasiconformal mapping of a domain G_1 containing $G \cup C$.*

Proof: Let f and g be quasiconformal mappings of certain neighbourhoods U and U' of C and C', respectively, such that C and C' are carried onto a line segment or a circle. It follows from the discussion on free boundary arcs in I.1.9 that there exists in $f(U)$ a domain D which is symmetric with respect to $f(C)$, contains $f(C)$ and is divided by $f(C)$ into two parts D_1 and D_2 such that $f^{-1}(D_1)$ lies in G and $f^{-1}(D_2)$ in $-G$. Let D', D_1' and D_2' be the corresponding sets for the mapping g. We can choose D so small that $w(f^{-1}(D_1))$ is contained in $g^{-1}(D_1')$. The reflection principle can then be applied to the mapping $g \circ w \circ f^{-1}$ in D. It follows that w can be extended to a quasiconformal mapping of the domain $G_1 = G \cup f^{-1}(D)$ and the theorem is proved.

Let us apply Theorem 8.2 to the special case when G and G' are n-tuply connected domains with quasiconformal boundary curves. By I.1.8 the boundary curves correspond pairwise under the mapping w. According to Theorem 8.2 w can therefore be extended to a quasiconformal mapping of a domain G_1 which contains the closure of G. If we combine this result with Theorem 8.1 we deduce (Springer [1]):

Theorem 8.3. *Let G and G' be two n-tuply connected domains whose boundary curves are quasiconformal. Then every quasiconformal mapping $w : G \to G'$ can be extended to a quasiconformal mapping of the whole plane.*

8.4. Quasiconformal reflection. The property of a Jordan curve of being quasiconformal can also be characterized in a simple way with the help of the anti-quasiconformal mappings mentioned in I.3.2, which are composed of a quasiconformal mapping and a reflection. Let C be a Jordan curve bounding the domains G_1 and G_2. An anti-quasiconformal mapping of G_1 onto G_2 which leaves the points of C invariant is called a *quasiconformal reflection* in C (Ahlfors [3]).

If C is quasiconformal there exists a quasiconformal mapping w of the plane which carries G_1 onto the upper half-plane H_1. Then the mapping $\Phi : G_1 \to G_2$, $\Phi(z) = w^{-1}\left(\overline{w(z)}\right)$, is a quasiconformal reflection in C.

If, conversely, C permits a quasiconformal reflection Φ, then, starting with a conformal mapping $f : H_1 \to G_1$, we define a mapping w by the formulae $w(z) = f(z)$ for $z \in H_1$, $w(z) = \Phi(f(\bar{z}))$ for $z \notin H_1$. Then w is a quasiconformal mapping of the plane which maps the real axis onto C, and C is therefore quasiconformal. Thus we have the following result: *A Jordan curve admits a quasiconformal reflection if and only if it is quasiconformal.*

8.5. Characterization of quasiconformal curves by module conditions. Let C be a Jordan curve bounding the domains G_1 and G_2. We choose four arbitrary points z_i on C such that the sequence z_1, z_2, z_3, z_4 defines the positive orientation with respect to G_1. The quadrilaterals $Q_1 = G_1(z_1, z_2, z_3, z_4)$ and $Q_2 = G_2(z_4, z_3, z_2, z_1)$ are called *conjugate* with respect to C.

If C is quasiconformal there exists a quasiconformal mapping w of the whole plane which carries C into the real axis. The images $w(Q_1)$ and $w(Q_2)$ of the above defined quadrilaterals are placed symmetrically with respect to the real axis and their canonical mappings are reflections of one another. Hence $M(w(Q_1)) = M(w(Q_2))$. Since w is quasiconformal it follows that the quotient $M(Q_2)/M(Q_1)$ has a fixed upper bound.

This necessary condition is also sufficient:

Theorem 8.4.[27] *A Jordan curve C is quasiconformal if and only if for all conjugate quadrilaterals Q_1, Q_2 with $M(Q_1) = 1$, the module of Q_2 has a fixed upper bound.*

Proof: The necessity follows from the above argument. To prove the sufficiency we denote by G_2^* the reflection of G_2 in the real axis. It follows from the hypothesis that the correspondence $z \to \bar{z}$ between the boundaries of G_1 and G_2^* satisfies a condition of type (6.2). Hence there exists a quasiconformal mapping $w : G_1 \to G_2^*$ with boundary values $w(z) = \bar{z}$. Then the mapping $\bar{w}$ is a quasiconformal reflection in C and the assertion follows from the result in 8.4.

8.6. Characterization of a quasiconformal curve by means of conformal mappings. Here we remind the reader of the function-theoretical sewing theorem proved in 7.5. It follows from the next theorem (Tienari [1]) that the sufficient condition given in the sewing theorem is also necessary under the additional assumption that the image of the real axis is a quasiconformal curve. As in 7.5 let $f_1 : H_1 \to G_1$ and $f_2 : H_2 \to G_2$ be conformal mappings with the same value at infinity.

---

[27] cf. Pfluger [3].

Theorem 8.5. *The Jordan curve C is quasiconformal if and only if $f_2^{-1} \circ f_1$ is quasisymmetric on the real axis.*

Proof: If C is quasiconformal then by Theorem 8.3, f_2 can be quasiconformally continued throughout the plane. Thus there exists a quasiconformal mapping $w_1 : H_1 \to G_1$ with the boundary values f_2. The composite function $w_1^{-1} \circ f_1$ is then a quasiconformal mapping of the upper half-plane onto itself which preserves the point at infinity and has the boundary values $f_2^{-1} \circ f_1$. Hence $f_2^{-1} \circ f_1$ is quasisymmetric on the real axis.

Conversely, if $f_2^{-1} \circ f_1$ is quasisymmetric on the real axis, there exists a quasiconformal mapping $w_1 : H_1 \to H_1$ with the boundary values $f_2^{-1} \circ f_1$. Hence $f_1 \circ w_1^{-1}$ has the same boundary values on the real axis as f_2, and these functions define a homeomorphism of the plane. Since an analytic curve is removable, this mapping is quasiconformal. It maps the real axis onto C, which is therefore quasiconformal.

8.7. Metric characterization of quasiconformal curves. Let us consider a Jordan curve C and two finite points z_1, z_2 on it. They divide C into two arcs, and we consider the one with the smaller euclidean diameter. If the quotient of this diameter and the distance $|z_1 - z_2|$ is bounded for all finite points z_1, z_2 on C, we say that C is of *bounded turning*.

This geometrically intuitive property turns out to be equivalent to the quasiconformality of C (cf. Ahlfors [3]):

Theorem 8.6. *A Jordan curve is quasiconformal if and only if it is of bounded turning.*

Proof: We suppose first that C is of bounded turning. The proof that it is then quasiconformal depends on Theorem 8.4. Let G_1 and G_2 be the domains bounded by C and $Q_1 = G_1(z_1, z_2, z_3, z_4)$ a quadrilateral of module one. We shall derive an upper bound for the module of the conjugate quadrilateral $Q_2 = G_2(z_4, z_3, z_2, z_1)$.

We denote by s_a and s_b the distances between the a- and b-sides of Q_1 measured in Q_1 (cf. I.4.3), and by d_a and d_b the corresponding distances in the plane. Then clearly $s_b \geqq d_b$. Since $M(Q_1) = 1$ the inequality (6.23) of 6.6 holds so that $s_a/s_b > 10^{-3}$. Hence

$$s_a > 10^{-3}\, d_b\,. \tag{8.1}$$

Since C is of bounded turning there exists a number h with the following property: If C is divided into two arcs by any two points z_1, z_2 then at least one of these arcs lies inside a circle of diameter $\leqq h\,|z_1 - z_2|$. In particular, a b-side of Q_1 (and of Q_2) lies inside a circle of diameter $h\,d_a$.

This fact, together with (8.1), yields

$$d_a > \frac{10^{-3}}{\pi h} d_b . \tag{8.2}$$

For if not, one b-side lies inside a circle of diameter $\leqq 10^{-3} d_b/\pi$. The other b-side, which is at a distance d_b from this one, must lie outside the circle. It follows that a-sides can be joined in Q_1 by a circular arc of length $\leqq 10^{-3} d_b$. This contradicts (8.1), and (8.2) is true.

The module of Q_2 can now be estimated by means of Theorem I.4.1. First we obtain as above a disc $|z - z_0| < h d_b/2$ which contains an a-side of Q_2. We define

$$\varrho(z) = \begin{cases} 1 & \text{for} \quad |z - z_0| < \frac{1}{2} h d_b + 10^{-3} \frac{d_b}{\pi h}, \\ 0 & \text{elsewhere} . \end{cases}$$

It follows from (8.2) that the ϱ-length of an arc joining the a-sides of Q_2 is at least $10^{-3} d_b/(\pi h)$. Hence by Theorem I.4.1 we have

$$M(Q_2) \leqq \frac{\pi (h d_b/2 + 10^{-3} d_b/(\pi h))^2}{10^{-6} d_b^2/(\pi h)^2} = 10^6 \pi \left(\frac{\pi h^2}{2} + 10^{-3}\right)^2 .$$

The quasiconformality of C follows from Theorem 8.4.

Now let us suppose that C is quasiconformal. Then for some K, there is a K-quasiconformal mapping w of the plane which maps C onto the real axis.

To prove that C is of bounded turning we choose two finite points z_1, z_2 which divide C into the subarcs C_1, C_2. We show the existence of a number h, independent of z_1 and z_2, such that one of the arcs C_i lies inside a circle of diameter $h |z_1 - z_2|$.

Let p_i be the point of C_i which is at the maximum distance from z_1. There is no loss of generality in assuming that the distances $|p_1 - z_1|$ and $|p_2 - z_1|$ are greater than $|z_2 - z_1|$, since otherwise the assertion holds trivially for every $h > 2$. If $d = \min (|p_1 - z_1|, |p_2 - z_1|)$ we can then construct the annulus $B = \{z \mid d > |z - z_1| > |z_2 - z_1|\}$; this has the module

$$M(B) = \log \frac{d}{|z_2 - z_1|} . \tag{8.3}$$

The image domain $B' = w(B)$ separates the points $z_1' = w(z_1)$, $z_2' = w(z_2)$ from the points $p_1' = w(p_1)$, $p_2' = w(p_2)$. All these points lie on the real axis and we can normalize w so that $z_1' = 0 < p_1' < z_2'$ and $p_2' = \infty$.

Thus B' separates 0, z_2' from p_1', ∞. Its module can therefore be estimated by Teichmüller's module theorem (cf. 1.3). This gives

$$M(B') \leqq 2\,\mu\left(\sqrt{\frac{z_2'}{z_2' + p_1'}}\right).$$

Since μ is decreasing $M(B') < 2\,\mu\left(1/\sqrt{2}\right)$ and it follows from (2.8) that

$$M(B') < \pi\,. \tag{8.4}$$

On the other hand, $M(B) \leqq K\,M(B')$, since w is K-quasiconformal. By (8.3) and (8.4) we thus have

$$d < e^{\pi K}\,|z_2 - z_1|\,.$$

Hence, if $h = 2\,e^{\pi K}$, one of the arcs C_i lies inside a circle of diameter $h\,|z_1 - z_2|$, and the second part of the theorem is proved.

8.8. Turning in the spherical metric. The property of a Jordan curve of being of bounded turning can also be expressed in the spherical metric. Of the arcs into which the curve C is divided by two (finite or infinite) points z_1, z_2 we consider the one having the smaller spherical diameter. Then the following result is true: *C is of bounded turning if and only if the quotient of this diameter and the spherical distance between z_1 and z_2 is bounded for all points $z_1, z_2 \in C$.*

This follows from Theorem 8.6: By a linear mapping which corresponds to a rotation of the Riemann sphere the curve C can always be transformed so as not to contain the point at infinity. Such a rotation preserves the spherical distances and the quasiconformality of a curve. Thus the curve C is quasiconformal and hence of bounded turning if and only if it satisfies the above spherical condition.

For a Jordan arc it is convenient to give the definition in the spherical metric: We say that the arc C is of bounded turning if the ratio of the spherical diameter of its subarc $\widehat{z_1 z_2}$ to the spherical distance between z_1 and z_2 is bounded for all pairs $z_1, z_2 \in C$.

If C is bounded, spherical distances can be replaced by euclidean ones, and we obtain the following result:

A bounded Jordan arc is of bounded turning if and only if there exists a finite number h such that

$$|z_1 - z_2| \leqq h\,|z_1 - z_3| \tag{8.5}$$

for all points z_1, z_2, z_3 lying in this order on C.

8.9. Local characterization of a quasiconformal curve. Let C be a curve or an arc of bounded turning and $h > 0$ a number such that the inequality

$$\Delta(\widehat{z_1, z_2}) \leqq h\, k(z_1, z_2) \tag{8.6}$$

holds for every pair of points z_1, z_2 on C, where $\Delta(\widehat{z_1, z_2})$ denotes the spherical diameter of a subarc $\widehat{z_1, z_2}$ of C bounded by z_1 and z_2[28]. Since $\Delta \leqq \pi/2$ in any case, (8.6) is automatically true if $k(z_1, z_2) \geqq \pi/2\, h$.

To express more clearly the local character of the quasiconformality of a curve, we first note: *Every closed subarc C_1 of a quasiconformal curve C is of bounded turning.*

In fact, if z_1 and z_2 lie on C_1, then (8.6) holds at least for one subarc $\widehat{z_1 z_2}$ of C. If $k(z_1, z_2) < \Delta\,(C - C_1)/h$, this arc must belong to C_1. By the above remark, (8.6) then holds for subarcs of C_1 for some h.

By 8.2 every closed quasiconformal arc is a subarc of a quasiconformal curve. We thus infer:

A closed quasiconformal arc is of bounded turning.

The converse is also true as has been proved by Rickman [1]. We restrict ourselves to the following result which is easier to establish: *If an open Jordan arc C is of bounded turning then its closed subarcs are quasiconformal.*

To prove this we may suppose that C lies in the finite plane. Every closed subarc C_1 of C lies in a closed subarc C_2 of C whose endpoints p_1, p_2 are at a positive distance $2r$ from C_1. Since C_2 does not separate the plane (I.1.3), we can choose a point q_i in each of the discs $|z - p_i| < r$, $i = 1, 2$, and join these points by a polygonal arc lying at a positive distance from C_2. If we also join q_1 and q_2 with C_2 by shortest possible line segments then (by discarding certain loops if necessary) we obtain a closed curve C_3 which contains C_1 as a subarc. This curve C_3 is of bounded turning. For if the points z_1, z_2 both lie on $C_2 \cap C_3$ or on $C_3 - C_2$, then the required condition is obviously satisfied. In the other cases its validity follows from the fact that the q_i are joined to C_2 by the shortest possible line segments. Thus by Theorem 8.6, C_3 is quasiconformal and hence so also is its subarc C_1.

The proposed local characterization of the quasiconformality of a Jordan curve runs as follows:

Theorem 8.7. *A Jordan curve C is quasiconformal if and only if every point of C belongs to an open subarc of C of bounded turning.*

---

[28] The smaller of two such arcs if C is a curve.

Proof: If C is quasiconformal, every closed subarc of C is of bounded turning as we have just seen. The same is true for open subarcs of C which are contained in some closed subarc of C. This means that the condition given in the theorem is necessary.

We now suppose that every point of C belongs to an open subarc of bounded turning. Then C can be covered by finitely many subarcs $C_1, \ldots, C_n$ of this type. Let z_1, z_2 be two points of C. If they lie on the same arc C_i then by hypothesis there exists an h satisfying (8.6). If z_1 and z_2 do not belong to the same C_i then $k(z_1, z_2)$ is greater than a positive bound r, and (8.6) holds for $h = \pi/2\, r$. Hence C is of bounded turning and so quasiconformal by Theorem 8.6.

8.10. Examples of quasiconformal arcs. From condition (8.5) we can deduce simple sufficient criteria for the quasiconformality of a Jordan arc C.[29] If for example there is a homeomorphism φ of the unit interval $0 < t < 1$ onto C, satisfying a Lipschitz condition

$$m\,|t_1 - t_2| \leqq |\varphi(t_1) - \varphi(t_2)| \leqq M\,|t_1 - t_2|\,,$$

then all compact subarcs of C are quasiconformal. In particular, every smooth closed Jordan arc is quasiconformal. The same holds if C consists of finitely many smooth arcs which meet pairwise at non-zero angles.

If the arc C has a zero angle at some point, it is not quasiconformal, for the condition (8.5) cannot hold in any neighbourhood of this point.

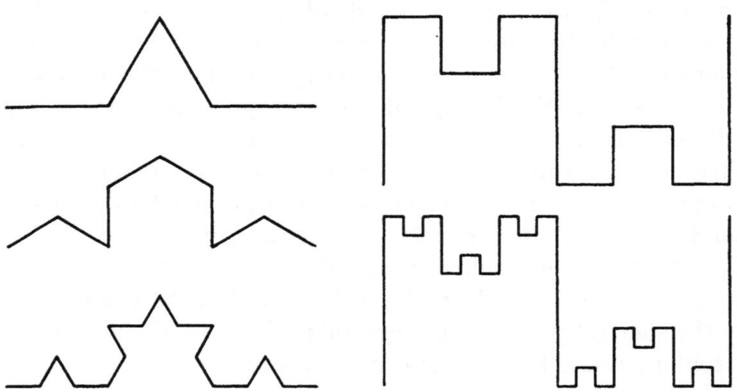

Fig. 11

On the other hand *there exist quasiconformal arcs which are not rectifiable.* Without giving a precise proof, we have shown two methods of construction in Fig. 11 which by unlimited repetition lead to non-rectifiable

[29] See also Pfluger [3], Tienari [1], Väisälä [2].

quasiconformal arcs.[30] The first of these does not even contain any rectifiable subarcs.

Once it was shown that quasiconformal arcs can be non-rectifiable, the question arose about their Hausdorff dimension (see III.1.7). This problem has recently been solved by Gehring and Väisälä [2]. They proved that while the Hausdorff dimension of a quasiconformal arc is always < 2 it can take any value λ, $1 \leqq \lambda < 2$.

§ 9. Circular Dilatation

9.1. Definition of the circular dilatation. In I.9.4 we defined the dilatation quotient D of a homeomorphism w at every regular point of w. The dilatation quotient admits a differential geometric interpretation as the ratio of the axes of the image ellipse of an infinitesimal disc. At a (not necessarily regular) point z it is therefore natural to measure the local dilatation of w not only by the dilatation of quadrilaterals in the neighbourhood of z but also by the distortion of the images of small circles with centre z.

For this purpose we write

$$H_w(z) = H(z) = \limsup_{r \to +0} \frac{\max_\alpha |w(z + r e^{i\alpha}) - w(z)|}{\min_\alpha |w(z + r e^{i\alpha}) - w(z)|}$$

for $z \neq \infty$, $w(z) \neq \infty$, and extend the definition by the formulae $H_w(\infty) = H_{\tilde{w}}(0)$ where $\tilde{w}(z) = w(1/z)$, and $H_w(z) = H_{1/w}(z)$ for $w(z) = \infty$. The function H is called the *circular dilatation* of the topological mapping w.

If $z \neq \infty$ is a regular point and $w(z) \neq \infty$, then

$$\begin{aligned} w(z + r e^{i\alpha}) - w(z) &= r\left(w_z(z)\, e^{i\alpha} + w_{\bar{z}}(z)\, e^{-i\alpha}\right) + o(r) \\ &= r\, e^{i\alpha}\, \partial_\alpha w(z) + o(r)\,. \end{aligned}$$

It follows that $H(z) = D(z)$ in this case. Because of the invariance of H with respect to linear mappings this relation also holds if $z = \infty$ or $w(z) = \infty$. In other words *the dilatation quotient D is the restriction of the circular dilatation H to regular points.*

9.2. Upper bounds for the circular dilatation. By Theorem I.9.4, for a regular quasiconformal mapping we have $F(z) = D(z) = H(z)$ at every point of its domain. We now examine the mutual relationship between F and H at non-regular points.

---

[30] The idea of such direct constructions is due to G. Piranian.

For the homeomorphism w,

$$w(z) = e^{-\frac{1}{|z|} + i \arg z},$$

of the finite plane onto the unit disc we have $H(0) = 1$, $F(0) = \infty$. Thus we cannot infer that $F(z)$ is finite if $H(z)$ is.

However, in the opposite direction the following result holds (Mori [2]):

Theorem 9.1. *For a K-quasiconformal mapping, H is bounded by a number depending only on K.*

Proof: If we do not attempt to find the best possible bound for H, the theorem can be proved as follows.

Let $w: G \to G'$ be a K-quasiconformal mapping; we may assume here that G and G' are domains in the finite plane. For $z_0 \in G$, $D_r = \{z \mid |z - z_0| < r\}$, $\overline{D}_r \subset G$, we write

$$m_1(r) = \max_{|z - z_0| = r} |w(z) - w(z_0)|, \quad m_2(r) = \min_{|z - z_0| = r} |w(z) - w(z_0)|$$

and denote by z_1 and z_2 the points at which the maximum and minimum are attained. Choose the number r so small that the disc $|w - w(z_0)| \leqq m_1(r)$ is contained in G'.

We consider the annulus $B' = \{w \mid m_2(r) < |w - w(z_0)| < m_1(r)\} \subset G'$. Its preimage B is a ring domain which separates the points z_0, z_2 from z_1, ∞. Since $|z_1 - z_0| = |z_2 - z_0|$ we have

$$M(B) \leqq 2\,\mu\left(\frac{1}{\sqrt{2}}\right) = \pi$$

by Teichmüller's module theorem (see 1.3) and by the formula (2.8).

On the other hand, $M(B') = \log\left(m_1(r)/m_2(r)\right)$ and $M(B') \leqq K\,M(B)$. It follows that

$$m_1(r)/m_2(r) \leqq e^{\pi K}. \tag{9.1}$$

Hence also $H(z_0) \leqq e^{\pi K}$.

9.3. Lower bound for the circular dilatation. For a K-quasiconformal mapping, $H(z)$ can in fact be greater than K. We make use of the distortion function $\lambda(K)$, introduced in 6.3, which has the asymptotic expansion

$$\lambda(K) = \frac{1}{16} e^{\pi K} - \frac{1}{2} + O(e^{-\pi K})$$

and prove (Lehto-Virtanen-Väisälä [1]):

Theorem 9.2. *For every $\varepsilon > 0$ there exists a K-quasiconformal mapping for which*

$$H(z) \geqq \lambda(K) - \varepsilon \tag{9.2}$$

at some point z.

Proof: By 2.2 and 6.3 the module of the quadrilateral consisting of the upper half-plane and the vertices $-\lambda(K)$, 0, 1, ∞ is

$$\frac{2}{\pi}\mu\left(\frac{1}{\sqrt{\lambda(K)+1}}\right) = K.$$

It follows from the remark made in I.3.4 that there exists a K-quasiconformal self-mapping w of the upper half-plane with $w(0) = 0$, $w(1) = 1$, $w(-1) = -\lambda(K)$, $w(\infty) = \infty$. Let this mapping be extended to the whole plane by reflection.

We write $B_n = \{z \mid 1/n < |z| < n\}$, $B_n' = w(B_n)$, $n = 2, 3, \ldots$, and map B_n' conformally onto an annulus $A_n = \{\zeta \mid a_n < |\zeta| < b_n\}$ by a function f_n with $f_n(1) = 1$. The composite K-quasiconformal mapping $w_n = f_n \circ w : B_n \to A_n$ can then be extended to the punctured plane $0 < |z| < \infty$ by repeated reflection. Since $z = 0$ and $z = \infty$ are removable singularities by Theorem I.8.1, w_n admits a K-quasiconformal continuation to the whole plane and we have $w_n(0) = 0$, $w_n(1) = 1$, $w_n(\infty) = \infty$.

It follows from Theorem 5.1 that $\{w_n\}$ is a normal family. By considering a subsequence and applying Theorem 5.3 we see that w_n converges to a quasiconformal mapping of the plane as $n \to \infty$. The mappings $f_n = w_n \circ w^{-1}$ are conformal in every compact subset of the punctured plane $0 < |z| < \infty$ from some n onwards, and converge to a conformal mapping f of the plane. Since $f(0) = 0$, $f(1) = 1$, $f(\infty) = \infty$, f is the identity mapping.

Hence to every $\varepsilon > 0$ there corresponds an N such that

$$|w_N(-1) + \lambda(K)| = |w_N(w^{-1}(-\lambda(K))) + \lambda(K)| < \varepsilon.$$

Since $w_N(1) = 1$ we have therefore on $|z| = 1$

$$\frac{\max |w_N(z)|}{\min |w_N(z)|} > \lambda(K) - \varepsilon.$$

This inequality also holds on every circle $|z| = N^{-2k}$, $k = 1, 2, \ldots$, obtained as images of $|z| = 1$ under the above reflections. Hence (9.2) holds at $z = 0$ for the K-quasiconformal mapping w_N.

9.4. Supremum of the circular dilatation. By Theorem 9.2 the supremum of the values which the circular dilatation can assume for K-quasiconformal mappings is at least $\lambda(K)$. In fact, equality holds:

Theorem 9.3. *In the family $\mathscr{W}$ of K-quasiconformal mappings w we have*

$$\sup_{w \in \mathscr{W}} H_w(z) = \lambda(K) .$$

The proof of this result requires function-theoretical tools different from the other methods we have used here. We shall therefore not prove the result but refer to the joint work of Väisälä and the authors [1].

III. Auxiliary Results from Real Analysis

Introduction to Chapter III

In Chapters I and II we have developed the theory of quasiconformal mappings from the geometric definition by using topological and function-theoretical methods. However, it is clear that some fundamental problems in the theory of quasiconformal mappings have a natural connection with real analysis. As examples we mention the questions concerning differentiability and preservation of null sets. For the further development of the theory it is therefore necessary to apply methods belonging to measure and integration theory.

While the auxiliary results from topology and complex function theory were only briefly dealt with in the first two sections of Chapter I, we have here preferred to devote a whole chapter to results from real analysis. This is not only because of the difficulty of giving precise references, but because some of the theorems are proved in a form in which they are directly applicable to quasiconformal mappings. For example, we deduce the results on differentiability and preservation of null sets in Chapter IV as special cases of theorems proved in the present chapter.

In § 1 we collect some notations and basic results on the concepts of measure and integral. For further particulars we refer the reader to the text-books of Munroe [1] or Saks [1]. Also § 2, in which we assign to every homeomorphism a completely additive set function, is of a general nature.

In § 3 we prove that a homeomorphism with almost everywhere finite partial derivatives is almost everywhere differentiable. This result will often find application later.

In § 4 we investigate the module of a curve family by applying methods developed in §§ 1—3. This concept was introduced in Chapter I in special cases.

Having presented in § 5 some methods of approximating a measurable function, we consider in § 6 functions with L^p-derivatives. Many theorems proved here will be applied directly to quasiconformal mappings in Chapter IV, for example, the result saying that a homeomorphism with L^2-derivatives is absolutely continuous with respect to the area measure.

Finally in § 7 we deal with the two-dimensional Hilbert transformation. As an application we derive a relation between the complex derivatives of a function possessing L^2-derivatives.

§ 1. *Measure and Integral*

1.1. Outer measure. We begin with some introductory remarks on the basic concepts of measure and integration theory. We assume that the reader is acquainted with these; our purpose is to establish the terminology to be used later.

In what follows a real set function is a mapping into the extended line; thus the values $+\infty$ and $-\infty$ are not excluded. A real set function μ^* defined for all subsets of a space Ω is called an *outer measure*[31] if it satisfies the following axioms: 1. $A \subset B$ implies that $\mu^*(A) \leqq \mu^*(B)$. 2. $\mu^*(\emptyset) = 0$. 3. For every sequence of sets A_n we have $\mu^*(\bigcup A_n) \leqq \sum \mu^*(A_n)$.

We consider only the case when Ω is a metric space (usually a point set in the plane). All outer measures which occur will satisfy not only 1—3 but also the following two axioms: 4. If the sets A and B are at a positive distance from one another, then $\mu^*(A \cup B) = \mu^*(A) + \mu^*(B)$. 5. For every set $A \subset \Omega$ there exists a sequence of open sets $\Omega \supset G_1 \supset G_2 \supset \cdots$ such that $\bigcap G_n \supset A$ and $\mu^*(\bigcap G_n) = \mu^*(A)$.

A set function which satisfies axioms 1—4 is called an *outer measure in the sense of Carathéodory*. It follows from axiom 5 that μ^* is *regular*.

Of the outer measures to be used we mention the Lebesgue outer linear and area measures. The latter is defined for an arbitrary subset A of the finite plane[32] as follows: We consider all sequences $R_1, R_2, \ldots$, where the elements R_n are open rectangles whose sides are parallel to the coordinate axes and whose union covers A. The Lebesgue outer measure is defined as $\inf \sum m(R_n) = m^*(A)$.

The Lebesgue outer linear measure l^* can be analogously defined in the case when Ω is a finite line. For a set $A \subset \Omega$ we define $l^*(A)$ as the

---

[31] Although μ^* is defined as a function we speak of the outer measure of a set A instead of the value of μ^* on A.

[32] We restrict ourselves here to the finite plane, since this is the union of countably many open sets of finite area measure (cf. 1.2). In 4.1 the area measure will be defined for sets of the whole plane.

greatest lower bound of the sum $\Sigma\, l(I_n)$, where $I_1, I_2, \ldots$ are open intervals whose union covers A, and $l(I_n)$ denotes the length of I_n.

It follows from the above definitions that m^* and l^* satisfy axioms 1—5.

The Lebesgue outer volume measure in an n-dimensional euclidean space R^n, $n > 2$, can be defined in an analogous way. We shall use this measure in the case of R^4.

1.2. Measurable sets. Let μ^* be an outer measure, defined on the subsets of the space Ω and satisfying axioms 1—5. A set $A \subset \Omega$ is called *measurable* with respect to μ^* if for every set $X \subset \Omega$ the relation $\mu^*(X) = \mu^*\,(X \cap A) + \mu^*\,(X - A)$ holds. If A is measurable, $\mu^*(A)$ is called the *measure* of A. To express the measurability of A we omit the star and write $\mu(A)$ instead of $\mu^*(A)$.

Measurable sets form a *completely additive class* $\mathcal{M}$, i.e. $\emptyset \in \mathcal{M}$, $A \in \mathcal{M}$ implies that $-A \in \mathcal{M}$ and $A_n \in \mathcal{M}$, $n = 1, 2, \ldots$, implies that $\bigcup A_n \in \mathcal{M}$. The significance of this class lies in the fact that equality

$$\mu\left(\bigcup_{n=1}^{\infty} A_n\right) = \sum_{n=1}^{\infty} \mu(A_n)\,,$$

holds in axiom 3 if the sets A_n belong to $\mathcal{M}$ and are nonintersecting. Thus the measure μ is completely additive in $\mathcal{M}$.

If $\mu^*(A) = 0$, then A is called a *null set*; such a set is always measurable, and by axiom 1 the same is true of any subset of A. As usual we shall say that a statement holds almost everywhere if it holds everywhere with the possible exception of a null set.

It follows from axioms 1—4 that every open and every closed set is measurable. There exists a uniquely determined smallest completely additive class $\mathcal{B}$ which contains all closed (and consequently also all open) sets of the space Ω. The elements of $\mathcal{B}$ are called *Borel sets*[33]; it follows from the definition that they are always measurable. By axiom 5 there corresponds to every set A a Borel set B, such that $A \subset B$ and $\mu^*(A) = \mu(B)$.

Since the outer measure μ^* is regular it possesses the following continuity property (Munroe [1], p. 95): If $A_1 \subset A_2 \subset \cdots$ is a non-decreasing sequence, then $\mu^*\,(\bigcup A_n) = \lim \mu^*(A_n)$. For a non-increasing sequence of measurable sets $A_1 \supset A_2 \supset \cdots$ the corresponding relation $\mu\,(\bigcap A_n) = \lim \mu(A_n)$ holds under the assumption that $\mu(A_1)$ is finite. Because

[33] We make the following remark on the dependence of the class $\mathcal{B}$ on the space Ω: If a subspace Ω_0 of Ω is a Borel set when considered as a subset of Ω, then the subsets of Ω_0 are simultaneously Borel sets with respect to Ω and Ω_0.

of this supplementary requirement it is not always possible, despite axiom 5, to associate with every measurable set A a sequence of open sets $G_n \supset A$ such that $\mu(A) = \lim \mu(G_n)$.

A set which is the union of countably many sets A_n with $\mu^*(A_n) < \infty$ is called *σ-finite* (with respect to the outer measure μ^*). It follows from axiom 5 that we can associate with every σ-finite measurable set $A \subset \Omega$ two Borel sets B_1 and B_2 such that $B_1 \subset A \subset B_2$ and $\mu(B_2 - B_1) = 0$. Furthermore, B_1 can be chosen as a countable union of closed sets. Thus there exists a sequence of closed sets $F_n \subset A$ such that $\lim \mu(F_n) = \mu(A)$ (cf. Munroe [1], pp. 108—110).

If the space Ω is the union of countably many open sets with finite measure (as for example in the case of the Lebesgue measure), then the following condition is necessary and sufficient for the measurability of a set $A \subset \Omega$: To every $\varepsilon > 0$ there corresponds a closed set $F \subset A$ and an open set $G \supset A$ such that $\mu(G - F) < \varepsilon$ (cf. Munroe [1], p. 111). The supremum of the measures $\mu(F)$ of closed sets $F \subset A$ is often called the *inner* measure of A and denoted by $\mu_*(A)$. If A is contained in a space Ω with finite measure then the above characterization of measurability coincides with the validity of the equation $\mu^*(A) = \mu_*(A)$. The relation $\mu(\Omega) = \mu^*(A) + \mu^*(\Omega - A)$ is also equivalent to this.

1.3. Measurable functions. A function f which is defined in a measurable set and maps this into a topological space is called measurable if the preimage $f^{-1}(G)$ of every open set G is measurable. The measurability of f depends on the choice of the outer measure μ^*; if the preimages of the open sets are Borel sets then f is measurable with respect to every μ^*. In this case f is called a *Borel function*.

All functions f considered are either real (i.e. mappings into the extended line) or complex-valued (mappings into the whole plane). In both cases there are various criteria for the measurability of f. A real function f, for example, is measurable if the set $\{z \mid f(z) > r\}$ is measurable for every rational number r. For the measurability of an almost everywhere finite complex-valued function $f = u + i\,v$ it is necessary and sufficient that the real functions u and v are measurable.

The characteristic function c_A of the set A is the real function which has the value unity at every point of A and is zero elsewhere. The function c_A is measurable if and only if A is measurable.

For measurable functions we have the following convergence theorem which is formulated here in the special case of the Lebesgue area measure:

Egoroff's Theorem. *Let f_n be a sequence of real or complex measurable functions which converges almost everywhere in a set A of finite measure to a finite function. Then for every $\varepsilon > 0$ there exists a closed set $F \subset A$, with $m(A - F) < \varepsilon$, where the convergence is uniform.*

Egoroff's theorem in this form follows almost immediately from the definition of a measurable function and from the characterization mentioned in 1.2 of the measure of A as the supremum of the measures of the closed sets $F \subset A$ (cf. Munroe [1], p. 157).

1.4. Integrable functions. Let A be a measurable set and f a real non-negative measurable function defined in A. We define

$$\int_A f\, d\mu = \sup \sum_{k=1}^{n} \mu(A_k) \inf_{z \in A_k} f(z) ,$$

where $A_1, \ldots, A_n$ are disjoint measurable subsets of A and where the product $\mu(A_k) \inf f(z) = 0$ if one of the factors vanishes even if the other is infinite. If the integral is finite we say that f is *integrable* in A.

As was remarked in 1.2, a measurable σ-finite set can be represented as the union of a Borel set and a null set. In the definition of the integral over a σ-finite set A we may therefore consider only Borel sets A_k instead of all measurable subsets of A.

A real or complex measurable function f is defined to be integrable if $|f|$ is integrable. If f is real, the non-negative functions $f^+ = (|f| + f)/2$, $f^- = (|f| - f)/2$ are integrable simultaneously with f, and the integral of f can be defined as the difference

$$\int_A f\, d\mu = \int_A f^+\, d\mu - \int_A f^-\, d\mu .$$

A complex integrable function $f = u + i\, v$ is finite almost everywhere. The real functions u, v are then integrable, no matter how they are defined at points where f is infinite. The integral of f is the sum

$$\int_A f\, d\mu = \int_A u\, d\mu + i \int_A v\, d\mu .$$

1.5. Lebesgue integrals. The integrals defined with respect to the Lebesgue linear and area measures are called Lebesgue integrals. For these we shall use the notation introduced in Chapter I. Thus, for example, we write $\iint f\, d\sigma$ instead of $\int f\, dm$ (see also 2.6).

The following result on the mutual dependence between the two Lebesgue integrals will be used frequently (cf. Munroe [1], p. 207).

Fubini's Theorem. *Let f be a real or complex valued measurable function defined in a rectangle $R = \{x + i\, y \mid x_1 < x < x_2,\ y_1 < y < y_2\}$.*

Then f is l-measurable as a function of x for almost all y, $y_1 < y < y_2$, and as a function of y for almost all x, $x_1 < x < x_2$. The integrals

$$\int_{y_1}^{y_2} |f(x, y)| \, dy \, , \qquad \int_{x_1}^{x_2} |f(x, y)| \, dx$$

are also measurable as functions of x and y, respectively, and we have

$$\iint_R |f| \, d\sigma = \int_{x_1}^{x_2} dx \int_{y_1}^{y_2} |f(x, y)| \, dy = \int_{y_1}^{y_2} dy \int_{x_1}^{x_2} |f(x, y)| \, dx \, . \tag{1.1}$$

If f is integrable in R, then the functions defined by the integrals

$$\int_{y_1}^{y_2} f(x, y) \, dy \, , \qquad \int_{x_1}^{x_2} f(x, y) \, dx$$ [34]

are integrable in $x_1 < x < x_2$ and $y_1 < y < y_2$, respectively, and

$$\iint_R f \, d\sigma = \int_{x_1}^{x_2} dx \int_{y_1}^{y_2} f(x, y) \, dy = \int_{y_1}^{y_2} dy \int_{x_1}^{x_2} f(x, y) \, dx \, . \tag{1.2}$$

Fubini's theorem can be analogously stated for arbitrary euclidean spaces R^m, R^n, R^{m+n}. We shall use the theorem also in the case $m = n = 2$. It then reads as follows:

Let $R_z = \{x + i\,y \mid x_1 < x < x_2,\ y_1 < y < y_2\}$, $R_\zeta = \{\xi + i\,\eta \mid \xi_1 < \xi < \xi_2,\ \eta_1 < \eta < \eta_2\}$ be rectangles in the $z = x + i\,y$ and $\zeta = \xi + i\,\eta$ planes and let $R = R_z \times R_\zeta = \{(z, \zeta) \mid z \in R_z,\ \zeta \in R_\zeta\}$ be their Cartesian product in R^4. If f is defined and measurable in R (with respect to the four-dimensional Lebesgue volume measure), then

$$\iint_{R_z} d\sigma_z \iint_{R_\zeta} |f(z, \zeta)| \, d\sigma_\zeta = \iint_{R_\zeta} d\sigma_\zeta \iint_{R_z} |f(z, \zeta)| \, d\sigma_z \, . \tag{1.1$'$}$$

If f is integrable in R then

$$\iint_{R_z} d\sigma_z \iint_{R_\zeta} f(z, \zeta) \, d\sigma_\zeta = \iint_{R_\zeta} d\sigma_\zeta \iint_{R_z} f(z, \zeta) \, d\sigma_z \, . \tag{1.2$'$}$$

As a second result on Lebesgue integrals we mention the following theorem, which can easily be proved with the help of Egoroff's theorem (see 1.3) (Munroe [1], p. 234).

Lebesgue convergence theorem. *Let f_n be a sequence of measurable functions which satisfy the following two conditions almost everywhere in a set A of finite Lebesgue area measure*: 1. $|f_n(z)| \leqq g(z)$, $n = 1, 2, \ldots$, *where g is integrable in A.* 2. $\lim f_n(z) = 0$. *Then*

$$\lim_{n \to \infty} \iint_A f_n \, d\sigma = 0 \, .$$

---

[34] These functions exist for almost all x and y, respectively, since the integrals $\int_{y_1}^{y_2} |f(x, y)| \, dy$ and $\int_{x_1}^{x_2} |f(x, y)| \, dx$ are finite for almost all x and y on account of (1.1).

1.6. Points of density of a set. Let A be a point set on the line. For a finite point x we consider all intervals I_x which contain x and set

$$S(x) = \liminf_{l(I_x) \to 0} \frac{l^*(A \cap I_x)}{l(I_x)}.$$

If $S(x) = 1$, then x is called a *point of* (linear) *density* of the set A. The lower limit can then be replaced by the ordinary limit.

Analogously we define points of density for a plane set A. For a finite $z \in A$ we consider all closed squares Q_z which contain z and have sides parallel to the coordinate axes. If

$$\liminf_{m(Q_z) \to 0} \frac{m^*(A \cap Q_z)}{m(Q_z)} = 1,$$

then z is called a point of density of A.

In both cases we have the following result on the existence of points of density (Munroe [1], p. 290, Saks [1], p. 129): *Almost every point of A is a point of density of A.*

For plane sets we need another more special concept of density. A finite point $z_0 = x_0 + i\,y_0$ of the $z = x + i\,y$-plane is called an *x-point of density* of the set A, if x_0 is a point of linear density of the one-dimensional set $\{x \mid x + i\,y_0 \in A\}$. Similarly, z_0 is defined to be a y-point of density if y_0 is a point of linear density of $\{y \mid x_0 + i\,y \in A\}$. A point which is both an x- and a y-point of density is called an *$x\,y$-point of density.*

In § 3 we shall need the following result on $x\,y$-points of density (Saks [1], p. 298).

Lemma 1.1. *Almost all points of a measurable set A in the finite plane are $x\,y$-points of density of A.*

1.7. Hausdorff measure. The Lebesgue linear measure is defined only for point sets which lie on a line. To determine the length of a more general plane set another type of measure must be introduced. Such a measure is obtained as a special case of the concept of α-dimensional measure, $\alpha > 0$, defined as follows:

Let A be a set of the finite plane. Consider all coverings of A by countably many open sets of diameter less than a positive number d. We associate with every such covering $\{G_1, G_2, \ldots\}$ the number $\sum d_n^\alpha$, where d_n denotes the diameter of the set G_n. The infimum of these numbers increases as $d \to 0$; thus the limit

$$\mu_\alpha^*(A) = \lim_{d \to 0} (\inf \sum_n d_n^\alpha),$$

exists. It is called the *α-dimensional Hausdorff outer measure of A*.

It follows immediately from the definition that for every $\alpha > 0$, μ_α^* is an outer measure satisfying the axioms 1–5. The restriction μ_α of μ_α^* to sets measurable with respect to μ_α^* is called the α-dimensional Hausdorff measure.

We conclude from the definition that for a fixed A, $\mu_\alpha^*(A)$ does not increase with increasing α and is positive and finite for at most one value of α. We define the *dimension* of A as the least upper bound

$$\dim A = \sup \{\alpha \mid \mu_\alpha^*(A) = \infty\};$$

then we have $\mu_\alpha^*(A) = 0$ for $\alpha > \dim A$ and $\mu_\alpha^*(A) = \infty$ for $\alpha < \dim A$.

In some later examples we shall use the following result on the dimension of the Cartesian product of two sets:

Lemma 1.2. *Let A and B be compact sets on the finite x- and y-axes. Then* $\dim (A \times B) = \dim A + \dim B$.

The inequality $\dim (A \times B) \leqq \dim A + \dim B$ is an almost immediate consequence of the definition of dimension. To prove the converse relation $\dim (A \times B) \geqq \dim A + \dim B$ we refer to Marstrand [1], p. 198.

1.8. Linear measure. Among all the above measures μ_α the one-dimensional Hausdorff measure μ_1 will most often be used here. If the set A lies on a line then this measure agrees with the Lebesgue linear measure. We shall therefore call $\mu_1(A)$ the linear measure of A and denote it by $l(A)$ also in the general case.

For a set A of finite linear measure the Lebesgue area measure vanishes. By Fubini's theorem the one-dimensional set $A(y_0) = \{z = x + i y_0 \mid z \in A\}$ is therefore of zero linear measure for almost all values y_0. In fact, a much sharper result is true: On every horizontal line $y = y_0$ there is only a finite number of points of A, with the possible exception of a null set of values y_0 (Gross [1]). This can be deduced from the following lemma, which will find an application in V.3.

Lemma 1.3. *Let A be a plane set, N a positive integer, and E the set of values y_0 for which $A(y_0)$ contains at least N points. Then* $l^*(E) \leqq l^*(A)/N$.

Proof: We denote by E_n the set of values y_0 with the following property: On the line $y = y_0$ there exist N points of A whose mutual distances are greater than $1/n$. Then we have $E_1 \subset E_2 \subset \cdots$ and $\bigcup E_n = E$. Thus $l^*(E) = \lim l^*(E_n)$, and we must now prove that $l^*(E_n) \leqq l^*(A)/N$ for every positive integer n.

For a fixed n we consider an arbitrary covering of A by open sets $G_1, G_2, \ldots$, with diameters $d_1, d_2, \ldots$ less than $1/n$. In view of the definition of the linear measure we have to prove that $\Sigma\, d_k \geqq N\, l^*(E_n)$.

For $y_0 \in E_n$ the line $y = y_0$ contains N points which lie in different sets G_k. If P_k denotes the projection of G_k onto the y-axis, then every point of E_n belongs to at least N sets P_k. Let $P_k = \bigcup I_{jk}$, where I_{jk}, $j = 1, 2, \ldots$, are disjoint open segments. Then every point of E_n belongs to at least N segments I_{jk}.

If the number of segments I_{jk} is finite, the relation $N\, l^*(E_n) \leqq \Sigma\,\Sigma\, l(I_{jk}) \leqq \Sigma\, d_k$ obviously follows. For a countably infinite covering we obtain the same result by a limiting process. In fact, if O_m denotes the set of points which belong to at least N of the first m segments in the sequence $I_{11}, I_{12}, I_{21}, \ldots$, then $O_N \subset O_{N+1} \subset \cdots$ and $E_n \subset \bigcup O_m$. It follows that

$$l^*(E_n) \leqq \lim_{m \to \infty} l(O_m) \leqq \frac{1}{N}\, \Sigma\, d_k\,,$$

and the lemma is proved.

§ 2. *Absolute Continuity*

2.1. Absolutely continuous additive set functions. Let τ be a completely additive set function defined in the class of Borel sets of the finite plane, and μ a Hausdorff or Lebesgue measure. The function τ is called *absolutely continuous* in a set E (with respect to the measure μ) if the following condition is satisfied: For every $\varepsilon > 0$ there exists $\delta > 0$ such that $|\tau(B)| < \varepsilon$ for every Borel set $B \subset E$ with $\mu(B) < \delta$. If τ is absolutely continuous in every compact subset of E, we say that τ is *locally absolutely continuous* in E.

If τ is absolutely continuous and B is a Borel set with $\mu(B) = 0$, then obviously $\tau(B) = 0$. Conversely it is easy to prove (Munroe [1], p. 191):

A bounded completely additive set function τ is absolutely continuous if it vanishes for every Borel null set.

As examples of completely additive absolutely continuous set functions we exhibit the integrals of functions integrable with respect to μ. They are bounded and vanish for every null set. On the other hand, the

Radon-Nikodym theorem (Munroe [1], p. 196) states that a completely additive absolutely continuous set function τ is always the integral of an integrable function, provided that τ is bounded and vanishes outside a σ-finite set.

2.2. Absolutely continuous homeomorphisms. Suppose that E and E' are Borel sets of the finite plane, $f: E \to E'$ is a homeomorphism, and μ is a Hausdorff or Lebesgue measure. Since a homeomorphism maps closed sets onto closed sets, Borel sets remain Borel sets (cf. the footnote in 1.2). The image of a Borel set lying in E is therefore measurable with respect to μ.

We associate with the mapping f a set function μ_f by the formula

$$\mu_f(B) = \mu\left(f(B \cap E)\right),$$

where B runs through all Borel sets. The function μ_f clearly has all the properties of a measure.

We define the homeomorphism f to be *absolutely continuous* or *locally absolutely continuous* (with respect to μ) if the set function μ_f is absolutely continuous or locally absolutely continuous in E. We shall consider only two measures here, the (Lebesgue) area measure and the (Hausdorff or Lebesgue) linear measure. In the first case E and E' will be chosen to be domains, in the second Jordan arcs. Before dealing with these special cases we shall, however, make some general remarks on absolutely continuous homeomorphisms.

Suppose that f is absolutely continuous and B is a Borel set with $\mu(B) = 0$. According to the above remarks $\mu_f(B) = 0$, i.e. Borel null sets are mapped onto null sets by f. Actually the following stronger result is true:

An absolutely continuous homeomorphism maps all null sets onto null sets.

In fact, every null set A is contained in a Borel null set B by 1.2, and $f(A)$ is therefore a subset of the null set $f(B)$.

If $A \subset E$ is a measurable σ-finite set, there exists by 1.2 a Borel set B, $A \subset B \subset E$ such that $B - A$ is a null set. If f is an absolutely continuous homeomorphism, then we have $\mu(f(B) - f(A)) = \mu(f(B - A)) = 0$, and hence $f(B) - f(A)$ is measurable. Since $f(B)$, being a Borel set, is measurable, we infer that $f(A)$ is also measurable. Thus we have shown:

An absolutely continuous homeomorphism maps every σ-finite measurable set onto a measurable set.

As a consequence of this result an absolutely continuous mapping is often called measurable in the case when E is a domain. We have preferred the term "absolutely continuous" to avoid confusion with the concept of measurable function and to point out the analogy with the one-dimensional case.

If E is σ-finite and f is an absolutely continuous homeomorphism, then μ_f is defined for all measurable sets and satisfies the following condition: For every $\varepsilon > 0$ there exists $\delta > 0$ such that $\mu_f(A) < \varepsilon$ for every *measurable* set A with $\mu(A) < \delta$.

2.3. Derivative of an additive set function. Retaining the above notation we now introduce the following special assumptions which are to remain in force throughout the rest of this section: *E is either a bounded or unbounded segment or a domain of the finite plane, and μ is to be the linear measure in the first case and the area measure in the second.*

In both cases E is the union of countably many compact sets. It follows that the results of 2.2 on the preservation of null sets and measurability remain true if the homeomorphism in question is only locally absolutely continuous.

Let τ be a non-negative completely additive set function which is defined for all Borel sets; at present we do not require it to be absolutely continuous.

The derivative of τ can be defined in various ways. The following procedure is analogous to the definition of a point of density (cf. 1.6).

Consider a fixed point $z \in E$. If E is a domain, let Q_z denote a square containing the point z and having sides parallel to the coordinate axes. If E is a segment, Q_z is to denote a closed interval containing z and lying on the line containing E. If the limit

$$\tau'(z) = \lim_{\mu(Q_z) \to 0} \frac{\tau(Q_z \cap E)}{\mu(Q_z)}$$

exists, we call it the *derivative* of the set function τ at z.

In 1.6 we have already mentioned a special result on the existence of the derivative. In fact, if c_B is the characteristic function of a measurable set B and

$$\tau(A) = \int_A c_B \, d\mu ,$$

then we conclude from the discussion of 1.6 that the set function τ has the derivative $\tau'(z) = c_B(z)$ almost everywhere.

In the general case we have the following result (Saks [1], pp. 115 and 119):

Theorem of Lebesgue: *A non-negative, completely additive set function τ, bounded in E, has almost everywhere in E a finite derivative $\tau'(z)$, which is measurable as a function of z. For every Borel set $B \subset E$ we have*

$$\tau(B) \geqq \int_B \tau' \, d\mu \,, \tag{2.1}$$

where equality holds for every B if and only if τ is locally absolutely continuous in E.

2.4. Derivative of the set function associated with a homeomorphism. We now consider the special case when τ is the set function μ_f associated with a homeomorphism $f: E \to E'$. If μ_f is bounded in every compact subset of E the finite derivative μ_f' exists almost everywhere in E by the above theorem of Lebesgue. This condition is always satisfied in the two-dimensional case, when E and E' are domains in the finite plane and μ is the Lebesgue area measure. In this case we call μ_f' the *areal derivative* of the mapping f.[35]

Lebesgue's theorem further states that in this case μ_f' is measurable and locally integrable in E (i.e. integrable over every compact subset of E) and that

$$\mu(f(B)) \geqq \int_B \mu_f' \, d\mu \tag{2.2}$$

for every Borel set $B \subset E$. The same holds in the one-dimensional case provided that μ_f is locally bounded.

If f is locally absolutely continuous, Lebesgue's theorem can be sharpened. By 2.2, μ_f is then defined for every measurable set, since E is σ-finite. Every σ-finite measurable set A is contained in a Borel set B with the property $\mu(B - A) = 0$ and we see that equality holds in (2.2) not only for Borel sets but for all measurable sets:

If f is a locally absolutely continuous homeomorphism, then

$$\mu(f(A)) = \int_A \mu_f' \, d\mu \tag{2.3}$$

for every measurable set A.

We need not assume here that μ_f is bounded: for a compact set $A \subset E$ a locally absolutely continuous μ_f is automatically bounded and for an unbounded A we obtain (2.3) by a limiting process.

---

[35] It should be noted that the areal derivative coincides almost everywhere with the Jacobian if the homeomorphism has partial derivatives almost everywhere (see 3.3).

2.5. Transformation of variables in integrals. Let $f: E \to E'$ be a locally absolutely continuous homeomorphism. We remind the reader that E is a segment on a finite line or a domain of the finite plane. In the latter case E' is trivially σ-finite; by the local absolute continuity of f the set E' has this property also in the first case.

Let g be a complex-valued function in E' and $h = g \circ f$. We conclude from the relation $g^{-1}(A) = f(h^{-1}(A))$, in view of 2.2, that the measurability of h implies that of g. Furthermore it follows from $h^{-1}(A) = f^{-1}(g^{-1}(A))$ that h and g are Borel functions simultaneously, and that h is measurable if g is measurable and f^{-1} is locally absolutely continuous.

If g is measurable and either integrable or non-negative, then the (finite or infinite) integral

$$\int_{A'} g \, d\mu$$

exists for every measurable set $A' \subset E$. We prove the following generalization of (2.3):

Lemma 2.1. *Let $f: E \to E'$ be a locally absolutely continuous homeomorphism and g an integrable or non-negative function in E', such that $h = g \circ f$ is measurable. If $A \subset E$ is a measurable set and $A' = f(A)$, then*

$$\int_{A'} g \, d\mu = \int_A h \, \mu_f' \, d\mu \, , \tag{2.4}$$

where we set $h \, \mu_f' = 0$ at the points of A where $h = \infty$ and $\mu_f' = 0$.

Proof: Without loss of generality we may assume that $g \geqq 0$. Since A' is σ-finite we may suppose that in the definition

$$\int_{A'} g \, d\mu = \sup_{\cup A_k' \subset A'} \Sigma \, g_k \, \mu(A_k') \, , \qquad g_k = \inf_{A_k'} g \, ,$$

the sets A_k' are Borel, by the remark made in 1.4. Then $A_k = f^{-1}(A_k')$ is measurable, and by (2.3)

$$\Sigma \, g_k \, \mu(A_k') = \Sigma \, g_k \int_{A_k} \mu_f' \, d\mu \leqq \int_{\cup A_k} h \, \mu_f' \, d\mu \leqq \int_A h \, \mu_f' \, d\mu \, .$$

Hence

$$\int_{A'} g \, d\mu \leqq \int_A h \, \mu_f' \, d\mu \, . \tag{2.5}$$

To prove the opposite inequality we consider next the set $h^{-1}(\infty) = A_\infty$. Since $g = \infty$ in $A_\infty' = f(A_\infty)$ we may suppose that

$$\mu(A_\infty') = \int_{A_\infty} \mu_f' \, d\mu = 0 \, .$$

Then μ_f' vanishes almost everywhere in A_∞ and, in view of our convention, so does $h\,\mu_f'$. The set A_∞ can therefore be disregarded in the integration.

It is thus enough to consider the case when h is everywhere finite. We choose an arbitrary $\varepsilon > 0$ and set $A_k = \{z \mid z \in A,\ (1+\varepsilon)^k \leqq h(z) < (1+\varepsilon)^{k+1}\}$, $A_k' = f(A_k)$. Then

$$\int_A h\,\mu_f'\,d\mu \leqq \sum_{k=-\infty}^{\infty} (1+\varepsilon)^{k+1}\,\mu(A_k') \leqq (1+\varepsilon) \sum_{k=-\infty}^{\infty} g_k\,\mu(A_k') \leqq (1+\varepsilon) \int_{A'} g\,d\mu\,.$$

This, together with (2.5), yields (2.4).

2.6. Rectifiable arcs. In 2.2 we considered a homeomorphism $f: E \to E'$ in the general case when E and E' were arbitrary Borel sets of the finite plane, and in 2.3—2.5 we made the special assumption that E was either a segment or a domain. We now look only at the case when E is a bounded or unbounded segment I on $-\infty < x < \infty$ and E' is a Jordan arc C of the finite plane. The measure μ is the Hausdorff linear measure l and the set function μ_f will now be denoted by V.

If $V(I) = l(C)$ is finite, then the arc C is called *rectifiable* and $l(C)$ is the *length* of C. The arc C is called *locally rectifiable* if every compact subarc of C is rectifiable. A Jordan curve is called rectifiable if all its subarcs are.

Since the Hausdorff linear measure has the continuity properties mentioned in 1.2, we have $\lim l(C_n) = l(\bigcap C_n)$ for every monotone sequence $C_1 \supset C_2 \supset \cdots$ of subarcs of C, provided that $l(C_1)$ is finite. If we choose the sequence such that $\bigcap C_n$ is a point, then $\lim l(C_n) = 0$. In the case of a locally rectifiable arc C, the set function V is therefore continuous in the following sense: If x is an arbitrary point of I, then for every $\varepsilon > 0$ there exists a neighbourhood Δ of x such that $V(\Delta) < \varepsilon$.

From now on, we suppose that C is locally rectifiable. To the set function V there corresponds a function s of x, determined up to an additive constant, such that the growth of s on every subinterval $[a, b] = \{x \mid a \leqq x \leqq b\}$ of I is equal to $V([a, b])$. The function s, called the *arc length*, is continuous at every point of continuity of V, and hence on the whole interval I.

If V has a finite derivative at a point $x \in I$, then s is evidently differentiable at x and $s'(x) = V'(x)$. It follows from Lebesgue's theorem that this derivative exists almost everywhere on I and satisfies the inequality

$$\int_I s'(x)\,dx \leqq V(I) = l(C)\,,$$

where equality holds for $l(C) < \infty$ if and only if the mapping f is locally absolutely continuous.

Since the function s is continuous and strictly increasing it maps I topologically onto a segment I_0. By the definition of arc length, we have $l(I_0) = V(I) = l(C)$ and so I_0 is bounded or unbounded according as C is rectifiable or not. We note that in the latter case C cannot be compact.

Denote by $\zeta : I_0 \to C$ the homeomorphism composed of f and the inverse of $s : I \to I_0$. We call ζ the *length preserving parametrization* of C. Among all parametrizations of C this one is distinguished by the fact that it maps every interval $\Delta \subset I_0$ onto an arc of length $l(\Delta)$. It follows that ζ is absolutely continuous and satisfies $l(\zeta(A)) = l(A)$ for every measurable set A. The inverse ζ^{-1} is also absolutely continuous.

Since a mapping composed of two absolutely continuous homeomorphisms is obviously absolutely continuous, it follows from $f = \zeta \circ s$, $s = \zeta^{-1} \circ f$, that f and s are simultaneously locally absolutely continuous. In this case and only then the equation

$$\int_A s' \, dx = l(f(A)) \tag{2.6}$$

holds for every measurable set $A \subset I$ (cf. (2.3)). Under the same assumptions Lemma 2.1 yields the more general formula

$$\int_A (g \circ f) \, s' \, dx = \int_{f(A)} g \, dl \tag{2.7}$$

for every non-negative or integrable Borel function g defined on C. In the integral on the right we write instead of dl also $|dz|$, $|dw|$, ..., according to the variable used.

In an obvious way the above considerations can be carried over to the case when the Jordan arc C consists of finitely many locally rectifiable arcs and their endpoints. The same is true if C is a Jordan curve; we only need to divide C into two subarcs.

2.7. Functions of bounded variation on an interval. The definition of the set function μ_f given in 2.2 assumes that $f : E \to E'$ is a homeomorphism. In the case when E is a segment, μ_f admits another characterization which can be applied to non-topological mappings of I.

This generalization depends on the fact that the length of an arc $f(\Delta)$, $\Delta \subset I$, can also be defined in the following way (see Saks [1], IV. § 8): Consider all finite sequences of non-intersecting intervals $a_i < x < b_i$ which, together with their endpoints, are contained in Δ. Then

$$l(f(\Delta)) = V(\Delta) = \sup \sum_i |f(b_i) - f(a_i)| \, . \tag{2.8}$$

This formula yields the required new characterization for the restriction of the set function V to the family of all segments $\Delta \subset I$. We use it as a definition in the case when f is an arbitrary continuous complex-valued function defined on I; the expression (2.8) is called the *variation* of f on Δ.

If $V(I)$ is finite we say that f is *of bounded variation* on I. If $V(\Delta) < \infty$ for every compact segment $\Delta \subset I$, then f is of *bounded local variation*; in the case of a homeomorphism this is equivalent to the local rectifiability of $f(I)$.

The variation V of f is defined by (2.8) for segments only. If V is locally bounded it can, however, be extended to a completely additive set function defined in the class of Borel subsets of I. To do this we first associate with every set $A \subset I$ an outer measure $V^*(A)$ by considering all coverings of A by sequences of open segments $\Delta_1, \Delta_2, \ldots$ and setting

$$V^*(A) = \inf \Sigma\, V\, (I \cap \Delta_k)\,.$$

Then V^* satisfies the axioms 1—5 of 1.1 and so the Borel sets are measurable with respect to V^*. Since $V^*(\Delta) = V(\Delta)$ for every compact segment $\Delta \subset I$ we obtain the required extension of V by setting $V(B) = V^*(B)$ for every Borel set B.

This definition of V agrees with the original one when f is a homeomorphism.

As a generalization of the arc length of 2.6, we associate with every continuous function f of bounded local variation a function s on I, defined by the formula $s(b) - s(a) = V([a, b])$. The function s is continuous and determined up to an additive constant. If f is real we write $f_1(x) = s(x) + x$, $f_2(x) = f_1(x) - f(x)$. Then

$$f = f_1 - f_2\,,$$

where f_1 and f_2 are strictly increasing and thus homeomorphisms of I.

A complex-valued function $f = u + i\,v$ is said to be of bounded variation or of bounded local variation if and only if the real functions u and v have the property in question. Hence, if f is of bounded local variation on I, it admits a representation $f = u_1 - u_2 + i\,(v_1 - v_2)$, where u_k, v_k are homeomorphisms of I.

By Lebesgue's theorem a monotone function on I has a finite derivative almost everywhere on I. It follows from the above that a function f of bounded local variation on I also has this property. The derivative f' is integrable over every compact subset of I.

The derivatives of a function f of bounded local variation and of the function s associated with f satisfy the equation

$$|f'(x)| = s'(x) \tag{2.9}$$

for almost all $x \in I$ (Saks [1], p. 123).

2.8. Absolutely continuous functions on an interval. As in the case of a homeomorphism, an arbitrary continuous function f on the interval I will be called absolutely continuous if its variation V is absolutely continuous on I. We see from the definition of V given in 2.7 that absolute continuity can also be characterized in the following way: The function f is absolutely continuous on I if to every $\varepsilon > 0$ there corresponds $\delta > 0$ such that $\sum |f(b_i) - f(a_i)| < \varepsilon$ for every finite sequence of non-intersecting intervals $a_i < x < b_i$ which together with their endpoints lie in I and have a total length $< \delta$.

Since an absolutely continuous function f on I is of bounded local variation it possesses a derivative which is integrable over compact subintervals of I. It follows from Lebesgue's theorem that

$$f(b) - f(a) = \int_a^b f'(x)\, dx$$

for every pair of points $a, b \in I$.

2.9. An example of a singular function. A function f of bounded variation on the interval I is called *singular* if the derivative f' vanishes almost everywhere on I. Since the increment of an absolutely continuous function equals the integral of its derivative, no non-constant function can be simultaneously singular and absolutely continuous.

For later applications we need an example of a non-constant continuous singular function. This is constructed with the help of a *Cantor set,* which will also appear later in connection with certain other questions.

A Cantor set is defined in the following way: Let a sequence of numbers p_ν, $0 < p_\nu < 1$, be given. We first remove from the middle of the closed interval $I = [0, 1]$ an open interval I_{11} of length p_1. From the middle of the two remaining closed segments we take away open intervals I_{21}, I_{22}, each of length $\frac{1}{2} p_2 (1 - p_1)$. At the nth step, we remove from the middle of the 2^{n-1} remaining closed segments open intervals I_{nk}, each of length $2^{1-n} p_n (1 - p_1) \cdots (1 - p_{n-1})$. The Cantor set $E = E(p_1, p_2, \ldots)$ is defined as the complement of the union of all the removed intervals:

$$E = I - \bigcup_{n=1}^{\infty} \bigcup_{k=1}^{2^{n-1}} I_{nk}\,.$$

The set E is perfect, that is it coincides with the set of all its limit points, and it has the linear measure

$$l(E) = \prod_{\nu=1}^{\infty} (1 - p_\nu) .$$

The proposed construction of a singular function can now be carried out as follows: We first set $g(x) = 2^{-n} (2k - 1)$ for $x \in I_{nk}$. Then g is defined as a non-decreasing function in $I - E$ whose values are everywhere dense in $[0, 1]$. Therefore the limit

$$f(x) = \lim_{t \to x} g(t)$$

exists for every $x \in I$ and defines a continuous function f on I which decreases nowhere and grows from 0 to 1.

By definition f is constant on every interval I_{nk}. Hence $f'(x) = 0$ for $x \in I - E$. If the numbers p_ν are so chosen that $l(E) = 0$, say $p_\nu = p =$ const., then $f'(x)$ vanishes almost everywhere on I. Thus the function f is singular and gives the required example.

2.10. Variation and absolute continuity on a Jordan arc. Let f be a continuous, complex-valued function defined on a locally rectifiable arc C. For all parametrizations $t: I \to C$, the variation $V(I)$ of the composite function $f \circ t$ is clearly the same. We call it the variation of f on C. If $V(I)$ is finite we say that f is of bounded variation on C.

The function f is said to be *absolutely continuous on the arc* C if it is absolutely continuous as a function of the arc length s of C. It should be noted that in contrast to the independence of $V(I)$ on the parametric representation $t: I \to C$, the absolute continuity of $f \circ t$ depends on the homeomorphism t.

2.11. Integrals over oriented arcs. Let C be a closed rectifiable Jordan arc and $\zeta: I_0 \to C$, $I_0 = \{s \mid a \leqq s \leqq b\}$ its length-preserving parametrization. Let f and g be continuous complex-valued functions on C, g being of bounded variation. We define a Stieltjes integral over the oriented arc $C^+ = (C, \alpha)$ by setting

$$\int_{C^+} f\, dg = \pm \int_a^b f(\zeta(s))\, dg(\zeta(s)) . \tag{2.10}$$

On the right hand side we have a classical Stieltjes integral, and the sign is chosen to be $+$ or $-$ according as ζ belongs to the orientation α or not (see I.1.4).

A homeomorphism w of C onto an arc C' induces a mapping of the orientation α of C onto an orientation α' of C' (see I.1.5), i.e. w carries

C^+ onto the oriented arc $w(C^+) = (C', \alpha')$. The definition (2.10) thus implies the following invariance property: If C' is rectifiable, then

$$\int_{C^+} f\, dg = \int_{w(C^+)} (f \circ w^{-1})\, d\, (g \circ w^{-1})\,. \tag{2.11}$$

§ 3. *Differentiability of Mappings of Plane Domains*

3.1. Existence of partial derivatives. As we have seen in 2.4 it follows from Lebesgue's theorem on the derivative of a set function that a homeomorphism of a plane domain has a finite areal derivative almost everywhere. The question of the differentiability of such a mapping in the usual sense is more difficult and cannot be answered directly by Lebesgue's theorem as in the one-dimensional case. We now consider this problem and begin with some remarks on the existence of the partial derivatives of a function of two real variables.

Let f be a complex-valued finite continuous function defined in a domain G of the finite $z = x + i\,y$-plane. If there exists points at which f_x and f_y exist and are finite, it is easy to see that the set of all such points is a Borel set and that f_x and f_y are Borel functions in this set (Saks [1], p. 170).

We seek conditions under which the partial derivatives of f exist almost everywhere in G. To connect this problem with the results mentioned in § 2 on the differentiability of functions of a real variable we give the following definition:

A function f, continuous in G, is *of bounded variation on lines in G* (*absolutely continuous on lines in G*) if the following condition is satisfied:

In every rectangle $R = \{x + i\,y \mid a < x < b,\ c < y < d\}$, $\overline{R} \subset G$, f is of bounded variation (is absolutely continuous) as a function of x on almost all segments $I_y = \{x + i\,y \mid a < x < b\}$, and as a function of y on almost all segments $I_x = \{x + i\,y \mid c < y < d\}$.[36]

Since absolute continuity on a segment implies bounded variation, f is of bounded variation on lines in G if it is absolutely continuous on lines in G.

By 2.7 a function of bounded local variation on a segment I has a finite derivative almost everywhere in I. The corresponding result for plane domains reads as follows:

---

[36] We could just as well use closed segments $\overline{I}_y$, $\overline{I}_x$ here.

Lemma 3.1. *A function f of bounded variation on lines in G possesses finite partial derivatives almost everywhere in G.*

Proof: We consider a rectangle $R = \{x + i\,y \mid a < x < b, c < y < d\}$, $\bar{R} \subset G$. Let E denote the set of points of R at which f_x exists and is finite. Since E is a Borel set its characteristic function c_E is measurable. Thus by Fubini's theorem,

$$m(E) = \iint_R c_E\, d\sigma = \int_c^d dy \int_a^b c_E\, dx = \int_c^d l\,(E \cap I_y)\, dy\,, \tag{3.1}$$

where I_y denotes the horizontal segment $\{x + i\,y \mid a < x < b\}$.

If f is of bounded variation on I_y, then f_x exists almost everywhere on I_y, and $l\,(E \cap I_y)$ thus equals the length $b - a$ of I_y. This holds for almost all $y \in (c, d)$, by hypothesis, and so (3.1) yields

$$m(E) = (b - a)\,(d - c) = m(R)\,.$$

The function f thus has a finite partial derivative f_x almost everywhere in R. In the same way we can show that f_y exists and is finite almost everywhere in R. Since G is a union of countably many rectangles with sides parallel to the coordinate axes we see that f has partial derivatives almost everywhere in G.

3.2. Differentiability of a homeomorphism. The existence of the partial derivatives of a function f at a point $z_0 = x_0 + i\,y_0$ does not necessarily imply the differentiability of f at z_0, i.e. the validity of an expansion $f(z) = f(z_0) + f_x(z_0)\,(x - x_0) + f_y(z_0)\,(y - y_0) + o\,(z - z_0)$.

However, in the case when f is a homeomorphism the following result is true (Gehring-Lehto [1]):

Theorem 3.1. *Let G and G' be domains in the finite plane and $f: G \to G'$ a homeomorphism having finite partial derivatives almost everywhere in G. Then f is differentiable almost everywhere in G.*

Proof: It is clearly enough to prove that f is almost everywhere differentiable in an arbitrary compact subset G_0 of G.

We consider the function F_h,

$$F_h(z) = \left|\frac{f(z+h) - f(z)}{h} - f_x(z)\right| + \left|\frac{f(z+i\,h) - f(z)}{h} - f_y(z)\right|,$$

for every point $z \in G_0$ at which f has finite derivatives f_x and f_y and for real numbers $h \neq 0$ such that $z + h$ and $z + i\,h$ lie in G. For sufficiently small $|h|$, F_h is a Borel function defined almost everywhere in G_0. It follows that the functions g_n,

$$g_n(z) = \sup_{0<h<1/n} F_h(z)\,, \qquad n = 1, 2, \ldots,$$

are also Borel functions from some $n = n_0$ onwards, since we can let h run through only the rational values in $(0, 1/n)$.

From the assumption on the partial derivatives of f we conclude that $g_n(z) \to 0$ as $n \to \infty$ almost everywhere in G_0. Hence, by Egoroff's theorem (see 1.3), there exists for every $\eta > 0$ a closed set $E \subset G_0$ with $m\,(G_0 - E) < \eta$ such that

$$f_x(z) = \lim_{h\to 0} \frac{f(z+h) - f(z)}{h}\,, \qquad f_y(z) = \lim_{h\to 0} \frac{f(z+i\,h) - f(z)}{h}$$

uniformly for $z \in E$. The restrictions of f_x and f_y to E are then continuous. If $\varepsilon \in (0, 1)$ is given, we can thus find $\delta > 0$ such that for z, $z_0 \in E$ the inequalities

$$|f_x(z) - f_x(z_0)| < \varepsilon\,, \quad |f_y(z) - f_y(z_0)| < \varepsilon \quad \text{for} \quad |z - z_0| < \delta \tag{3.2}$$

and

$$|F_h(z)| < \varepsilon \quad \text{for} \quad 0 < |h| < \delta \tag{3.3}$$

hold.

Suppose that $z_0 = x_0 + i\,y_0 \in E$ is an $x\,y$-point of density of E. By Lemma 1.1 almost every point of E has this property. According to the above the theorem will therefore be proved if we can show that f is differentiable at z_0. For this purpose we shall prove that f satisfies the inequality

$$|f(z) - f(z_0) - f_x(z_0)\,(x - x_0) - f_y(z_0)\,(y - y_0)| < M\,\varepsilon\,|z - z_0| \tag{3.4}$$

in a neighbourhood of z_0, where M is finite and depends neither on z nor on ε.

To simplify the notation we suppose that $z_0 = 0$ and write $L(z) = f(0) + f_x(0)\,x + f_y(0)\,y$. Using the triangle inequality we obtain

$$|f(z) - L(z)| \leqq |f(x + i\,y) - f(x) - f_y(x)\,y| + |f(x) - f(0) - f_x(0)\,x| + |f_y(x) - f_y(0)|\,|y|\,.$$

From (3.2) and (3.3) it then follows that

$$|f(z) - L(z)| < 3\,\varepsilon\,|z| \tag{3.5}$$

for $|z| < \delta$, $x \in E$. Analogously we see that (3.5) also holds for $|z| < \delta$, $i\,y \in E$.

To prove that (3.4) (for $z_0 = 0$) also holds without the special hypotheses $x \in E$ and $i\,y \in E$, we first use our assumption that $z = 0$ is an

xy-point of density of E. It follows that there exists a square $Q = \{x + iy \mid |x| < d, |y| < d\}$ whose length of side is at most δ and which satisfies the following condition: For every closed segment $J \subset Q$ lying on either coordinate axis and containing the origin we have

$$l(J \cap E) > \frac{l(J)}{1+\varepsilon}. \tag{3.6}$$

Now take an arbitrary point $z = x + iy \neq 0$ of the square $Q_0 = \{x + iy \mid |x| < d/2, |y| < d/2\}$. By (3.6) each of the open segments $((1-\varepsilon)x, x)$, $(x, (1+\varepsilon)x)$, $((1-\varepsilon)iy, iy)$, $(iy, (1+\varepsilon)iy)$ contains at least one point of E. Thus we can choose four points x_1, x_2, iy_1, iy_2 of E which satisfy the conditions $x_1 < x < x_2$, $y_1 < y < y_2$, $x_2 - x_1 < 2\varepsilon|x|$, $y_2 - y_1 < 2\varepsilon|y|$. Let R denote the rectangle with corners $x_1 + iy_1$, $x_2 + iy_1$, $x_2 + iy_2$, $x_1 + iy_2$. Then every boundary point $z^* = x^* + iy^*$ of R has the property that either x^* or iy^* belongs to E. Since $|z^*| < \delta$, (3.5) then holds everywhere on the boundary of R.

We now make use of the assumption that f is a homeomorphism. This implies that f satisfies the maximum principle in R. The expression $|f(\zeta) - L(z)|$, considered as a function of ζ, therefore takes its maximum at a boundary point z^* of R. But (3.5) holds at z^* and so

$$\begin{aligned} |f(z) - L(z)| &\leqq |f(z^*) - L(z)| \leqq |f(z^*) - L(z^*)| + |L(z^*) - L(z)| \\ &\leqq 3\varepsilon|z^*| + |f_x(0)|\,|x^* - x| + |f_y(0)|\,|y^* - y|. \end{aligned}$$

By the above $|z^* - z| < 2\varepsilon|z| < 2|z|$. If we now replace the origin by a general point z_0, we obtain (3.4) for $z - z_0 \in Q_0$ with $M = 9 + 2\,(|f_x(z_0)| + |f_y(z_0)|)$, and the theorem is proved.

Remark. The above proof is based on a method due to Stepanoff (Saks [1], p. 300—303) as well as on the maximum principle. Since the maximum principle holds not only for homeomorphisms but also for all continuous open mappings, Theorem 3.1 is also true for these mappings.

3.3. Areal derivative and Jacobian of a homeomorphism. Let f be a homeomorphism between the domains G and G' of the finite plane; henceforth we shall assume that f is sense-preserving. If f has finite partial derivatives at a point of G, then the Jacobian

$$J = \frac{i}{2}(f_x \bar{f}_y - \bar{f}_x f_y) = |f_z|^2 - |f_{\bar{z}}|^2,$$

also exists at that point. It is known that for sufficiently regular homeomorphisms J equals the areal derivative μ_f' defined in 2.4. The following result of this type will be needed later:

Lemma 3.2. *Let the homeomorphism f be differentiable at a point $z_0 \in G$. Then f has a finite areal derivative at z_0 and $\mu_f'(z_0) = J(z_0)$.*

Proof: For the sake of simplicity we suppose that $z_0 = 0$ and $f(0) = 0$. Let $Q \subset G$ be a closed square with length of side r and let $0 \in Q$. We write $L(z) = f_x(0)\, x + f_y(0)\, y$ and

$$\varepsilon(r) = \sup_{0<|z|<2r} \left| \frac{f(z) - L(z)}{z} \right| .$$

Since f is differentiable at the origin we have $\varepsilon(r) \to 0$ as $r \to 0$.

The linear function L maps the square Q onto either a parallelogram, or a segment, or a point. In any case, we have for the image $L(Q)$ the formulae $m(L(Q)) = r^2 J(0)$ for the area and $d(L(Q)) \leqq M r$ for the diameter, where $M = 2\,(|f_x(0)| + |f_y(0)|)$.

In Q we have $|f(z) - L(z)| \leqq |z|\, \varepsilon(r) < 2\, r\, \varepsilon(r)$, and so every boundary point of $f(Q)$ lies within this distance of the boundary of $L(Q)$. An elementary geometrical argument then yields

$$r^2 J(0) - 8\, M\, r^2\, \varepsilon(r) \leqq m(f(Q)) \leqq r^2 J(0) + 8\, M\, r^2\, \varepsilon(r) + 4\, \pi\, r^2 (\varepsilon(r))^2 .$$

Hence

$$\lim_{r \to 0} \frac{m(f(Q))}{r^2} = \mu_f'(0) = J(0) ,$$

as required.

3.4. Integration of the Jacobian. If the homeomorphism $f: G \to G'$ has finite partial derivatives almost everywhere in G then by Theorem 3.1 it is differentiable at almost all points of G. By Lemma 3.2 we then have $J(z) = \mu_f'(z)$ almost everywhere in G. Applying Lebesgue's theorem to the areal derivative of a topological mapping (see 2.4) we infer:

Lemma 3.3. *If the homeomorphism $f: G \to G'$ possesses finite partial derivatives almost everywhere in G, then*

$$\iint_B J\, d\sigma \leqq m(f(B))$$

for every Borel set $B \subset G$. Equality holds for every B if and only if f is locally absolutely continuous in G. Then

$$\iint_A J\, d\sigma = m(f(A))$$

even for every measurable set $A \subset G$.

By 2.1 and 2.2 the inverse f^{-1} of f is absolutely continuous in every compact subset of G' if and only if f maps no set of positive measure onto a null set. Thus from the second part of Lemma 3.3 we obtain:

Lemma 3.4. *Let the homeomorphism f be locally absolutely continuous in G and have finite partial derivatives almost everywhere in G. The*

Jacobian J of f is positive almost everywhere in G if and only if the inverse $f^{-1}: G' \to G$ is locally absolutely continuous in G'.

We remark that the transformation formula (2.4) of 2.5 can in the case considered here be expressed in the following form:

Lemma 3.5. *Let the homeomorphism $f: G \to G'$ and its inverse f^{-1} be locally absolutely continuous in their respective domains and let f have finite partial derivatives almost everywhere in G. Let g be a complex-valued, measurable function in G' which is non-negative or integrable. Then*

$$\iint_A g(f(z))\, J(z)\, d\sigma = \iint_{f(A)} g\, d\sigma\,,$$

where $A \subset G$ is an arbitrary measurable set.

§ 4. *Module of a Family of Arcs or Curves*

4.1. Generalization of the notion of module. The module introduced in Chapter I for quadrilaterals and ring domains can be characterized by means of a family of Jordan arcs or curves. In fact the concept of module can be generalized to arbitrary families of arcs or curves as we shall now show (cf. Ahlfors-Beurling [1], Fuglede [1]).

As in Chapter I we consider in this section point sets of the whole compact plane. All set functions τ (measures and integrals) must therefore be extended to sets containing the point at infinity. This we do simply by writing $\tau(A) = \tau(A - \{\infty\})$. In the case when τ is the Lebesgue area measure the extended concept of measure differs from the usual one in that the measure of a set need no longer be the infimum of the measures of its open coverings.

An arc C containing ∞ is called locally rectifiable if it can be transformed into a locally rectifiable arc by a linear conformal mapping. If we remove the point at infinity from such an arc we obtain two locally rectifiable subarcs and the integration methods developed in § 2 can be applied to these.

Let $\mathscr{C}$ be a family of Jordan arcs or curves in the plane Ω. A non-negative Borel function ϱ defined in Ω is called *admissible for the family* $\mathscr{C}$ if the relation

$$\int_C \varrho\, |dz| \geqq 1 \tag{4.1}$$

holds for every locally rectifiable $C \in \mathscr{C}$. We write

$$m_\varrho(\Omega) = \iint_\Omega \varrho^2\, d\sigma \tag{4.2}$$

(cf. I.4.2) for every admissible ϱ and call the infinum

$$M(\mathscr{C}) = \inf_{\varrho} m_{\varrho}(\Omega) \tag{4.3}$$

the module of the family $\mathscr{C}$. The number $1/M(\mathscr{C})$ is called *the extremal length of* $\mathscr{C}$.

Note that condition (4.1) does not concern arcs and curves which are not locally rectifiable so that these do not affect the module of $\mathscr{C}$.

Let B be a Borel set of the plane such that all curves $C \in \mathscr{C}$ lie in B. Then clearly the infimum (4.3) does not change if, instead of all admissible functions ϱ, we consider only those which vanish outside B. Thus for sets B of this type we have

$$M(\mathscr{C}) = \inf_{\varrho} m_{\varrho}(B) . \tag{4.4}$$

4.2. Properties of the generalized module. We infer from the above definition that the generalized module $M(\mathscr{C})$ is *monotonic* (cf. I.4.6). It follows first from

$$\mathscr{C}_1 \supset \mathscr{C}_2$$

that every function ϱ admissible for $\mathscr{C}_1$ is also admissible for $\mathscr{C}_2$. Hence

$$M(\mathscr{C}_1) \geqq M(\mathscr{C}_2) .$$

In order to obtain a more general form of monotonicity we consider families $\mathscr{C}_1$, $\mathscr{C}_2$ with the following property: To every $C_2 \in \mathscr{C}_2$ there corresponds a $C_1 \in \mathscr{C}_1$ which is a subarc of C_2. Then every function admissible for $\mathscr{C}_1$ is admissible for $\mathscr{C}_2$, and so we have again

$$M(\mathscr{C}_1) \geqq M(\mathscr{C}_2) .$$

The module $M(\mathscr{C})$ is also *subadditive*: If $\mathscr{C} = \bigcup_{n=1}^{\infty} \mathscr{C}_n$, then

$$M(\mathscr{C}) \leqq \sum_{n=1}^{\infty} M(\mathscr{C}_n) . \tag{4.5}$$

To prove this we remark that (4.5) always holds if the right hand side is infinite. If it is finite, then given $\varepsilon > 0$, we can for every n choose a function ϱ_n admissible for $\mathscr{C}_n$ such that

$$m_{\varrho_n}(\Omega) \leqq M(\mathscr{C}_n) + 2^{-n}\varepsilon .$$

The function $\varrho = (\Sigma\, \varrho_n^2)^{1/2}$ is admissible for $\mathscr{C}$. Consequently, we have

$$M(\mathscr{C}) \leqq m_{\varrho}(\Omega) = \sum_{n=1}^{\infty} m_{\varrho_n}(\Omega) \leqq \sum_{n=1}^{\infty} M(\mathscr{C}_n) + \varepsilon ,$$

and hence (4.5).

From the first monotonicity property and the subadditivity it follows in particular that

$$M(\mathscr{C} \cup \mathscr{C}_0) = M(\mathscr{C}) \tag{4.6}$$

if $M(\mathscr{C}_0) = 0$.

4.3. Families of zero module. The next theorem contains a useful criterion for the module to vanish (cf. Fuglede [1]).

Lemma 4.1. *A necessary and sufficient condition for $M(\mathscr{C})$ to vanish is the existence of a non-negative Borel function ϱ_0 which is square integrable over the plane and satisfies*

$$\int_C \varrho_0 |dz| = \infty$$

for every locally rectifiable $C \in \mathscr{C}$.

Proof: If there exists a function ϱ_0 with the above properties then all the functions ϱ_0/n, $n = 1, 2, \ldots$, are admissible for $\mathscr{C}$, and we have

$$M(\mathscr{C}) \leqq \lim_{n\to\infty} \frac{1}{n^2} \iint_\Omega \varrho_0^2 \, d\sigma = 0 .$$

Conversely, if $M(\mathscr{C}) = 0$ then there exists a sequence $\varrho_1, \varrho_2, \ldots$ of functions which are admissible for $\mathscr{C}$ and satisfy

$$\iint_\Omega \varrho_n^2 \, d\sigma < 4^{-n}, \qquad n = 1, 2, \ldots .$$

Then $\varrho_0 = \Sigma \, \varrho_n$ is a Borel function, and

$$\iint_\Omega \varrho_0^2 \, d\sigma = \iint_\Omega \left(\sum_{n=1}^\infty 2^{-n/2} \, 2^{n/2} \varrho_n \right)^2 d\sigma \leqq \iint_\Omega \sum_{n=1}^\infty 2^{-n} \sum_{n=1}^\infty 2^n \varrho_n^2 \, d\sigma$$

$$= \sum_{n=1}^\infty 2^n \iint_\Omega \varrho_n^2 \, d\sigma < \sum_{n=1}^\infty 2^{-n} = 1 .$$

Since each ϱ_n is admissible for $\mathscr{C}$, we have further

$$\int_C \varrho_0 \, |dz| = \sum_{n=1}^\infty \int_C \varrho_n \, |dz| = \infty$$

for every locally rectifiable $C \in \mathscr{C}$, and the lemma is proved.

From this lemma we deduce the following result: *The module of a family $\mathscr{C}$ vanishes if all arcs and curves of $\mathscr{C}$ possess a common point z_0. If z_0* is finite, we see this by setting

$$\varrho_0(z) = - \frac{1}{|z - z_0| \log |z - z_0|}$$

in a neighbourhood of z_0 and $\varrho_0(z) = 0$ elsewhere. In the case $z_0 = \infty$ we choose

$$\varrho_0(z) = \frac{1}{1 + |z| \log |z|} .$$

The latter function ϱ_0 can also be used to prove the following result: *The family of all non-rectifiable arcs and curves has module zero.*

4.4. Modules and L^2-convergence. A measurable function f belongs to the class L^p, $1 \leqq p < \infty$, in the measurable set A if $|f|^p$ is integrable over A; we then write $f \in L^p(A)$. In what follows we use the Lebesgue area measure.

The class $L^p(A)$ becomes a metric space if we define

$$||f - g||_p = \left(\iint\limits_A |f - g|^p \, d\sigma\right)^{1/p}$$

as the distance between f and g, and identify functions which are equal almost everywhere in A. The metric is called the L^p *metric* and the expression $||f||_p$ the L^p *norm* of f.

We shall later need the following theorem of Fuglede [1] on convergence in the L^2 metric.

Lemma 4.2. *Let f_n, $n = 1, 2, \ldots$, be a sequence of Borel functions belonging to the class L^2 in a Borel set B and converging to a Borel function f in the L^2 metric. Then there exists a subsequence f_{n_k} of the sequence f_n such that*

$$\lim_{k \to \infty} \int\limits_C |f_{n_k} - f| \, |dz| = 0 \tag{4.7}$$

for all locally rectifiable arcs $C \subset B$, with the exception of a family of module zero.

Proof: We choose that subsequence f_{n_k} such that

$$\iint\limits_B |f_{n_k} - f|^2 \, d\sigma < 2^{-3k} \tag{4.8}$$

and set $|f_{n_k} - f| = g_k$. Let $\mathscr{C}_0$ be the family of those locally rectifiable arcs $C \subset B$ for which (4.7) does not hold.

If $C \in \mathscr{C}_0$, there exist arbitrarily large values of k such that

$$\int\limits_C g_k |dz| > 2^{-k} .$$

Writing $\mathscr{C}_k = \left\{C \mid C \subset B, \int\limits_C g_k \, |dz| > 2^{-k}\right\}$ we therefore have

$$\mathscr{C}_0 \subset \bigcup_{k \geqq n} \mathscr{C}_k \tag{4.9}$$

for every n.

On the other hand, since the function $2^k g_k$ is admissible for $\mathscr{C}_k$ it follows from the definition of module and from (4.8) that

$$M(\mathscr{C}_k) \leqq 2^{2k} \iint_B g_k^2 \, d\sigma < 2^{-k} .$$

Since the module is subadditive we have

$$M\left(\bigcup_{k \geqq n} \mathscr{C}_k\right) \leqq \sum_{k=n}^{\infty} M(\mathscr{C}_k) < 2^{-n+1} ,$$

and $M(\mathscr{C}_0) = 0$ follows from (4.9).

§ 5. *Approximation of Measurable Functions*

5.1. Preparatory remarks. In this section we discuss two types of approximation theorems. In 5.2 and 5.3, we approximate *pointwise* a given measurable function by continuous functions or step functions. The second type of problem is dealt with in 5.4—5.8, where the approximation is measured *in the L^p metric*.

We use the Lebesgue area measure, and the functions to be approximated are defined and measurable in the whole finite plane. If necessary we then have to extend functions which are measurable in a set A by assigning them the value zero outside A.

Let f be a complex-valued finite measurable function. We first construct a sequence of measurable functions g_n, $n = 1, 2, \ldots,$ converging to f, each g_n taking only finitely many values. To this end for every n we divide the square $\{u + i\,v \mid -n < u \leqq n, \ -n < v \leqq n\}$ into $4\,n^4$ half-open squares Q_k, $k = 1, 2, \ldots, 4\,n^4$, with length of side $1/n$. If $f^{-1}(Q_k) = A_k$ and q_k is the corner of Q_k nearest to the origin, then

$$|q_k| \leqq \inf_{z \in A_k} |f(z)| , \qquad \sup_{z \in A_k} |f(z) - q_k| \leqq \frac{\sqrt{2}}{n} . \tag{5.1}$$

We now write

$$g_n = \sum_{k=1}^{4\,n^4} q_k \, c_{A_k} ,$$

where c_A is the characteristic function of A. By the measurability of f, g_n is measurable, and by (5.1) $|g_n(z)| \leqq |f(z)|$ and $\lim g_n(z) = f(z)$ for every z. If f is bounded, the convergence is uniform.

5.2. Pointwise approximation by continuous functions. By a small modification of the functions g_n we obtain for f an approximating sequence of *continuous* functions f_n. However, we can only conclude that f_n

converges to f almost everywhere. On the other hand, f then needs to be only almost everywhere finite.

To construct an approximating sequence of continuous functions we again consider a function g_n having the constant value q_k in the set A_k, $k = 1, 2, \ldots, 4n^4$, and vanishing in $A_0 = - \bigcup A_k$. By 1.2, every set $A_k \cap D_n$, $k = 0, 1, 2, \ldots, 4n^4$, where $D_n = \{z \mid |z| < n\}$, contains a compact subset F_k such that

$$\sum_{k=0}^{4n^4} m\,(A_k \cap D_n - F_k) = m\left(D_n - \bigcup_{k=0}^{4n^4} F_k\right) < 2^{-n}.$$

We denote by $d(z, F_k)$ the distance between the point z and the set F_k and by d the smallest of the mutual distances between the sets F_k, $k = 0, 1, 2, \ldots, 4n^4$. If we write $q_0 = 0$,

$$h_k(z) = \left(1 - \frac{2\,d(z, F_k)}{d}\right) q_k, \qquad k = 0, 1, \ldots, 4n^4,$$

for $d(z, F_k) \leqq d/2$, and $h_k(z) = 0$ for $d(z, F_k) > d/2$, then every function h_k is uniformly continuous in the finite plane. The sum

$$f_n = \sum_{k=0}^{4n^4} h_k$$

coincides with g_n in the set $\bigcup F_k$. Thus it differs from g_n inside D_n at most in a set of measure less than 2^{-n}.

Since the discs D_n exhaust the finite plane as $n \to \infty$ we conclude that for almost every z the equation $f_n(z) = g_n(z)$ holds from some h onwards. Hence $\lim f_n(z) = \lim g_n(z) = f(z)$ almost everywhere, and we have constructed the required approximation to f by continuous functions.

The values $|f_n(z)|$ can be greater than $|f(z)|$ at certain points z, but they cannot exceed $\sup\limits_z |f(z)|$.

Remark. Consider the restriction of the above function f to a set A of finite measure. By Egoroff's theorem (see 1.3), for every $\varepsilon > 0$ there is a closed set $F \subset A$ with $m\,(A - F) < \varepsilon$, such that the approximating sequence f_n converges uniformly in F. Since the limit function of a uniformly convergent sequence of continuous functions is continuous we have proved the following result on the relationship between a measurable and a continuous function.

Lusin's Theorem. *Let f be a complex-valued, almost everywhere finite, measurable function in a set A of finite measure. Then for every $\varepsilon > 0$ there exists a closed set $F \subset A$ with $m\,(A - F) < \varepsilon$ such that the restriction of f to F is finite and continuous.*

5.3. Special approximating sequences. We call the set of open squares $Q_{hk} = \{x + iy \mid (h-1)\delta < x < h\delta,\ (k-1)\delta < y < k\delta\}$, $h, k = 0, \pm 1, \pm 2, \ldots$, a *net* and say that the net becomes finer when δ is made smaller. A function which takes a constant value in every square of some net is called a *step function*.

As above suppose that f is an almost everywhere finite, measurable function in the finite plane. To approximate f by step functions we consider the sequence f_n constructed in 5.2; this converges to f almost everywhere. Since every f_n is uniformly continuous, we can construct a net N_n for every n, so fine that $|f_n(z_1) - f_n(z_2)| < 1/n$ if z_1 and z_2 lie in the same square $Q_{hk} \in N_n$. For every Q_{hk} we denote by z_{hk} the point of $\overline{Q}_{hk}$ where $|f_n|$ assumes its minimal value. We then set $\varphi_n(z) = f_n(z_{hk})$ if $z \in Q_{hk}$ and $\varphi_n(z) = f_n(z)$ if z is a boundary point of Q_{hk}. The function φ_n is then a step function which satisfies the inequalities $|\varphi_n| \leqq |f_n|$, $|\varphi_n - f_n| \leqq 1/n$ everywhere in the finite plane. We thus infer:

For any almost everywhere finite, measurable function f we can construct a sequence of step functions φ_n such that $\sup\limits_z |\varphi_n(z)| \leqq \sup\limits_z |f(z)|$ *and* $f(z) = \lim\limits_{n\to\infty} \varphi_n(z)$ *almost everywhere.*

Finally we remark that an almost everywhere finite, measurable function f can also be approximated by polynomials. To prove this we again consider the above approximating sequence of continuous functions f_n. By *Weierstrass's approximation theorem*, there exists for every f_n a sequence of polynomials $P_{nk}(z, \bar{z})$ such that $P_{nk}(z, \bar{z}) \to f_n(z)$ as $k \to \infty$, uniformly in every bounded set. Such polynomials are, for instance,

$$P_{nk}(z, \bar{z}) = \frac{1}{c_k} \iint\limits_{|\zeta| < \log k} f_n(\zeta) \left[1 - \left(\frac{|\zeta - z|}{\log k}\right)^2\right]^k d\sigma\,,$$

where

$$c_k = \iint\limits_{|\zeta| < \log k} \left[1 - \left(\frac{|\zeta|}{\log k}\right)^2\right]^k d\sigma = \frac{\pi \log^2 k}{k+1}\,.$$

For sufficiently large $k = k_n$ we have for $P_{nk_n} = P_n$ the inequality $|P_n(z) - f_n(z)| < 1/n$ in $|z| < n$. Then $\lim P_n(z) = \lim f_n(z) = f(z)$ almost everywhere.

In what follows it is significant that from some n onwards the absolute value of the polynomials P_n in an arbitrary given bounded set is not larger than the supremum of $|f|$ in the plane. Since $|P_n(z)| - 1/n \leqq |f_n(z)| \leqq \sup |f(z)|$ for $|z| < n$, this can be achieved by replacing P_n by $(1 - 1/(n \sup |f|))\, P_n$.

We remark that P_n is a polynomial in z and $\bar{z}$ (or x and y) and should not be confused with an analytic function of $z = x + iy$.

5.4. Smoothing of integrable functions. Let f be a measurable function which is integrable in every compact subset of the finite plane Ω. Such a function is called *locally integrable* in Ω. Since f is finite except perhaps on a null set it can be approximated by the methods set out in 5.2 and 5.3. We now study the approximation of f relative to the L^p metric.

Let ϑ be a finite continuous function which vanishes outside a disc. The function $f * \vartheta$ defined by

$$f * \vartheta(z) = \iint_{\Omega} f(z - \zeta)\, \vartheta(\zeta)\, d\sigma = \iint_{\Omega} f(\zeta)\, \vartheta\, (z - \zeta)\, d\sigma \qquad (5.2)$$

is then finite in the finite plane.

The convolution process (5.2) regularizes f: *the function* $f * \vartheta$ *is continuous.* This follows immediately from

$$f * \vartheta(z) - f * \vartheta(z_0) = \iint_{\Omega} f(\zeta)\, [\vartheta(z - \zeta) - \vartheta(z_0 - \zeta)]\, d\sigma\,,$$

since f is locally integrable and $\vartheta(z - \zeta) - \vartheta(z_0 - \zeta)$ vanishes outside a disc and tends to zero uniformly as $z \to z_0$.

If ϑ is continuously differentiable, then the mean value theorem, applied to the real and imaginary parts of ϑ, implies that the expression

$$\frac{\vartheta(z + h) - \vartheta(z)}{h} - \vartheta_x(z)\,,$$

where $h \neq 0$ is real, tends to zero uniformly as $h \to 0$. It follows that

$$\lim_{h \to 0} \frac{f * \vartheta(z + h) - f * \vartheta(z)}{h} - f * \vartheta_x(z)$$

$$= \lim_{h \to 0} \iint_{\Omega} f(\zeta) \left[\frac{\vartheta(z + h - \zeta) - \vartheta(z - \zeta)}{h} - \vartheta_x(z - \zeta)\right] d\sigma = 0\,.$$

Thus $f * \vartheta$ has the partial derivative

$$(f * \vartheta)_x = f * \vartheta_x\,, \qquad (5.3)$$

which is continuous everywhere in the finite plane since ϑ_x is continuous. In the same way we can show that the partial derivative $(f * \vartheta)_y$ exists and equals $f * \vartheta_y$.

We say that a function belongs to the *class* C^n if it is n times (in the case $n = \infty$ infinitely many times) continuously differentiable. By repeating the above argument we deduce in general that $f * \vartheta \in C^n$ if $\vartheta \in C^n$.

The *support* of a function f is the closure of the set of all points z at which $f(z) \neq 0$. The class of functions belonging to C^n in the finite plane Ω and having bounded support will be denoted by C_0^n. If f has bounded support in Ω then it follows from the definition (5.2) that $f * \vartheta$ also has bounded support.

5.5. A lemma on convergence in the L^p metric. The importance of the above regularization process lies in the fact that for suitably chosen ϑ the function $f * \vartheta$ can be used to approximate f in the L^p metric.

First we prove the following lemma:

If f belongs to L^p in the finite plane Ω, then

$$\lim_{\lambda \to 0} \iint_{\Omega} |f(z + \lambda) - f(z)|^p \, d\sigma = 0 . \tag{5.4}$$

Proof: For every given $\varepsilon > 0$ we can find a number R such that

$$\iint_{|z| \geqq R-1} |f|^p \, d\sigma < \frac{\varepsilon}{2^{p+2}} .$$

Since $|f(z + \lambda) - f(z)|^p \leqq 2^{p-1} \left(|f(z + \lambda)|^p + |f(z)|^p\right)$, the integral

$$I_\lambda(A) = \iint_A |f(z + \lambda) - f(z)|^p \, d\sigma ,$$

then satisfies

$$I_\lambda(\{z \mid |z| \geqq R\}) < \frac{\varepsilon}{4} \tag{5.5}$$

for $|\lambda| < 1$.

The integral of $|f|^p$ over A is absolutely continuous as a function of the set A (see 2.1). Thus we can find an $\eta > 0$ such that this integral is less than $\varepsilon/2^{p+2}$ for every A with $m(A) < \eta$, and consequently

$$I_\lambda(A) < \frac{\varepsilon}{4} . \tag{5.6}$$

By the theorem of Lusin proved in 5.2 the disc $D_{R+1} = \{z \mid |z| < R + 1\}$ contains a closed subset F with $m\,(D_{R+1} - F) < \eta$, such that the restriction of f to F is continuous, and since F is compact, even uniformly continuous. Thus there exists $\delta \in (0, 1)$ such that

$$|f(z + \lambda) - f(z)|^p < \frac{\varepsilon}{4\, m(F)} \qquad \text{for} \quad |\lambda| < \delta , \tag{5.7}$$

provided that $z \in F$, $z + \lambda \in F$.

We shall prove that $I_\lambda(\Omega) < \varepsilon$ for $|\lambda| < \delta$. To this end we consider the following three sets: $A_1 = \{z \mid z \in F, z + \lambda \in F\}$, $A_2 = \{z \mid z + \lambda \in D_{R+1} - F\}$, $A_3 = D_{R+1} - F$. Since $|\lambda| < 1$, $z + \lambda$ belongs to D_{R+1} if $|z| < R$. The set $A_1 \cup A_2 \cup A_3$ therefore covers the disc D_R and so

$$I_\lambda(\Omega) \leqq \sum_{k=1}^{3} I_\lambda(A_k) + I_\lambda(\{z \mid |z| \geqq R\}) .$$

The last term on the right was estimated in (5.5). From (5.7) we see that $I_\lambda(A_1) < \varepsilon/4$. Since the measures of A_2 and A_3 are less than η we deduce from (5.6) that $I_\lambda(A_k) < \varepsilon/4$, for $k = 2, 3$ also. Thus $I_\lambda(\Omega) < \varepsilon$ and the assertion (5.4) is proved.

5.6. Approximation in the L^p metric. By applying the above result (5.4) we can carry out the required approximation to f in the L^p metric. Denote by ϑ_n a finite non-negative continuous function which vanishes outside the square $Q_n = \{x + iy \mid -1/n < x < 1/n, -1/n < y < 1/n\}$ and satisfies the condition

$$\iint_\Omega \vartheta_n \, d\sigma = 1 . \tag{5.8}$$

The following lemma states that the functions $f * \vartheta_1, f * \vartheta_2, \ldots$, constructed from the sequence $\vartheta_1, \vartheta_2, \ldots$ by (5.2), approximate f in the L^p metric.

Lemma 5.1. *Let $f \in L^p$ in the finite plane Ω. Then*

$$\lim_{n\to\infty} \iint_\Omega |f - f * \vartheta_n|^p \, d\sigma = 0 . \tag{5.9}$$

Proof: Because

$$f(z) - f * \vartheta_n(z) = \iint_\Omega [f(z) - f(z-\zeta)] \, \vartheta_n(\zeta) \, d\sigma$$

it follows from Hölder's inequality[37] (or immediately if $p = 1$) that

$$\begin{aligned} |f(z) - f * \vartheta_n(z)|^p &= \left| \iint_\Omega [f(z) - f(z-\zeta)] \, (\vartheta_n(\zeta))^{\frac{1}{p}} \, (\vartheta_n(\zeta))^{1-\frac{1}{p}} \, d\sigma \right|^p \\ &\leqq \iint_\Omega |f(z) - f(z-\zeta)|^p \, \vartheta_n(\zeta) \, d\sigma . \end{aligned} \tag{5.10}$$

Now integrate both sides with respect to z. In order to be able to apply Fubini's theorem as quoted in 1.5 to the right hand side we must know that the integrand is measurable with respect to the Lebesgue volume measure in the four-dimensional $z\zeta$-space. This follows from the fact that $f(z - \zeta)$ can be transformed into a function of z by a rotation of the $z\zeta$-space and that the Cartesian product $A \times \Omega = \{(z, \zeta) \mid z \in A, \zeta \in \Omega\}$ is measurable in the four-dimensional volume measure if A is measurable with respect to the area measure.

We can thus apply Fubini's theorem and obtain from (5.10)

$$\iint_\Omega |f(z) - f * \vartheta_n(z)|^p \, d\sigma_z \leqq \iint_\Omega \left(\iint_\Omega |f(z) - f(z-\zeta)|^p \, d\sigma_z \right) \vartheta_n(\zeta) \, d\sigma_\zeta .$$

---

[37] Hölder's inequality

$$\iint F G \, d\sigma \leqq \left(\iint F^p \, d\sigma\right)^{1/p} \left(\iint G^q \, d\sigma\right)^{1/q}$$

holds for $F, G \geqq 0$, $p > 1$, $1/p + 1/q = 1$.

Since ϑ_n vanishes outside Q_n it is sufficient to take the outer integral on the right hand side over Q_n. Then $|\zeta| < \sqrt{2}/n$. Since the inner integral tends to zero as $\zeta \to 0$ by (5.4) and ϑ_n is normalized by (5.8), it follows that the whole double integral has limit zero as $n \to \infty$. This proves the lemma.

5.7. L^p-approximation by C_0^∞-functions. Write

$$\vartheta_n(z) = a_n\, e^{\frac{1}{|z|^2 - n^{-2}}} \quad \text{for} \quad |z| < \frac{1}{n}, \qquad \vartheta_n(z) = 0 \quad \text{for} \quad |z| \geqq \frac{1}{n}, \tag{5.11}$$

where a_n is to be chosen so that (5.8) is satisfied. Then $\vartheta_n \in C^\infty$, and by 5.4 the functions $f * \vartheta_n$ also belong to C^∞. A function $f \in L^p$ can thus be approximated in the L^p metric by infinitely differentiable functions.

In fact, a stronger result holds: *A function $f \in L^p(\Omega)$ can be approximated in the L^p metric by functions of the class C_0^∞.*

To see this we write $\varphi_k(z) = f(z)$ for $z \in D_k = \{\zeta \mid |\zeta| < k\}$, and $\varphi_k(z) = 0$ for $z \in -D_k$, $k = 1, 2, \ldots$. If ϑ_n is the function defined by (5.11), then by Lemma 5.1 we have

$$\iint_\Omega |\varphi_k - \varphi_k * \vartheta_n|^p \, d\sigma < k^{-p}$$

for a sufficiently large $n = n_k$. Every approximating function $f_k = \varphi_k * \vartheta_{n_k}$ belongs to C^∞ and vanishes outside the disc D_{k+1/n_k}. The triangle inequality now yields

$$\left(\iint_\Omega |f - f_k|^p \, d\sigma\right)^{1/p} \leqq \left(\iint_{-D_k} |f|^p \, d\sigma\right)^{1/p} + \frac{1}{k},$$

and the desired result follows.

5.8. Pointwise convergence of an L^p-approximation. For a continuous function $f \in L^p$ the sequence $f * \vartheta_n$ constructed above converges not only in the L^p metric but also pointwise.

Lemma 5.2. *Let the integrable function f be finite and continuous in an open set G. Then*

$$f(z) = \lim_{n \to \infty} f * \vartheta_n(z), \tag{5.12}$$

the convergence being uniform in every compact subset F of G.

Proof: We choose a compact set F_0 such that $F \subset F_0 \subset G$ and F lies at a positive distance d from the complement of F_0.

It follows from the definition of $f * \vartheta_n$ that

$$|f(z) - f * \vartheta_n(z)| \leqq \iint_{Q_n} |f(z) - f(z - \zeta)|\, \vartheta_n(\zeta)\, d\sigma. \tag{5.13}$$

If $n > \sqrt{2}/d$ and z belongs to F, then the point $z - \zeta$ lies in F_0 if $\zeta \in Q_n$. Since f is uniformly continuous in the compact set F_0, the assertion (5.12) follows immediately from (5.13).

§ 6. Functions with L^p-derivatives

6.1. Definition of a function with L^p-derivatives. We now introduce a class of functions which will play an important role in Chapters IV to VI.

In what follows G will always denote a domain of the finite plane. We say that a finite complex-valued function f defined in G has *L^p-derivatives, $p \geqq 1$,* if it satisfies the following two conditions:

1. f is absolutely continuous on lines in G (cf. 3.1).

2. The partial derivatives f_x, f_y belong to L^p in every compact subset of G.

A function with L^1-derivatives is also called *absolutely continuous in the sense of Tonelli.*

It follows from Hölder's inequality that a function with L^p-derivatives also has L^q-derivatives for $q \leqq p$.

6.2. Integral transformations for functions with L^p-derivatives. We shall give two more characterizations of the functions with L^p-derivatives, both expressing important properties of this class of functions. The first of these characterizations is based on the fact that a transformation formula from classical analysis between curve and surface integrals holds for functions with L^p-derivatives and only for these. We remaind the reader that ∂D denotes the boundary of the Jordan domain D, positively oriented with respect to D (cf. I.1.4).

Lemma 6.1. *A function f which is finite and continuous in G possesses L^p-derivatives in G if and only if there exist two functions g and h belonging to L^p in every compact subset of G such that the following condition holds: If R, $\overline{R} \subset G$, is an arbitrary rectangle with sides parallel to the axes, then*

$$\int_{\partial R} f\,dx = -\iint_R g\,d\sigma\,, \qquad \int_{\partial R} f\,dy = \iint_R h\,d\sigma\,. \tag{6.1}$$

In this case $f_x = h$, $f_y = g$ almost everywhere in G.

Proof: We first suppose that f has L^p-derivatives in G. Let $R = \{x + iy \mid a < x < b,\ c < y < d\}$, $\overline{R} \subset G$. Since the restriction of f to the vertical segments $\{x + i\,y \mid c \leqq y \leqq d\}$ is an absolutely continuous function of y for almost all x, $a < x < b$, we have

$$\int_{\partial R} f\,dx = \int_a^b \big(f(x + i\,c) - f(x + i\,d)\big)\,dx = -\int_a^b dx \int_c^d f_y(x + i\,y)\,dy\,.$$

Here f_y belongs to $L^p \subset L^1$ in $\bar{R}$, and the double integral can therefore be transformed into a surface integral over R by Fubini's theorem. We thus obtain the first of the relations (6.1) with $g = f_y$. The second relation is proved analogously and the necessity of (6.1) follows.

To prove that the condition is also sufficient we start with two functions g and h locally in L^p in G and satisfying (6.1). We apply the first relation (6.1) to the rectangle $R_{xy} \subset R$ with vertices $a + i\,c$, $x + i\,c$, $x + i\,y$, $a + i\,y$. This gives

$$\int_a^x \big(f(\xi + i\,c) - f(\xi + i\,y)\big)\,d\xi = -\iint_{R_{xy}} g\,d\sigma = -\int_a^x d\xi \int_c^y g(\xi + i\,\eta)\,d\sigma\,. \tag{6.2}$$

Let y_n, $n = 1, 2, \ldots$, be a sequence everywhere dense on $c < y < d$. If we differentiate (6.2) with respect to x for a fixed y_n and remember that the derivative of an integral is almost everywhere equal to the integrand, we obtain

$$f(x + i\,y_n) - f(x + i\,c) = \int_c^{y_n} g(x + i\,\eta)\,d\eta \tag{6.3}$$

for every x, $a < x < b$, with the exception of a set E_n of linear measure zero. If x does not belong to the null set $E = \bigcup E_n$, then (6.3) holds for every y_n. Since both sides of (6.3) depend continuously on y_n, we have

$$f(x + i\,y) - f(x + i\,c) = \int_c^y g(x + i\,\eta)\,d\eta$$

everywhere in R except for the vertical segments $x \in E$.

Being an integral, f is an absolutely continuous function of y for almost every $x \in (a, b)$. Furthermore, we see that $f_y(x + i\,y) = g(x + i\,y)$ for almost all y provided that $x \notin E$. Since f_y and g are measurable with respect to the area measure it follows from Fubini's theorem that $f_y(z)$ and $g(z)$ coincide except on a set of zero area measure.

If we exchange the roles of x and y we see that f is also absolutely continuous on almost all horizontal segments and has the partial derivative $f_x = h$ almost everywhere. The proof is now complete.

6.3. Approximation of functions with L^p-derivatives. A further characterization of the functions with L^p-derivatives depends on the fact that they admit a particular C^∞-approximation. The approximating sequence can be chosen such that the functions converge pointwise and their derivatives converge in the L^p metric.

Lemma 6.2. *Let f, g, h be three functions defined in the domain G and suppose that g and h belong to L^p in every compact subset of G. Further let f_n be a sequence of continuously differentiable functions in G such that the following conditions hold for every compact subset F of G:*

1. $\lim\limits_{n\to\infty} (f_n - f) = 0$ *uniformly in F,*

2. $$\lim_{n\to\infty} \iint_F \left|\frac{\partial f_n}{\partial x} - h\right|^p d\sigma = 0, \quad \lim_{n\to\infty} \iint_F \left|\frac{\partial f_n}{\partial y} - g\right|^p d\sigma = 0 .$$

Then f has L^p-derivatives in G and $f_x = h$, $f_y = g$ almost everywhere in G.

Conversely, if f has L^p-derivatives in G there exists a sequence of functions $f_n \in C_0^\infty$ satisfying conditions 1 and 2 with $h = f_x$, $g = f_y$.

If the support of f is a compact subset of G, then the functions f_n can be chosen so that the conditions 1 and 2 hold with F replaced by G.

Proof: We start by proving the first assertion. It follows from condition 1 that f is continuous. If R, $\overline{R} \subset G$, denotes a rectangle with sides parallel to the axes, then

$$\int_{\partial R} f\,dx = \lim_{n\to\infty} \int_{\partial R} f_n\,dx = -\lim_{n\to\infty} \iint_R \frac{\partial f_n}{\partial y}\,d\sigma .$$

From the second equation in 2 and Hölder's inequality we deduce that

$$\lim_{n\to\infty} \iint_R \frac{\partial f_n}{\partial y}\,d\sigma = \iint_R g\,d\sigma .$$

Hence

$$\int_{\partial R} f\,dx = -\iint_R g\,d\sigma .$$

In the same way we can prove that

$$\int_{\partial R} f\,dy = \iint_R h\,d\sigma .$$

Lemma 6.1 now shows that f has L^p-derivatives in G and that $f_x = h$, $f_y = g$ almost everywhere in G.

Conversely, let us assume that f has L^p-derivatives in G. In order to construct a sequence of C_0^∞-functions with properties 1 and 2 we make use of the regularization process introduced in 5.4.

It is enough to carry out the construction so that conditions 1 and 2 are satisfied with $h = f_x$, $g = f_y$ for a fixed compact $F \subset G$. The required approximation of f in every compact subset of G is then obtained by exhausting G by an increasing sequence of compact sets, constructing the corresponding sequence of C_0^∞-functions and choosing the diagonal sequence.

We choose a compact set F_0, $F \subset F_0 \subset G$, such that F is at a positive distance d from the complement of F_0. The function $\tilde{f}$ defined by the formulae $\tilde{f}(z) = f(z)$ for $z \in F_0$, $\tilde{f}(z) = 0$ for $z \notin F_0$, is integrable in the plane. We can choose F_0 such that its boundary has zero area measure (cf. II.8.1). Then the derivatives $\tilde{f}_x$ and $\tilde{f}_y$ belong to L^p in the finite plane. Using the functions (5.11) introduced in 5.7 we construct the sequence

$$f_n = \tilde{f} * \vartheta_n\,, \qquad n = 1, 2, \ldots, \tag{6.4}$$

where every f_n then belongs to C_0^∞. By Lemma 5.2 we have $\lim f_n = \tilde{f} = f$ uniformly in F, and so condition 1 is fulfilled. If the support of f lies in F, then $f_n \to f$ uniformly even in G.

To prove that condition 2 holds we use Lemma 5.1. Since $\tilde{f}_x$ belongs to L^p in the finite plane, we obtain

$$\lim_{n \to \infty} \iint_F |f_x - \tilde{f}_x * \vartheta_n|^p \, d\sigma = 0\,. \tag{6.5}$$

If the support of f lies in F, then this relation holds with G replacing F. The derivative f_y is approximated in the same way by $\tilde{f}_y * \vartheta_n$.

To complete the proof we must show that $\tilde{f}_x * \vartheta_n = (\tilde{f} * \vartheta_n)_x$ and $\tilde{f}_y * \vartheta_n = (\tilde{f} * \vartheta_n)_y$ from some n onwards, everywhere in F. It will appear from our proof that these relations are valid everywhere in G if the support of f lies in F.

Choose $n > \sqrt{2}/d$ and, for brevity, write ϑ instead of ϑ_n. Since $\tilde{f}_x(z - \zeta) = f_x(z - \zeta)$ for $z \in F$, $\zeta \in Q_n$ (with Q_n as defined in 5.6) we first obtain

$$\tilde{f}_x * \vartheta(z) = \iint_{Q_n} f_x(z - \zeta)\, \vartheta(\zeta)\, d\sigma = \int_{-1/n}^{1/n} d\eta \int_{-1/n}^{1/n} f_x(z - \xi - i\,\eta)\, \vartheta(\xi + i\,\eta)\, d\xi\,, \tag{6.6}$$

by Fubini's theorem.

By the hypothesis, the function φ defined by $\varphi(\zeta) = f(z - \zeta)$ is absolutely continuous on almost all horizontal segments $I_\eta = \{\zeta = \xi + i\,\eta \mid -1/n < \xi < 1/n\}$, $I_\eta \subset Q_n$. For every such value of η the inner integral on the right hand side of (6.6) can be transformed by integrating by parts. Since ϑ vanishes at the endpoints of I_η we obtain

$$\int_{-1/n}^{1/n} f_x(z - \xi - i\,\eta)\, \vartheta(\xi + i\,\eta)\, d\xi = \int_{-1/n}^{1/n} f(z - \xi - i\,\eta)\, \vartheta_x(\xi + i\,\eta)\, d\xi\,.$$

This equation holds for almost all η and it follows from (6.6) that

$$\tilde{f}_x * \vartheta(z) = \int_{-1/n}^{1/n} d\eta \int_{-1/n}^{1/n} f(z - \xi - i\,\eta)\, \vartheta_x(\xi + i\,\eta)\, d\xi$$

$$= \iint_{Q_n} f(z - \zeta)\, \vartheta_x(\zeta)\, d\sigma = \tilde{f} * \vartheta_x(z)\,.$$

By (5.3) we have $\tilde{f} * \vartheta_x = (\tilde{f} * \vartheta)_x$, and consequently $\tilde{f}_x * \vartheta(z) = (\tilde{f} * \vartheta)_x\,(z)$ as required.

The proof that $\tilde{f}_y * \vartheta(z) = (\tilde{f} * \vartheta)_y\,(z)$ is similar, and the lemma is now proved.

6.4. Approximation in the finite plane. We now consider the special case when $p = 2$ and G is the finite plane Ω. If we suppose that the partial derivatives of f are not merely locally square integrable but belong to $L^2(\Omega)$, then Lemma 6.2 can be complemented as follows.

Lemma 6.3. *Let f be absolutely continuous on lines in the finite plane Ω and suppose that its partial derivatives belong to $L^2(\Omega)$. Then there exists a sequence of functions f_n, $n = 1, 2, \ldots$, in the class C_0^∞ for which the following conditions hold:*

1. $\lim_{n\to\infty} f_n(z) = f(z)$ *uniformly in every bounded set,*

2. $\lim_{n\to\infty} \iint_\Omega |f_x - (f_n)_x|^2\, d\sigma = \lim_{n\to\infty} \iint_\Omega |f_y - (f_n)_y|^2\, d\sigma = 0.$

Proof: If the support of f is bounded, then the assertion follows immediately from Lemma 6.2. Hence it is sufficient to approximate f in the above way by functions $\tilde{f}_n$ with L^2-derivatives and bounded support.

We set $\tilde{f}_n(z) = \psi_n(|z|)\, f(z)$, $n = 2, 3, \ldots$, where the function ψ_n is continuous for all $|z| = r \geqq 0$, takes the values $\psi_n(r) = 1$ for $0 \leqq r \leqq n$, $\psi_n(r) = 0$ for $r \geqq e^n$, and has derivative $\psi_n'(r) = c_n/(r \log r)$ for $n < r < e^n$. Then

$$\int_n^{e^n} \frac{c_n}{r \log r}\, dr = -1\,, \tag{6.7}$$

and so

$$c_n = \frac{1}{\log\log n - \log n}\,. \tag{6.8}$$

Each function $\tilde{f}_n$ is absolutely continuous on lines in Ω and has bounded support. We assert that the above conditions 1 and 2 hold with $\tilde{f}_n$ replacing f_n. It then also follows that $\tilde{f}_n$ has L^2-derivatives.

Condition 1 clearly holds. We denote by $f_x \psi_n$ the function with the value $f_x(z)\,\psi_n(|z|)$ at z. Since $f_x \psi_n \to f_x$ in the L^2 metric, the first relation in 2 will follow from the triangle inequality if we show that

$$\lim_{n\to\infty} \iint_\Omega |f_x \psi_n - (\tilde{f}_n)_x|^2\, d\sigma = 0 . \tag{6.9}$$

The corresponding relation for derivatives with respect to y can then be proved in the same way.

We have

$$\iint_\Omega |f_x \psi_n - (\tilde{f}_n)_x|^2\, d\sigma \leqq \int_n^{e^n} dr \int_0^{2\pi} |\psi_n'(r)\, f(r\, e^{i\varphi})|^2\, r\, d\varphi$$

$$= \int_n^{e^n} \frac{c_n^2}{r \log^2 r} \left(\int_0^{2\pi} |f(r\, e^{i\varphi})|^2\, d\varphi \right) dr . \tag{6.10}$$

If f is continuously differentiable in the disc $|z| \leqq r$, then

$$|f(r\, e^{i\varphi}) - f(e^{i\varphi})|^2 = \left| \int_1^r \frac{\partial f(r\, e^{i\varphi})}{\partial r}\, dr \right|^2 \leqq \int_1^r \frac{dr}{r} \int_1^r \left| \frac{\partial f(r\, e^{i\varphi})}{\partial r} \right|^2 r\, dr ,$$

and consequently

$$\int_0^{2\pi} |f(r\, e^{i\varphi}) - f(e^{i\varphi})|^2\, d\varphi \leqq \log r \iint_{|z|<r} (|f_x|^2 + |f_y|^2)\, d\sigma .$$

From Lemma 6.2 we see that this also holds without the assumption of differentiability. For all sufficiently large values of r we thus have

$$\int_0^{2\pi} |f(r\, e^{i\varphi})|^2\, d\varphi \leqq (1 + 2\,\|f_x\|_2^2 + 2\,\|f_y\|_2^2) \log r = A \log r .$$

From (6.10), using (6.7) and (6.8), we deduce that

$$\iint_\Omega |f_x \psi_n - (\tilde{f}_n)_x|^2\, d\sigma \leqq A\, c_n \int_n^{e^n} \frac{c_n}{r \log r}\, dr = -A\, c_n = \frac{A}{\log n - \log\log n}$$

from some n onwards. This implies (6.9), and so $\tilde{f}_n$ satisfies conditions 1 and 2.

6.5. Generalized Green's formula. By aid of Lemma 6.2 the classical Green's formula can be generalized:

Green's formula. *Let f and g be functions having respectively L^p- and L^q-derivatives in the domain G, where $1/p + 1/q = 1$. If $D, \overline{D} \subset G$, is a Jordan domain with a rectifiable boundary on which g is of bounded variation, then*

$$\int_{\partial D} f\, dg = \iint_D (f_x\, g_y - f_y\, g_x)\, d\sigma . \tag{6.11}$$

The same relation holds if f has L^1-derivatives and g belongs to the class C^2.[38]

Proof: If f and g have L^p- and L^q-derivatives, Lemma 6.2 implies the existence of functions $f_n, g_n \in C^\infty$, $n = 1, 2, \ldots$, such that $\lim_{n\to\infty} |f_n - f| = \lim_{n\to\infty} |g_n - g| = 0$ uniformly in $\overline{D}$ and

$$\lim_{n\to\infty} \iint_D \left|\frac{\partial(f_n - f)}{\partial x}\right|^p d\sigma = \lim_{n\to\infty} \iint_D \left|\frac{\partial(f_n - f)}{\partial y}\right|^p d\sigma$$
$$= \lim_{n\to\infty} \iint_D \left|\frac{\partial(g_n - g)}{\partial x}\right|^q d\sigma = \lim_{n\to\infty} \iint_D \left|\frac{\partial(g_n - g)}{\partial y}\right|^q d\sigma = 0. \quad (6.12)$$

If $g \in C^2$ and f has L^1-derivatives, then we choose the sequence f_n as above (with $p = 1$) and set $g_n = g$ for $n = 1, 2, \ldots$.

We assume here the validity of Green's formula (6.11) for C^2-functions (see for example Apostol [1], p. 289). Then

$$\int_{\partial D} f_n\, dg_m = \iint_D \left(\frac{\partial f_n}{\partial x}\frac{\partial g_m}{\partial y} - \frac{\partial f_n}{\partial y}\frac{\partial g_m}{\partial x}\right) d\sigma \quad (6.13)$$

for every pair f_n, g_m in both cases above.

Now let $m \to \infty$. Then g_m converges uniformly to g on the boundary of D. Since f_n, g_m and g are continuous and of bounded variation, we obtain by integrating by parts

$$\lim_{m\to\infty} \int_{\partial D} f_n\, dg_m = -\lim_{m\to\infty} \int_{\partial D} g_m\, df_n = -\int_{\partial D} g\, df_n = \int_{\partial D} f_n\, dg\,. \quad (6.14)$$

To carry out the limiting process $m \to \infty$ on the right hand side of (6.13) we note first that the integrals

$$\iint_D \left|\frac{\partial f_n}{\partial x}\right|^p d\sigma\,, \qquad \iint_D \left|\frac{\partial f_n}{\partial y}\right|^p d\sigma$$

are uniformly bounded, by (6.12). Since we only need to deal with the case $p > 1$, Hölder's inequality yields

$$\left|\iint_D \left[\frac{\partial f_n}{\partial x}\left(\frac{\partial g_m}{\partial y} - \frac{\partial g}{\partial y}\right) - \frac{\partial f_n}{\partial y}\left(\frac{\partial g_m}{\partial x} - \frac{\partial g}{\partial x}\right)\right] d\sigma\right|$$
$$\leqq \left(\iint_D \left|\frac{\partial f_n}{\partial x}\right|^p d\sigma\right)^{1/p} \left(\iint_D \left|\frac{\partial(g_m - g)}{\partial y}\right|^q d\sigma\right)^{1/q}$$
$$+ \left(\iint_D \left|\frac{\partial f_n}{\partial y}\right|^p d\sigma\right)^{1/p} \left(\iint_D \left|\frac{\partial(g_m - g)}{\partial x}\right|^q d\sigma\right)^{1/q}.$$

[38] The requirement $g \in C^2$ can clearly be replaced by a much weaker one.

By (6.12) the right hand side tends to zero as $m \to \infty$. It follows from this and (6.14) that (6.13) remains valid if g_m is replaced by g.

The uniform convergence of f_n on the boundary of D implies that

$$\lim_{n\to\infty} \int_{\partial D} f_n\, dg = \int_{\partial D} f\, dg . \tag{6.15}$$

In the case $p > 1$ we apply Hölder's inequality as above and infer that

$$\lim_{n\to\infty} \iint_D \left(\left(\frac{\partial f_n}{\partial x} - \frac{\partial f}{\partial x} \right) \frac{\partial g}{\partial y} - \left(\frac{\partial f_n}{\partial y} - \frac{\partial f}{\partial y} \right) \frac{\partial g}{\partial x} \right) d\sigma = 0 . \tag{6.16}$$

For $p = 1$, $g \in C^2$, we have

$$\left| \iint_D \left(\left(\frac{\partial f_n}{\partial x} - \frac{\partial f}{\partial x} \right) \frac{\partial g}{\partial y} - \left(\frac{\partial f_n}{\partial y} - \frac{\partial f}{\partial y} \right) \frac{\partial g}{\partial x} \right) d\sigma \right|$$

$$\leqq \max_{z\in\overline{D}} |g_y(z)| \iint_D \left| \frac{\partial (f_n - f)}{\partial x} \right| d\sigma + \max_{z\in\overline{D}} |g_x(z)| \iint_D \left| \frac{\partial (f_n - f)}{\partial y} \right| d\sigma ,$$

and we see that (6.16) holds in this case too.

By (6.15) and (6.16), we can replace f_n by f in (6.13), as we have replaced g_m by g. The generalized Green's formula is now proved.

Example: In the two simple cases $g(z) = z$ and $g(z) = \bar{z}$, we obtain the following result from Green's formula: *If f has L^1-derivatives in G, then*

$$\int_{\partial D} f\, dz = 2i \iint_D f_{\bar{z}}\, d\sigma , \qquad \int_{\partial D} f\, d\bar{z} = -2i \iint_D f_z\, d\sigma \tag{6.17}$$

for every Jordan domain D, $\overline{D} \subset G$, with a rectifiable boundary.

6.6. Absolute continuity of homeomorphisms with L^2-derivatives. Of the functions with L^p-derivatives, the case $p = 2$ is the most important to us. In fact, we show in IV.1 that every quasiconformal mapping has L^2-derivatives. The following result can therefore later be applied directly to quasiconformal mappings.

Theorem 6.1. *A homeomorphism w of the domain G, having L^2-derivatives in G, is locally absolutely continuous in G.*

Proof: Without loss of generality we may assume that w is sense-preserving.

By Lemma 3.3, the Jacobian J of w satisfies the inequality

$$\iint_B J\, d\sigma \leqq m(w(B)) \tag{6.18}$$

for every Borel set $B \subset G$, where equality holds for every B if and only if w is locally absolutely continuous in G. We show first that equality holds in (6.18) if B is the closure of a rectangle R with sides parallel to the axes.

Since w is absolutely continuous on lines in G and the integral (6.18) depends continuously on B, we can, by enlarging R if necessary, ensure that w is absolutely continuous on Fr R. Then w is also of bounded variation there and the image under w of the boundary of R is consequently rectifiable. Applying Green's formula (6.11) to the domains R and $w(R) = R'$, we thus obtain

$$\int_{\partial R} u\, dv = \iint_R J\, d\sigma\,, \qquad \int_{\partial R'} u\, dv = \iint_{R'} d\sigma\,,$$

where in the left hand equation u and v denote the real and imaginary parts of w, and in the right they are the coordinates of the $w = u + i\,v$-plane.

By (2.11) the left hand sides of the above equations are equal. Thus equality holds in (6.18) if B is the closure of a rectangle R with sides parallel to the axes.

The same is then true for every open $B \subset G$, since an open set of the finite plane consists of countably many non-intersecting half-open rectangles with sides parallel to the axes.

To demonstrate equality in (6.18) for general Borel sets B, we need consider only sets B with $\overline{B} \subset G$. If B is such a set, then for every $\varepsilon > 0$ there exists an open set D_ε, $B \subset D_\varepsilon$, $\overline{D}_\varepsilon \subset G$, with $m\,(D_\varepsilon - B) < \varepsilon$. Since the integral (6.18) is locally absolutely continuous in G as a set function, we then have

$$\iint_B J\, d\sigma = \lim_{\varepsilon \to 0} \iint_{D_\varepsilon} J\, d\sigma = \lim_{\varepsilon \to 0} m\big(w(D_\varepsilon)\big) \geqq m\big(w(B)\big)\,.$$

By (6.18) equality holds, and Theorem 6.1 is proved.

6.7. Composition of functions with L^p-derivatives. As a second application of Green's formula we establish the following result:

Lemma 6.4. *Let $w = u + i\,v : G \to G'$ and its inverse $w^{-1} = x + i\,y$ be homeomorphisms which have L^p-derivatives with $p \geqq 2$ in G and G'. Further, suppose that f is a function with L^q-derivatives in G', where $1/p + 1/q = 1$. Then the composite function $f \circ w$ has L^1-derivatives in G and its partial derivatives satisfy the chain rule*

$$(f \circ w)_x = (f_u \circ w)\, u_x + (f_v \circ w)\, v_x\,, \quad (f \circ w)_y = (f_u \circ w)\, u_y + (f_v \circ w)\, v_y \quad (6.19)$$

almost everywhere in G.

Proof: By Theorem 6.1, w and w^{-1} are locally absolutely continuous in G and G', respectively. Furthermore it follows from Theorem 3.1 and Lemma 3.4 that w and w^{-1} are differentiable and have a positive Jacobian except on certain null sets $E_0 \subset G$, $E_0' \subset G'$. Since a locally absolutely continuous homeomorphism maps null sets onto null sets, the set $E = E_0 \cup w^{-1}(E_0')$ is also of zero area measure.

If the point $\zeta \in G$ does not belong to E, then w is differentiable at ζ and w^{-1} at $w(\zeta) = \zeta'$. A simple calculation then shows that

$$u_y(\zeta) = -x_v(\zeta')\, J(\zeta)\,, \qquad v_y(\zeta) = x_u(\zeta')\, J(\zeta)\,, \tag{6.20}$$

where J denotes the Jacobian of w. Thus these equations hold almost everywhere in G.

Having this preliminary result, we want to show that $f \circ w$ satisfies the condition (6.1) given in Lemma 6.1. To this end we consider a rectangle R, $\overline{R} \subset G$, with sides parallel to the axes. First we suppose that w is absolutely continuous on the boundary of R. Then $w(\mathrm{Fr}\, R)$ is rectifiable. Green's formula (6.11) can now be applied to the functions f and x in the domain $R' = w(R)$. This gives

$$\int_{\partial R} (f \circ w)\, dx = \int_{\partial R'} f\, dx = \iint_{R'} (f_u x_v - f_v x_u)\, d\sigma\,. \tag{6.21}$$

Furthermore Lemma 3.5 tells us that

$$\iint_{R'} (f_u x_v - f_v x_u)\, d\sigma = \iint_{R} \big((f_u \circ w)\,(x_v \circ w) - (f_v \circ w)\,(x_u \circ w)\big)\, J\, d\sigma\,.$$

Hence, in view of (6.20) and (6.21),

$$\int_{\partial R} (f \circ w)\, dx = -\iint_{R} \big((f_u \circ w)\, u_y + (f_v \circ w)\, v_y\big)\, d\sigma\,. \tag{6.22}$$

In the same way we prove the formula

$$\int_{\partial R} (f \circ w)\, dy = \iint_{R} \big((f_u \circ w)\, u_x + (f_v \circ w)\, v_x\big)\, d\sigma\,. \tag{6.23}$$

These relations are proved under the assumption that w is absolutely continuous on the boundary of R. However, since w is absolutely continuous on lines in G by hypothesis, it follows from the continuity of f and the surface integrals appearing in (6.22) and (6.23) that (6.22) and (6.23) are valid for every rectangle R, $\overline{R} \subset G$, with sides parallel to the axes. Lemma 6.1 now gives the required result that $f \circ w$ has L^1-derivatives in G and that the relations (6.19) are true almost everywhere in G.

Remark. If the mappings w and w^{-1} are continuously differentiable in G and G', respectively, it follows from (6.19) that $f \circ w$ has L^q-derivatives simultaneously with f.

6.8. Absolute continuity on arbitrary arcs. It follows from the above that a function with L^p-derivatives is absolutely continuous not only on horizontal and vertical segments, but on much more general arcs. To express this fact in an exact form we use the concept of the module of a family of arcs introduced in § 4. In the case $p = 2$ one can prove the following (cf. Väisälä [1]):

Theorem 6.2. *Let f be a function with L^2-derivatives in G. Then f is absolutely continuous on all arcs C, $\bar{C} \subset G$, with the exception of a family of module zero.*

Proof: Since the family of all non-rectifiable arcs has zero module by 4.3 we may restrict ourselves to rectifiable arcs.

Let F be a compact subset of G. By Lemma 6.2 there exists a sequence of functions $f_n \in C^\infty$, $n = 1, 2, \ldots$, such that $f_n \to f$ uniformly in F and the partial derivatives of f_n converge to f_x and f_y in $L^2(F)$. Being derivatives of a continuous function, f_x and f_y are Borel functions. Hence, by Lemma 4.2, there exists a subsequence of f_n, also denoted by f_n, such that the relations

$$\lim_{n\to\infty} \int_C \left| \frac{\partial f_n}{\partial x} - \frac{\partial f}{\partial x} \right| |dz| = \lim_{n\to\infty} \int_C \left| \frac{\partial f_n}{\partial y} - \frac{\partial f}{\partial y} \right| |dz| = 0 \qquad (6.24)$$

hold for all rectifiable arcs $C \subset F$, apart from a family of module zero. We show that f is absolutely continuous on every arc C for which (6.24) holds.

Let $\zeta = x + iy : I \to C$ be a length-preserving parametrization of C, choose two points s_1 and s_2, $s_1 < s_2$, on I, and set $z_1 = \zeta(s_1)$, $z_2 = \zeta(s_2)$. If the partial derivatives of f_n are considered as functions of s, then

$$f(z_2) - f(z_1) = \lim_{n\to\infty} (f_n(z_2) - f_n(z_1)) = \lim_{n\to\infty} \int_{s_1}^{s_2} \left[\frac{\partial f_n}{\partial x} x' + \frac{\partial f_n}{\partial y} y' \right] ds, \qquad (6.25)$$

where $|x'| \leqq 1$, $|y'| \leqq 1$. Using the inequality

$$\left| \int_{s_1}^{s_2} \left[\left(\frac{\partial f_n}{\partial x} - \frac{\partial f}{\partial x} \right) x' + \left(\frac{\partial f_n}{\partial y} - \frac{\partial f}{\partial y} \right) y' \right] ds \right| \leqq \int_C \left(\left| \frac{\partial f_n}{\partial x} - \frac{\partial f}{\partial x} \right| + \left| \frac{\partial f_n}{\partial y} - \frac{\partial f}{\partial y} \right| \right) |dz|$$

and the limit relations (6.24), we then obtain from (6.25)

$$f(z_2) - f(z_1) = \int_{s_1}^{s_2} \left(\frac{\partial f}{\partial x} x' + \frac{\partial f}{\partial y} y' \right) ds .$$

Being an integral f is thus absolutely continuous on C.

Until now we have considered only arcs C in the compact set F. Let $F_1, F_2, \ldots,$ be compact sets which exhaust G. By the above $M(\mathscr{C}_n) = 0$ for every n, where $\mathscr{C}_n$ is the family of those arcs $C \subset F_n$ on which f is not absolutely continuous. Since the module is subadditive by 4.2, we have

$$M(\cup \mathscr{C}_n) \leqq \Sigma M(\mathscr{C}_n) = 0\,,$$

and the theorem is proved.

§ 7. *Hilbert Transformation*

7.1. Generalization of Cauchy's integral formula. As an application of Green's formula we shall give an integral representation for functions with L^1-derivatives, which coincides with Cauchy's formula in the case of an analytic function. This result in the differentiated form and combined with certain general properties of L^p-spaces leads to far-reaching consequences.

Let f be a function having L^1-derivatives in a domain G and D, $\bar{D} \subset G$, a Jordan domain with rectifiable boundary. Further, let z be a point of D and denote by D_r the disc $\{\zeta \mid |\zeta - z| < r\}$, $\bar{D}_r \subset D$. The function ψ defined by the formulae

$$\psi(\zeta) = f(\zeta)\,(\zeta - z)^{-1} \qquad \text{for } \zeta \in G - D_r\,,$$

$$\psi(\zeta) = r^{-2} f(\zeta)\,(\bar{\zeta} - \bar{z}) \quad \text{for } \zeta \in D_r\,,$$

then has L^1-derivatives in G. Applying the first formula (6.17) to ψ in the domains D and D_r and subtracting gives

$$\int\limits_{\partial D} \frac{f(\zeta)}{\zeta - z}\, d\zeta - \int\limits_{\partial D_r} \frac{f(\zeta)}{\zeta - z}\, d\zeta = 2\,i \iint\limits_{D - D_r} \frac{f_{\bar{\zeta}}(\zeta)}{\zeta - z}\, d\sigma\,.$$

Now let $r \to 0$. Since f is continuous, the second integral on the left hand side has the limit $2\pi i\, f(z)$, and it follows that

$$f(z) = \frac{1}{2\pi i} \int\limits_{\partial D} \frac{f(\zeta)}{\zeta - z}\, d\zeta - \frac{1}{\pi} \lim_{r \to 0} \iint\limits_{D - D_r} \frac{f_{\bar{\zeta}}(\zeta)}{\zeta - z}\, d\sigma\,. \tag{7.1}$$

The first integral

$$W_D f(z) = \frac{1}{2\pi i} \int\limits_{\partial D} \frac{f(\zeta)}{\zeta - z}\, d\zeta \tag{7.2}$$

defines an analytic function $W_D f$ in D. For the second integral we write

$$-\frac{1}{\pi} \lim_{r \to 0} \iint\limits_{D - D_r} \frac{\omega(\zeta)}{\zeta - z}\, d\sigma = -\frac{1}{\pi} \int\limits_{D} \frac{\omega(\zeta)}{\zeta - z}\, d\sigma = T_D\, \omega(z)\,,$$

even if the integral exists only as a Cauchy principal value. Using this notation we can express (7.1) as follows:

Lemma 7.1. *A function f having L^1-derivatives in G has the representation*

$$f = W_D f + T_D f_{\bar{z}}$$

in D where $W_D f$ is analytic.

If f is identically zero on the boundary of D, then $W_D f$ vanishes in D and the above formula reduces to the simpler form

$$f = T_D f_{\bar{z}} . \tag{7.3}$$

In the special case when D is the disc $|\zeta| < R$ and $f(\zeta) = \bar{\zeta}$, the analytic part of f also vanishes. For then

$$\int_{\partial D} \frac{\bar{\zeta}\, d\zeta}{\zeta - z} = R^2 \int_{\partial D} \frac{d\zeta}{\zeta(\zeta - z)} = \frac{R^2}{z} \int_{\partial D} \left(\frac{1}{\zeta - z} - \frac{1}{\zeta}\right) d\zeta = 0$$

for $z \in D$ and so

$$\bar{z} = -\frac{1}{\pi} \iint_{|\zeta| < R} \frac{d\sigma}{\zeta - z} . \tag{7.4}$$

7.2. Definition of the Hilbert transformation. The question of differentiating (7.3) leads to the so-called Hilbert transformation which will now be defined. We restrict ourselves first to C_0^∞-functions (see 5.4) in the finite plane Ω. For $\omega \in C_0^\infty$ we write

$$T\,\omega(z) = -\frac{1}{\pi} \iint_{\Omega} \frac{\omega(\zeta)}{\zeta - z}\, d\sigma .$$

Fix a point $z_0 \in \Omega$ and define a function ψ by the formula

$$\psi(z) = T\,\omega(z) - \omega(z_0)\, \bar{z} . \tag{7.5}$$

We show that ψ has at z_0 a derivative $\psi'(z_0) = \psi_z(z_0)$ which is independent of direction, and that $\psi_{\bar{z}}(z_0)$ therefore vanishes. This leads to the Hilbert transformation of ω and simultaneously yields rules for the differentiation of $T\,\omega$.

In view of the definition of $T\,\omega$ and (7.4) we have

$$\frac{\psi(z_0 + h) - \psi(z_0)}{h} = -\frac{1}{\pi} \iint_D \frac{\omega(\zeta) - \omega(z_0)}{(\zeta - z_0 - h)(\zeta - z_0)}\, d\sigma , \tag{7.6}$$

where $D = \{\zeta \mid |\zeta| < R\}$ is an arbitrary disc containing the point z_0 and the support of ω. We let $h \to 0$ and show that this limiting process commutes with the integration.

To do this we note that there exists a finite number M such that $|\omega(\zeta) - \omega(z_0)| \leqq M\,|\zeta - z_0|$ in D. Thus the integral

$$\iint_D \frac{|\omega(\zeta) - \omega(z_0)|}{|\zeta - z_0|^2}\,d\sigma$$

exists and

$$\left|\iint_D \frac{\omega(\zeta) - \omega(z_0)}{(\zeta - z_0 - h)\,(\zeta - z_0)}\,d\sigma - \iint_D \frac{\omega(\zeta) - \omega(z_0)}{(\zeta - z_0)^2}\,d\sigma\right|$$

$$= \left|\iint_D \frac{\omega(\zeta) - \omega(z_0)}{(\zeta - z_0)^2\,(\zeta - z_0 - h)}\,h\,d\sigma\right| \leqq M|h| \iint_D \frac{d\sigma}{|\zeta - z_0|\,|\zeta - z_0 - h|}\,. \quad (7.7)$$

For sufficiently small $|h|$, the discs $|\zeta - z_0| < |h|/2$ and $|\zeta - z_0 - h| < |h|/2$ lie in D. Outside these discs

$$|\zeta - z_0|\,|\zeta - z_0 - h| \geqq \frac{1}{3}\,|\zeta - z_0|^2$$

and we have

$$|h| \iint_D \frac{d\sigma}{|\zeta - z_0|\,|\zeta - z_0 - h|} < 2 \iint_{|\zeta - z_0| < \frac{1}{2}|h|} \frac{d\sigma}{|\zeta - z_0|} + 2 \iint_{|\zeta - z_0 - h| < \frac{1}{2}|h|} \frac{d\sigma}{|\zeta - z_0 - h|}$$

$$+ 3\,|h| \iint_{\frac{1}{2}|h| \leqq |\zeta - z_0| < 2R} \frac{d\sigma}{|\zeta - z_0|^2} = 4\pi\,|h| + 6\pi\,|h| \log \frac{4R}{|h|}\,.$$

This expression tends to zero with h, and so by (7.6) and (7.7) the derivative

$$\psi'(z_0) = \psi_z(z_0) = -\frac{1}{\pi} \iint_D \frac{\omega(\zeta) - \omega(z_0)}{(\zeta - z_0)^2}\,d\sigma \quad (7.8)$$

exists.

The term $\omega(z_0)$ can here be omitted. In fact, in view of the second formula (6.17), we have (cf. 7.1)

$$2i \iint_{D - D_r} \frac{d\sigma}{(\zeta - z_0)^2} = \int_{\partial D} \frac{d\bar{\zeta}}{\zeta - z_0} - \int_{\partial D_r} \frac{d\bar{\zeta}}{\zeta - z_0} = -R^2 \int_{\partial D} \frac{d\zeta}{\zeta^2\,(\zeta - z_0)} = 0$$

for every disc $D_r = \{\zeta \mid |\zeta - z_0| < r\}$, $\bar{D}_r \subset D$, and the Cauchy principal value

$$S\,\omega(z) = -\frac{1}{\pi} \iint_\Omega \frac{\omega(\zeta)}{(\zeta - z)^2}\,d\sigma \quad (7.9)$$

thus exists. The linear operator S in (7.9) is called the *Hilbert transformation*, and $S\,\omega$ is the *Hilbert transform* of ω.

7.3. Differentiation of the Hilbert transform. By (7.5) and (7.8) we have

$$(T\,\omega)_z = S\,\omega\,, \qquad (T\,\omega)_{\bar z} = \omega\,. \tag{7.10}$$

Applying the second formula (6.17) we further obtain (cf. 7.1)

$$\begin{aligned} S\,\omega(z) - T\,\omega_z(z) &= \frac{1}{\pi}\iint\limits_D \frac{\partial}{\partial\zeta}\left(\frac{\omega(\zeta)}{\zeta - z}\right) d\sigma \\ &= -\frac{1}{2\pi i}\int\limits_{\partial D} \frac{\omega(\zeta)}{\zeta - z}\, d\bar\zeta + \frac{1}{2\pi i}\lim_{r\to 0}\int\limits_{\partial D_r} \frac{\omega(\zeta)}{\zeta - z}\, d\bar\zeta\,, \end{aligned}$$

where $D_r = \{\zeta \mid |\zeta - z| < r\}$. On the right hand side the first integral vanishes since ω vanishes on Fr D, and it is clear that the second term on the right is also zero. Consequently,

$$S\,\omega = T\,\omega_z\,, \tag{7.11}$$

and we deduce from (7.10) that

$$(S\,\omega)_{\bar z} = \omega_z\,. \tag{7.12}$$

It follows from (7.5) that $T\,\omega$ is everywhere continuous. Thus by (7.11) $S\,\omega$ is also continuous. From (7.10) we then infer that $T\,\omega$ is continuously differentiable and (7.11) shows that $S\,\omega$ has the same property. Repeating this argument we see that $T\,\omega$ and $S\,\omega$ belong to the class C^∞ in Ω. Furthermore, it follows from (7.10) and (7.12) that $T\,\omega$ and $S\,\omega$ are analytic outside the support of ω.

7.4. L^2-norm of the Hilbert transform of a C_0^∞-function. We wish to extend the definition of the Hilbert transformation to functions which belong to $L^2(\Omega)$. This extension is based on the fact that *the Hilbert transformation preserves the L^2-norms of C_0^∞-functions*:

$$\iint\limits_\Omega |S\,\omega|^2\, d\sigma = \iint\limits_\Omega |\omega|^2\, d\sigma\,. \tag{7.13}$$

To prove this we use the identities

$$\overline{(f_{\bar z})} = (\bar f)_z\,, \qquad \overline{(f_z)} = (\bar f)_{\bar z} \tag{7.14}$$

and obtain from (7.10)

$$\iint\limits_\Omega |\omega|^2\, d\sigma = \iint\limits_\Omega \omega\,\overline{(T\,\omega)}_z\, d\sigma = \iint\limits_{D_R} (\omega\,\overline{T\,\omega})_z\, d\sigma - \iint\limits_{D_R} \omega_z\,\overline{T\,\omega}\, d\sigma \tag{7.15}$$

for every disc $D_R = \{z|\,|z| < R\}$ containing the support of ω. On the other hand, it follows from (7.10), (7.12) and the second identity (7.14) that

$$\iint_{\Omega} |S\,\omega|^2\, d\sigma = \lim_{R\to\infty} \iint_{D_R} S\,\omega(\overline{T\,\omega})_{\bar z}\, d\sigma$$
$$= \lim_{R\to\infty} \left(\iint_{D_R} (S\,\omega\,\overline{T\,\omega})_{\bar z}\, d\sigma - \iint_{D_R} \omega_z\, \overline{T\,\omega}\, d\sigma \right). \quad (7.16)$$

Applying (6.17) gives

$$\iint_{D_R} (\omega\,\overline{T\,\omega})_z\, d\sigma = -\frac{1}{2i} \int_{\partial D_R} \omega\,\overline{T\,\omega}\, d\bar z\,,$$

$$\iint_{D_R} (S\,\omega\,\overline{T\,\omega})_{\bar z}\, d\sigma = \frac{1}{2i} \int_{\partial D_R} S\,\omega\,\overline{T\,\omega}\, dz\,.$$

The integrals in the first equation are equal to zero since ω vanishes on the boundary of D_R. If the support of ω lies in $|z| < R_0$ we obtain for $T\,\omega$ and $S\,\omega$ at a boundary point z of $D_R \supset D_{R_0}$ the estimates

$$|T\,\omega(z)| \leqq \frac{1}{\pi\,(|z| - R_0)} \iint_{\Omega} |\omega|\, d\sigma\,, \qquad |S\,\omega(z)| \leqq \frac{1}{\pi\,(|z| - R_0)^2} \iint_{\Omega} |\omega|\, d\sigma\,.$$

Thus

$$\lim_{R\to\infty} \iint_{D_R} (S\,\omega\,\overline{T\,\omega})_{\bar z}\, d\sigma = \frac{1}{2i} \lim_{R\to\infty} \int_{\partial D_R} S\,\omega\,\overline{T\,\omega}\, dz = 0\,,$$

and the required equation (7.13) follows from (7.15) and (7.16).

7.5. Completeness of L^p-spaces. With the help of (7.13) the Hilbert transformation S can now be extended to the class $L^2(\Omega)$. To this end we first make some general remarks on the spaces L^p, $p \geqq 1$.

A sequence f_n, $n = 1, 2, \ldots$, in L^p (cf. 4.4) is called a Cauchy sequence if for every $\varepsilon > 0$ we can find an n_ε such that $\|f_m - f_n\|_p < \varepsilon$ for m, $n > n_\varepsilon$. A convergent sequence in L^p is certainly a Cauchy sequence. The converse is also true: *If f_n is a Cauchy sequence in L^p then there exists $f \in L^p$ such that* $\lim \|f_n - f\|_p = 0$ (Munroe [1], p. 243). This property of L^p is called *completeness.*

In general it does not follow from the convergence of a sequence f_n in the L^p-metric that the functions f_n converge pointwise. However, we have (Munroe [1], pp. 225 and 229): *If the functions f_n converge to f in the space L^p, then there is a subsequence f_{n_k} of f_n such that*

$$\lim_{k\to\infty} f_{n_k}(z) = f(z)$$

for almost all z.

7.6. Extension of the Hilbert transformation to L^2. Using (7.13) we can now extend the Hilbert transformation to the space L^2 in the following way.

Let ω be an arbitrary function which belongs to L^2 in the finite plane Ω. By 5.7 there exists a sequence of functions $\omega_n \in C_0^\infty$ such that

$$\lim_{n\to\infty} \|\omega_n - \omega\|_2 = 0 .$$

By (7.13) we have

$$\|S\,\omega_m - S\,\omega_n\|_2 = \|\omega_m - \omega_n\|_2$$

for every m and n. Thus $S\,\omega_n$, $n = 1, 2, \ldots$, is a Cauchy sequence. In view of the completeness of L^2 there exists an L^2-function, which we denote by $S\,\omega$, with the property

$$\lim_{n\to\infty} \|S\,\omega_n - S\,\omega\|_2 = 0 .$$

This function $S\,\omega$, which is a uniquely determined element of L^2, is called the Hilbert transform of ω. It defines the Hilbert transformation S as a continuous linear operator in the whole space $L^2(\Omega)$, having the integral representation (7.9) in the subclass C_0^∞.[39] The equation

$$\|S\,\omega\|_2 = \|\omega\|_2$$

holds for every $\omega \in L^2(\Omega)$.

7.7. Application to functions with L^2-derivatives. We now return to formula (7.3) of 7.1. Let f be a function which is absolutely continuous on lines in the finite plane Ω and whose partial derivatives are L^2-integrable not merely locally but also over the whole of Ω. By Lemma 6.3 there then exists a sequence of functions $f_n \in C_0^\infty$ such that

$$\lim_{n\to\infty} \|(f_n)_z - f_z\|_2 = \lim_{n\to\infty} \|(f_n)_{\bar z} - f_{\bar z}\|_2 = 0 . \tag{7.17}$$

[39] The following remark can be made on the integral representation of the Hilbert transform of an arbitrary function $\omega \in L^2(\Omega)$.

If we write

$$S_\varrho\,\omega(z) = -\frac{1}{\pi} \iint\limits_{|\zeta - z| > \varrho} \frac{\omega(\zeta)}{(\zeta - z)^2}\,d\sigma ,$$

then

$$\lim_{\varrho\to 0} \|S_\varrho\,\omega - S\,\omega\|_2 = 0 .$$

(The result is due to Beurling; a proof can be found in Ahlfors [2].) Considered as an element of L^2, the integral defined by (7.9) exists therefore as a Cauchy principal value (with respect to the L^2-metric) for every $\omega \in L^2(\Omega)$ and coincides with the above defined Hilbert transform of ω. In fact, $S_\varrho\,\omega(z)$ even converges pointwise almost everywhere as $\varrho \to 0$ (Calderón-Zygmund [1]). We shall not make use of these results in the sequel.

Applying formulae (7.3) and (7.10) we deduce that

$$(f_n)_z = S(f_n)_{\bar{z}} \tag{7.18}$$

for every n. Since

$$\|S(f_n)_{\bar{z}} - S\,f_{\bar{z}}\|_2 = \|(f_n)_{\bar{z}} - f_{\bar{z}}\|_2$$

it follows from (7.17) that $S(f_n)_{\bar{z}} \to S\,f_{\bar{z}}$ in L^2. Thus by 7.5 the sequence f_n has a subsequence for which both terms in (7.18) converge pointwise almost everywhere. We have proved the following result:

Lemma 7.2. *If the function f is absolutely continuous on lines in the finite plane Ω and has square integrable derivatives over Ω, then*

$$f_z(z) = S\,f_{\bar{z}}(z)$$

almost everywhere.

IV. Analytic Characterization of a Quasiconformal Mapping

Introduction to Chapter IV

The methods developed in the previous chapter will now be applied to quasiconformal mappings. To bridge the gap between the theorems of Chapter III and the theory of quasiconformal mappings, we prove first of all that a quasiconformal mapping is absolutely continuous on lines.

Once we have this result we obtain immediately a series of properties of quasiconformal mappings. Thus it follows from Lemma III.3.1 and Theorem III.3.1 that a quasiconformal mapping is differentiable almost everywhere. This and Theorem I.9.3 imply that the dilatation condition $\max_\alpha |\partial_\alpha w| \leqq K \min_\alpha |\partial_\alpha w|$ holds almost everywhere for a K-quasiconformal mapping, a result which sheds new light on the connection between the general and the classical regular quasiconformal mappings.

It follows from the dilatation condition and Lemma III.3.3 that a quasiconformal mapping has L^2-derivatives. Consequently all the results proved in III.6 can be carried over to quasiconformal mappings. For example, we see that a quasiconformal mapping preserves null sets and possesses a positive Jacobian almost everywhere.

In § 2 we investigate the converse problem of how far the properties established in § 1 are sufficient to characterize a quasiconformal mapping. With the help of a counterexample we show first that the quasiconformality of a homeomorphism w cannot be characterized by prescribing the local behaviour of w only almost everywhere. We also deduce from this example that not only the dilatation condition but also the absolute continuity on lines must be required. We therefore obtain the *Analytic definition*, fundamental in our theory, under minimal conditions, when we prove the following result: A sense-preserving homeomorphism w, which is absolutely continuous on lines and satisfies the dilatation condition $\max_\alpha |\partial_\alpha w| \leqq K \min_\alpha |\partial_\alpha w|$ almost everywhere, is K-quasiconformal.

Using the analytic definition we obtain in § 3 various modifications of the original geometric definition. For example, we show that a K-

quasiconformal mapping w of the domain G satisfies the module condition $M(w(\mathscr{C})) \leqq K\, M(\mathscr{C})$ for every family $\mathscr{C}$ of arcs lying in G.

In § 4 we shall prove, again by the aid of the analytic definition, that a sense-preserving homeomorphism is K-quasiconformal precisely when its circular dilatation is everywhere finite and exceeds K in at most a null set.

Finally in § 5, as a preparation for Chapter V, we introduce the complex dilatation of a quasiconformal mapping w; its domain of definition consists of the regular points of w, where it agrees with $w_{\bar{z}}/w_z$. We prove the uniqueness theorem which states that the complex dilatation determines a quasiconformal mapping up to a conformal transformation. Finally the convergence properties of the complex dilatation of quasiconformal mappings w_n of a domain G are examined, under the hypothesis that the sequence w_n converges uniformly in compact subsets of G.

§ 1. *Analytic Properties of a Quasiconformal Mapping*

1.1. Absolute continuity on lines. In this section we give a series of theorems on the continuity and differentiability properties of quasiconformal mappings. As was mentioned in the introduction to this chapter, most of these theorems follow immediately from the results of the previous chapter once we have proved that a quasiconformal mapping is absolutely continuous on lines (cf. III.3.1). This result is due to Strebel [1] and Mori [2]. In our proof we follow the method of Pfluger [2] who obtains the result directly from the definition of absolute continuity using Rengel's inequality.

The definition of absolute continuity on lines in III.3.1 refers to a domain of the finite plane and to a finite-valued function. Here we extend the definition as follows: A homeomorphism f is called absolutely continuous on lines in a domain G if it is absolutely continuous on lines in the former sense in the domain $G - \{\infty\} - \{f^{-1}(\infty)\}$. Analogously we say that f has L^p-derivatives in G if it has them in $G - \{\infty\} - \{f^{-1}(\infty)\}$.

Lemma 1.1. *A quasiconformal mapping w of a domain G is absolutely continuous on lines in G.*

Proof: Let $R = \{x + i\, y \mid a < x < b,\ c < y < d\}$ be a rectangle such that $\overline{R} \subset G - \{f^{-1}(\infty)\}$, and I_y the horizontal segment $a < x < b$ with ordinate y. We have to prove that w is absolutely continuous on almost all I_y for $c < y < d$.

For this purpose we consider the subrectangle R_y of R lying below I_y, and its image $w(R_y)$. The area $A(y)$ of $w(R_y)$ is an increasing function of y and thus has a finite derivative $A'(y)$ for all y, $c < y < d$, except for a set of zero linear measure. We show that w is absolutely continuous on I_{y_0}, $c < y_0 < d$, if $A'(y_0)$ exists and is finite.

Let (x_k, x_k^*), $k = 1, \ldots, n$, be an arbitrary system of non-intersecting open subintervals of the segment $a < x < b$. To prove that w is absolutely continuous on I_{y_0}, it is enough to show that the sum $\Sigma\, |w_k^* - w_k|$, where $w_k^* = w\,(x_k^* + i\, y_0)$, $w_k = w\,(x_k + i\, y_0)$, has an upper bound which tends to zero with $\Sigma\, (x_k^* - x_k)$.

We first choose a positive δ such that $y_0 + \delta < d$. Let R_k, $k = 1, \ldots, n$, denote the quadrilateral consisting of the rectangle $\{x + i\, y \mid x_k < x < x_k^*,\ y_0 < y < y_0 + \delta\}$ and its corners. The horizontal sides will be regarded as a-sides for R_k (cf. I.2.3). Then

$$M(R_k) = \frac{1}{\delta}\,(x_k^* - x_k)\,. \tag{1.1}$$

For the module of the image of R_k, Rengel's inequality (see I.4.3) gives the estimate

$$M\bigl(w(R_k)\bigr) \geqq \frac{(d_k(\delta))^2}{m(w(R_k))}\,, \tag{1.2}$$

where $d_k(\delta)$ denotes the euclidean distance between the b-sides of $w(R_k)$, measured in the plane. These sides converge to the points w_k and w_k^* as $\delta \to 0$, and so

$$\lim_{\delta \to 0} d_k(\delta) = |w_k^* - w_k|\,. \tag{1.3}$$

From (1.1) and (1.2) we get the inequalities

$$K(x_k^* - x_k) \geqq \frac{\delta(d_k(\delta))^2}{m(w(R_k))}\,, \qquad k = 1, \ldots, n\,,$$

where K denotes the maximal dilatation of w. If we add these over k and use Schwarz's inequality we obtain

$$K \sum_{k=1}^{n} (x_k^* - x_k) \geqq \delta\, \frac{\left(\sum\limits_{k=1}^{n} d_k(\delta)\right)^2}{\sum\limits_{k=1}^{n} m(w(R_k))}\,. \tag{1.4}$$

Now we make use of the assumption that the function A introduced above has a finite derivative at $y = y_0$. Since

$$A(y_0 + \delta) - A(y_0) \geqq \sum_{k=1}^{n} m\bigl(w(R_k)\bigr)$$

it follows from (1.4) that

$$\left(\sum_{k=1}^{n} d_k(\delta)\right)^2 \leqq K \frac{A(y_0+\delta) - A(y_0)}{\delta} \sum_{k=1}^{n} (x_k^* - x_k) .$$

If we let $\delta \to 0$ this becomes by (1.3)

$$\left(\sum_{k=1}^{n} |w_k^* - w_k|\right)^2 \leqq K\, A'(y_0) \sum_{k=1}^{n} (x_k^* - x_k) ,$$

and the absolute continuity of w on I_{y_0} is proved. In the same way we can show that w is absolutely continuous on almost all vertical segments $c < y < d$, $x =$ constant, lying in R.

Remark. A quadrilateral consisting of a rectangle and its comens is called a *horizontal rectangle* if its a-sides are parallel to the x-axis. From the above proof it follow that *Lemma* 1.1 *holds for every homeomorphism* w *which satisfies the dilatation condition* $M(R)/K \leqq M(w(R)) \leqq K\, M(R)$ *for all horizontal rectangles.*
On the other hand, a K-quasiconformal mapping w of G need not be absolutely continuous on *every* closed segment in G. By II.8.10 the image of a segment I_y can indeed be non-rectifiable and then w is not even of bounded variation on I_y.

1.2. Differentiability and local dilatation conditions. It follows from the above Lemma 1.1 and from Lemma III.3.1 that a quasiconformal mapping w of the domain G has finite partial derivatives almost everywhere in G. Thus by Theorem III.3.1 w is differentiable almost everywhere in G.

From this result, important in itself, we can deduce more about the local behaviour of a K-quasiconformal mapping. According to Theorem I.9.3 the inequality $\max_\alpha |\partial_\alpha w(z)| \leqq K \min_\alpha |\partial_\alpha w(z)|$ holds at every point where w is differentiable. By the above this is true for almost every $z \in G$, and we obtain the following result:

Theorem 1.1. *A K-quasiconformal mapping w of the domain G is differentiable and satisfies the dilatation condition*

$$\max_\alpha |\partial_\alpha w(z)| \leqq K \min_\alpha |\partial_\alpha w(z)| \tag{1.5}$$

at almost all points z of G.

We remind the reader that for a regular K-quasiconformal mapping (1.5) holds at *all* points of G by definition.[40]

[40] A general quasiconformal mapping naturally need not be differentiable at all points of G.

Theorem 1.1 is due to Mori [2]. He proved the differentiability almost everywhere using the distortion theorem II.9.1 and the Rademacher-Stepanoff theorem (Saks [1], pp. 310—311). We prefer to replace this method by our Theorem III.3.1, which we shall also need later.

1.3. L^2-derivatives of a quasiconformal mapping. The Jacobian of the mapping w satisfies

$$J(z) = \max_\alpha |\partial_\alpha w(z)| \min_\alpha |\partial_\alpha w(z)|$$

at every point z where w is differentiable. The inequality (1.5) is thus equivalent to

$$\max_\alpha |\partial_\alpha w(z)|^2 \leqq K\, J(z)\,. \tag{1.6}$$

From this and Theorem 1.1 it follows that the partial derivatives of a K-quasiconformal mapping w of G satisfy the inequalities

$$|w_x(z)|^2 \leqq K\, J(z)\,, \qquad |w_y(z)|^2 \leqq K\, J(z) \tag{1.7}$$

almost everywhere in G.

We deduce from Lemma III.3.3 that J is integrable over every compact subset of $G - \{w^{-1}(\infty)\}$. By (1.7) the same is true of the functions $|w_x|^2$ and $|w_y|^2$. By Lemma 1.1 w is absolutely continuous on lines in G, and we obtain the following result (Bers [1]):

Theorem 1.2. *A quasiconformal mapping of a domain G has L^2-derivatives in G.*

This theorem will play a central role in the sequel, for it enables us to transfer the results of III.6 to quasiconformal mappings.

1.4. Absolute continuity with respect to the area measure. By combining the above Theorem 1.2 with Theorem III.6.1 we obtain the following result (cf. Morrey [1]):

Theorem 1.3. *A quasiconformal mapping w of the domain G is locally absolutely continuous in $G - \{\infty\} - \{w^{-1}(\infty)\}$.*

In view of III.2.2 we deduce from this theorem: *A quasiconformal mapping of G transforms every area-measurable subset of G into an area-measurable set, and preserves null sets.*

Further, by Lemma III.3.3, we have

$$\iint_A J\, d\sigma = m(w(A)) \tag{1.8}$$

for every measurable set $A \subset G$.

Remark. From the examples constructed in II.8.10 we see that a quasiconformal mapping need not be absolutely continuous with respect to the *linear measure.* As far as sets of zero linear measure are concerned, Beurling and Ahlfors [1] have shown that under a quasiconformal mapping such a set can have an image with positive linear measure. As an example we can take a quasiconformal mapping of the plane, whose restriction to the real axis is a real-valued singular function. The existence of such a mapping follows from the fact that there exist singular functions satisfying the boundary condition (6.11) of Theorem II.6.3.

1.5. Regular points of a quasiconformal mapping. If a mapping $w: G \to G'$ is K-quasiconformal, then its inverse w^{-1} is K-quasiconformal, and by Theorem 1.3 locally absolutely continuous in $G' - \{\infty\} - \{w(\infty)\}$. Hence by Lemma III.3.4 the Jacobian J of w is positive almost everywhere in G.

By the definition given in I.1.6 a sense-preserving homeomorphism w of G is regular at a point $z \in G - \{\infty\} - \{w^{-1}(\infty)\}$ if w is differentiable at z and $J(z) > 0$. From the above and Theorem 1.1 we obtain:

Theorem 1.4. *A quasiconformal mapping w of the domain G is regular at almost every point of G.*

We refer to the result of I.9.6 which states that at a regular point the conditions

$$\min_{\alpha} |\partial_\alpha w(z)| > 0\,, \qquad w_z(z) \neq 0\,, \tag{1.9}$$

are satisfied. Thus these relations hold for a quasiconformal mapping w almost everywhere in G.

It further follows from Theorem 1.4 that the dilatation quotient D (cf. I.9.4) which we introduced at regular points, is defined almost everywhere in G. The inequality (1.5) can be written in the form

$$D(z) \leqq K$$

for almost all $z \in G$.

§ 2. *Analytic Definition of Quasiconformality*

2.1. Statement of the problem. The results of the previous section show that the definition of quasiconformality given in I.3.2 strongly restricts the local behaviour of the mapping. Conversely, we know that the regular quasiconformal mappings can be characterized by local analytic properties. We shall now give a similar analytic characterization for general quasiconformal mappings.

We assume that w is a sense-preserving homeomorphism which satisfies the dilatation condition (1.5) at almost every point. This condition makes sense only if w is differentiable almost everywhere. By Theorem 1.1 these conditions are necessary for w to be quasiconformal. On the other hand, they are not sufficient. In fact the following counterexample shows that a sense-preserving homeomorphism need not be quasiconformal even if it is conformal except in a null set.

2.2. A counterexample. To construct the example we consider a Cantor set $E(p_1, p_2, \ldots)$ with linear measure zero and the corresponding singular function f (see III.2.9). The formula

$$w(z) = z + i\, f(x)$$

defines a homeomorphism w of the square $Q = \{z = x + i\,y \mid 0 < x < 1, 0 < y < 1\}$ which leaves the x-coordinate of every point invariant and adds the increment $f(x)$ to the y-coordinate. If, as in III.2.9, I_{nk} denote the open intervals on which f is constant, then the restriction of w to each rectangle $R_{nk} = \{x + i\,y \mid x \in I_{nk}, 0 < y < 1\}$ is a translation. Since $\bigcup R_{nk}$ contains almost every point of Q, the dilatation condition (1.5) holds with $K = 1$ almost everywhere in Q.

In spite of this, the mapping w is not quasiconformal in Q. This follows from Lemma 1.1, since the singular function f is not constant on $\{x \mid \delta \leqq x \leqq 1 - \delta\}$, $\delta = (1 - p_1)/4$, and consequently the mapping w cannot be absolutely continuous on any of the segments $\{x + i\,y_0 \mid \delta \leqq x \leqq 1 - \delta\}, 0 < y_0 < 1$.

We note that, with respect to the area measure, the mapping w is not only absolutely continuous, but is in fact measure-preserving: $m(w(A)) = m(A)$ for every measurable set $A \subset Q$. Thus this property, even if combined with conformality outside a null set, is not enough to ensure the quasiconformality of the mapping.

In particular, if we consider the Cantor set $E(p_1, p_2, \ldots)$ with $p_n = 1 - 2^{-n}$, $n = 1, 2, \ldots$, then an elementary calculation shows that $\dim E = 0$ (cf. III.1.7). The above exceptional set $Q - \bigcup R_{nk}$ is then so small that it is not only of zero area but of dimension 1, as follows from Lemma III.1.2.[41]

2.3. Analytic definition. It follows from the above example that we cannot characterize the quasiconformality of a homeomorphism by the

[41] By Theorem V.3.2 an exceptional set of this type cannot be essentially smaller.

properties in § 1 without taking into account the concept of absolute continuity on lines (Lemma 1.1). If a homeomorphism $w: G \to G'$ possesses this property, then finite partial derivatives w_x and w_y exist almost everywhere in G, and by Theorem III.3.1 w is differentiable almost everywhere in G. The hypothesis that w is absolutely continuous on lines in G can therefore be combined with the dilatation condition (1.5) without any assumptions about the differentiability of w. We shall show that these properties are sufficient to characterize quasiconformality.

Although we present this result as a theorem, we call it the *Analytic definition* of quasiconformality. Indeed, since the converse result is also true we could use the statement of the theorem as a new definition equivalent to our geometric definition.

Analytic definition. *Let a sense-preserving homeomorphism w of the domain G satisfy the following two conditions:*

1. *w is absolutely continuous on lines in G.*
2. *The dilatation condition*

$$\max_{\alpha} |\partial_{\alpha} w(z)| \leqq K \min_{\alpha} |\partial_{\alpha} w(z)|$$

holds almost everywhere in G.

Then w is a K-quasiconformal mapping of G.

Proof: Since an isolated point is a removable singularity of a quasiconformal mapping (Theorem I.8.1), there is no loss of generality in assuming that G and its image G' are domains of the finite plane.

We show first that w has L^2-derivatives in G. By Lemma III.3.1 and Theorem III.3.1, a homeomorphism with property 1 is differentiable almost everywhere in G. By Lemma III.3.3 the Jacobian J of w is then locally integrable in G. From condition 2 it follows that the functions $|w_x|^2$ and $|w_y|^2$ are majorized by $K\,J$ almost everywhere in G (cf. 1.3). Thus w possesses L^2-derivatives in G.

To prove that w is K-quasiconformal we consider a quadrilateral Q, $\overline{Q} \subset G$, and its image $w(Q)$. We have to show that the modules $M = M(Q)$ and $M' = M(w(Q))$ satisfy the inequality

$$M' \leqq K\,M\,.$$

Let f_2 be the canonical mapping of $w(Q)$ onto the rectangle $R_2 = \{u + i\,v \mid 0 < u < M',\ 0 < v < 1\}$, and f_1 the inverse of the canonical mapping of Q onto the rectangle $R_1 = \{\xi + i\,\eta \mid 0 < \xi < M,\ 0 < \eta < 1\}$. The composed mapping $w^* = f_2 \circ w \circ f_1$ is a sense-preserving homeomorphism of R_1 onto R_2, which can be extended topologically to the boundary. Our next step will be to show that w^* satisfies conditions 1 and 2 in R_1.

Since w has L^2-derivatives in G and f_1 is conformal in R_1, we can use Lemma III.6.4. This tells us that the mapping $w \circ f_1$ is absolutely continuous on lines in R_1, and that the equation

$$\partial_\alpha(w \circ f_1)(\zeta) = f_1'(\zeta)\, \partial_{\alpha+\beta} w(f_1(\zeta)),$$

where $\beta = \arg f_1'(\zeta)$, holds almost everywhere in R_1 for every direction α.

Since f_2 is conformal, $w^* = f_2 \circ w \circ f_1$ is also absolutely continuous on lines in R_1. Further we have

$$\partial_\alpha w^*(\zeta) = f_2'(w(f_1(\zeta)))\, f_1'(\zeta)\, \partial_{\alpha+\beta} w(f_1(\zeta))$$

for almost all $\zeta \in R_1$. Since the derivatives f_2' and f_1' are independent of the direction α, and w satisfies condition 2, this condition is also satisfied by w^*.

The inequality $M' \leqq K\, M$ can now be proved as follows. Since w^* satisfies condition 2, we have $|w_\xi^*|^2 \leqq K\, J^*$ almost everywhere in R_1, where J^* is the Jacobian of w^*. It follows from Lemma III.3.3 that

$$\iint_{R_1} |w_\xi^*|^2\, d\sigma \leqq K \iint_{R_1} J^*\, d\sigma \leqq K\, m(R_2) = K\, M'. \tag{2.1}$$

By Fubini's theorem (see III.1.5)

$$\iint_{R_1} |w_\xi^*|^2\, d\sigma = \int_0^1 d\eta \int_0^M |w_\xi^*|^2\, d\xi, \tag{2.2}$$

and by Schwarz's inequality

$$\int_0^M |w_\xi^*|^2\, d\xi \geqq \frac{1}{M}\left(\int_0^M |w_\xi^*|\, d\xi\right)^2. \tag{2.3}$$

For further estimation, condition 1 is needed. Since w^* is continuous on the boundary of R_1, we have

$$\int_0^M |w_\xi^*(\xi + i\,\eta)|\, d\xi \geqq |w^*(M + i\,\eta) - w^*(i\,\eta)| \geqq M' \tag{2.4}$$

for every η for which w^* is absolutely continuous on every closed subsegment of $I = \{\xi + i\,\eta \mid 0 < \xi < M\}$. Since w^* satisfies condition 1 in R_1, (2.4) holds for almost all η, $0 < \eta < 1$. Our inequality $M' \leqq K\, M$ now follows from (2.1)—(2.4).

2.4. Earlier forms of the analytic definition. As was mentioned in § 1, Mori [2] proved that a K-quasiconformal mapping satisfies the conditions of the Analytic definition. The first versions of the converse are due to Yûjôbô and Bers. Yûjôbô [1] proved that a homeomorphism w is K-quasiconformal if it has L^1-derivatives, is differentiable almost everywhere, and satisfies the conditions 1 and 2. Bers [1] replaced the first two of these conditions by the assumption that w has L^2-derivatives (cf. also Bers [2]). Pfluger [2] noted that it is enough to assume the existence of the L^1-derivatives. In the above formulation the Analytic definition is to be found in the paper of Gehring and Lehto [1].

The counterexample in 2.2 shows that conditions 1 and 2 are in a certain sense the weakest possible. The disadvantage of these minimal conditions is that condition 1 is stated in a form which depends on the coordinate system.

§ 3. *Variants of the Geometric Definition*

3.1. Quasiconformality and module conditions. By definition a K-quasiconformal mapping magnifies the module of quadrilaterals by at most a factor K, and in I.7.1 we proved that the same holds for ring domains. Since these modules are special cases of the general concept of module introduced in III.4, it is natural to ask whether the same dilatation condition holds for the module of *every* family of arcs. The answer is affirmative, as we shall show in this section.

On the other hand, one can seek the weakest possible module conditions which imply quasiconformality. In fact it was shown in I.5.3 and I.9.2 that if the dilatations of analytic quadrilaterals or of small quadrilaterals are bounded, then the homeomorphism is quasiconformal. Further results of this type are given in 3.5 and 3.6 of this section.

3.2. Absolute continuity on arcs. We begin with the following result, which will be applied in deriving the above-mentioned general module condition.

Theorem 3.1. *A quasiconformal mapping $w: G \to G'$ is absolutely continuous on all closed arcs $C \subset G$ with the exception of a family of module zero.*

Proof: By Theorem 1.2, w has L^2-derivatives in G. If G and G' lie in the finite plane then the assertion follows from Theorem III.6.2. In the case when G or G' contains the point at infinity we only need remark that the family of all arcs passing through a fixed point has module zero by III.4.3.

Theorem 3.1 contains Lemma 1.1 on absolute continuity on lines as a special case. To see this we consider a rectangle $R = \{x + i\,y \mid a < x < b,\ c < y < d\}$, $\overline{R} \subset G$, and denote by I_y the horizontal segment $a < x < b$ with ordinate y lying in R. It then follows immediately from Lemma III.4.1 that a family of segments I_y is of module zero if and only if the corresponding y-values in $c < y < d$ form a set of zero linear measure.

We do not know whether it is possible to find a more precise estimate for a family on whose arcs a quasiconformal mapping is not absolutely continuous or in what manner the exceptional family depends on the maximal dilatation.

3.3. General module condition. We now prove the above-mentioned general result on the magnification of the module of a family of arcs (Väisälä [1]).

Theorem 3.2. *Let $w: G \to G'$ be a K-quasiconformal mapping, $\mathscr{C}$ a family of arcs $C \subset G$ and $\mathscr{C}'$ the image family of $\mathscr{C}$ under w. Then*

$$M(\mathscr{C})/K \leqq M(\mathscr{C}') \leqq K\, M(\mathscr{C})\,. \tag{3.1}$$

Proof: Since the inverse of w is also K-quasiconformal it is enough to prove one of the above inequalities, for example that $M(\mathscr{C}) \leqq K\, M(\mathscr{C}')$.

Because the module is monotonic and subadditive we may delete from $\mathscr{C}$ an arbitrary family of module zero without changing the value of the module of $\mathscr{C}$ (cf. formula (4.6) in III.4.2). In view of Theorem 3.1 and the monotonicity of the module (III.4.2) we may therefore assume that no arc $C \in \mathscr{C}$ contains a closed subarc on which w is not absolutely continuous. Further it follows from the remark at the end of III.4.3 that we need only consider rectifiable arcs $C \in \mathscr{C}$. The image arcs $C' \in \mathscr{C}'$ are then locally rectifiable.

By Theorem 1.4 there exists a Borel set E of zero area such that $G - E$ consists of regular points of w only. The partial derivatives of w and so also the derivatives $\partial_\alpha w$ are Borel functions in $G - E$. The same holds for the function $\max_\alpha |\partial_\alpha w|$ since $\partial_\alpha w$ is continuous as a function of α, and so $\max_\alpha |\partial_\alpha w| = \sup_\beta |\partial_\beta w|$, where β assumes only rational values.

Let ϱ^* be a non-negative Borel function defined in the w-plane and admissible with respect to $\mathscr{C}'$, i.e.

$$\int_{C'} \varrho^*\, |dw| \geqq 1\,. \tag{3.2}$$

for every arc $C' \in \mathscr{C}'$.

If

$$\varrho(z) = \begin{cases} \varrho^*(w(z)) \max_\alpha |\partial_\alpha w(z)| & \text{for } z \in G - E, \\ \infty & \text{for } z \in E, \\ 0 & \text{for } z \in -G, \end{cases}$$

then ϱ is also a Borel function. We shall show that ϱ is admissible with respect to $\mathscr{C}$.

Let $\zeta : I \to C$ be a parametrization of C with the arc length s as parameter. The composed mapping $w \circ \zeta : I \to C'$ is a parametrization of C'. Let s^* be the arc length of C' increasing with s. It then follows from equation (2.9) of III.2.7 that

$$\left|\frac{d(w \circ \zeta)}{ds}\right| = \frac{ds^*}{ds} \tag{3.3}$$

almost everywhere on I.

On the other hand, for every $s_0 \in I$ for which the derivatives in (3.3) exist and $z_0 = \zeta(s_0)$ is a regular point of w, we have

$$\left|\frac{d(w \circ \zeta)}{ds}\right|_{s=s_0} = \lim_{s \to s_0} \left|\frac{w(\zeta(s)) - w(\zeta(s_0))}{s - s_0}\right| \leqq \liminf_{s \to s_0} \left|\frac{w(\zeta(s)) - w(\zeta(s_0))}{\zeta(s) - \zeta(s_0)}\right|$$

$$\leqq \max_\alpha |\partial_\alpha w(z_0)| . \tag{3.4}$$

Since the function ϱ defined above assumes the value ∞ at every non-regular point of w, we deduce from (3.3) and (3.4) that

$$\varrho(\zeta(s)) \geqq \varrho^*(w(\zeta(s))) \frac{ds^*}{ds} \tag{3.5}$$

for almost all $s \in I$.

The absolute continuity of w on every compact subset of C is by definition equivalent to s^* being absolutely continuous as a function of s on compact subsets of I. It follows therefore from (3.5), in view of the transformation formula (2.7) in III.2.6, that

$$\int_C \varrho\, |dz| = \int_I \varrho(\zeta(s))\, ds \geqq \int_I \varrho^*(w(\zeta(s))) \frac{ds^*}{ds} ds = \int_{C'} \varrho^*\, |dw| .$$

Hence by (3.2)

$$\int_C \varrho\, |dz| \geqq 1 ,$$

i.e. ϱ is admissible with respect to $\mathscr{C}$.

It follows that the module of $\mathscr{C}$ satisfies the inequality

$$M(\mathscr{C}) \leqq \iint_G \varrho^2\, d\sigma . \tag{3.6}$$

By the definition of ϱ we have

$$(\varrho(z))^2 = (\varrho^*(w(z)))^2 \max_\alpha |\partial_\alpha w(z)|^2 = (\varrho^*(w(z)))^2 D_w(z)\, J(z) \tag{3.7}$$

almost everywhere in G, where D_w denotes the dilatation quotient and J the Jacobian of w. At every regular point $z \in G$ we have $D_w(z) = D_{w^{-1}}(w(z))$. Therefore, by Theorem 1.3 and Lemma III.3.5,

$$\iint_G (\varrho^*(w(z)))^2 D_w(z) J(z) d\sigma = \iint_{G'} (\varrho^*)^2 D_{w^{-1}} d\sigma .$$

Consequently, it follows from (3.6) and (3.7) that

$$M(\mathscr{C}) \leqq \iint_{G'} (\varrho^*)^2 D_{w^{-1}} d\sigma . \tag{3.8}$$

Since this inequality holds for every ϱ^* admissible with respect to $\mathscr{C}'$ we finally obtain

$$M(\mathscr{C}) \leqq \inf_{\varrho^*} \iint_{G'} (\varrho^*)^2 D_{w^{-1}} d\sigma \leqq K \inf_{\varrho^*} \iint_{G'} (\varrho^*)^2 d\sigma = K\, M(\mathscr{C}') ,$$

as required.

3.4. Conformal invariance of the module. For $K = 1$, Theorem 3.2 yields: *The module of a family of arcs is invariant under conformal mappings.*

From this we deduce that the general module condition (3.1) contains the conditions for quadrilaterals and ring domains as special cases. Indeed, if Q is a quadrilateral and $\mathscr{C}$ denotes the family of all Jordan arcs which join the a-sides of Q inside Q, then

$$M(\mathscr{C}) = M(Q) .$$

If Q is a rectangle this follows immediately from Fubini's theorem. For general quadrilaterals it is a consequence of the conformal invariance of the module.[42]

Similarly for a ring domain B we have

$$M(\mathscr{C}') = \frac{M(B)}{2\pi} , \qquad M(\mathscr{C}'') = \frac{2\pi}{M(B)} ,$$

where $\mathscr{C}'$ is the family of Jordan curves separating the components of the complement of B, while $\mathscr{C}''$ consists of the arcs which join them.

3.5. Characterization of quasiconformality by means of rectangles. As we have already mentioned, the quasiconformality of a homeomorphism $w: G \to G'$ is guaranteed as soon as the dilatation condition $M(w(Q)) \leqq K\, M(Q)$ is valid for a suitable subclass of the quadrilaterals Q, $\overline{Q} \subset G$. We shall consider some classes of quadrilaterals which are sufficient for this purpose.

---

[42] Cf. the footnote 7 on p. 21.

As a first result we prove the following:

Theorem 3.3. *If a sense-preserving homeomorphism $w: G \to G'$ satisfies the module condition $M(w(R)) \leqq K\, M(R)$ for every rectangle R, $\bar{R} \subset G$, then w is a K-quasiconformal mapping of G.*

Proof: We show that w satisfies conditions 1 and 2 of the Analytic definition.

At the end of 1.1, as a supplement to Lemma 1.1, we remarked that a homeomorphism $w: G \to G'$ is absolutely continuous on lines if it satisfies the module condition $M(R)/K \leqq M(w(R)) \leqq K\, M(R)$ for every horizontal rectangle[43] R. Thus condition 1 is satisfied. It further follows from Lemma III.3.1 and Theorem III.3.1 that w is differentiable almost everywhere in G.

By the remark in I.9.5 we have

$$\max_\alpha |\partial_\alpha w(z)| \leqq K \min_\alpha |\partial_\alpha w(z)|$$

at every point of G at which w is differentiable, assuming that the dilatation of every square R, $\bar{R} \subset G$, is at most K. Hence condition 2 is also satisfied, and w is K-quasiconformal by the Analytic definition.

The above proof shows that Theorem 3.3 can be sharpened slightly. In fact, w is K-quasiconformal if the module condition $M(R)/K \leqq M(w(R)) \leqq K\, M(R)$ is satisfied for all horizontal rectangles and all squares.

3.6. Module condition for horizontal rectangles. It is natural to ask whether the validity of the module condition only for horizontal rectangles implies quasiconformality. This is in fact true, although the maximal dilatation of the mapping can be greater than the dilatation of horizontal rectangles (Gehring-Väisälä [1]).

Theorem 3.4. *A sense-preserving homeomorphism $w: G \to G'$ satisfying the module condition $M(R)/K \leqq M(w(R)) \leqq K\, M(R)$ for all horizontal rectangles R, $\bar{R} \subset G$, is $\left(K + \sqrt{K^2 - 1}\right)$-quasiconformal. The bound is best possible.*

Proof: As in the proof of Theorem 3.3 we can deduce that w is absolutely continuous on lines and almost everywhere differentiable in G. Hence,

[43] Note that the double inequality is necessary here since the a-sides are horizontal by definition.

by the Analytic definition, w is $\left(K + \sqrt{K^2 - 1}\right)$-quasiconformal if the condition

$$\max_{\alpha} |\partial_\alpha w(z_0)| \leqq \left(K + \sqrt{K^2 - 1}\right) \min_{\alpha} |\partial_\alpha w(z_0)|$$

holds at every point $z_0 \in G$ where w is differentiable. Using complex derivatives we write this inequality in the form (cf. I.9.4)

$$|w_z(z_0)| + |w_{\bar{z}}(z_0)| \leqq \left(K + \sqrt{K^2 - 1}\right) (|w_z(z_0)| - |w_{\bar{z}}(z_0)|) . \tag{3.9}$$

We may confine ourselves to the case when $w_z(z_0) \neq 0$, since otherwise both sides of (3.9) vanish (cf. I.9.4). Further, by means of a translation of the z-plane and a similarity transformation of the w-plane, we can achieve the normalizations $z_0 = w(z_0) = 0$, $w_z(z_0) = 1$. Denoting $w_{\bar{z}}(0)$ by $\varkappa$ we then have

$$w(z) = z + \varkappa \bar{z} + o(z) , \tag{3.10}$$

with $0 \leqq |\varkappa| \leqq 1$. The required inequality (3.9) becomes

$$1 + |\varkappa| \leqq \left(K + \sqrt{K^2 - 1}\right) (1 - |\varkappa|) . \tag{3.11}$$

Let R be the horizontal rectangle with corners 0, a, $a + i\,b$, $i\,b$, where $a > 0$, $b > 0$. If we denote the module a/b of R by M, then the module of the image quadrilateral $R' = w(R)$ satisfies

$$M/K \leqq M(R') \leqq K\,M \tag{3.12}$$

by hypothesis. By Rengel's inequality (see I.4.3) we have further

$$\frac{s_b^2}{m(R')} \leqq M(R') \leqq \frac{m(R')}{s_a^2} ,$$

where s_a and s_b denote the distances, measured in R', between the a- and the b-sides, respectively, and $m(R')$ is the area of R'. From (3.12) we deduce that

$$s_b^2 \leqq K\,M\,m(R') , \qquad s_a^2 \leqq \frac{K}{M} m(R') . \tag{3.13}$$

By (3.10) we have on the vertical sides of R

$$w\,(a + i\,y) = a\,(1 + \varkappa) + i\,y\,(1 - \varkappa) + o\,(a + b) ,$$
$$w\,(i\,y) = i\,y\,(1 - \varkappa) + o\,(a + b) .$$

This gives the estimate

$$s_b \geqq a\,|1 + \varkappa| - b\,|1 - \varkappa| + o\,(a + b) .$$

Correspondingly, for s_a we have

$$s_a \geqq b\,|1 - \varkappa| - a\,|1 + \varkappa| + o\,(a + b) ,$$

and a simple calculation gives

$$m(R') = a\, b\, (1 - |\varkappa|^2) + o(a^2 + b^2)\,.$$

If we now let $a \to 0$ with $a/b = M$ fixed, then these estimates, together with (3.13), yield the following result: For $M\,|1 + \varkappa| \geqq |1 - \varkappa|$, we have

$$\left(|1 + \varkappa| - \frac{|1 - \varkappa|}{M}\right)^2 \leqq K\,(1 - |\varkappa|^2)\,,$$

and for $M\,|1 + \varkappa| < |1 - \varkappa|$,

$$(|1 - \varkappa| - M\,|1 + \varkappa|)^2 \leqq K\,(1 - |\varkappa|^2)\,.$$

It follows that the values $\varkappa = \pm 1$ cannot occur. As $M \to \infty$ or $M \to 0$, we obtain

$$|1 + \varkappa|^2 \leqq K\,(1 - |\varkappa|^2)\,, \qquad |1 - \varkappa|^2 \leqq K\,(1 - |\varkappa|^2)\,.$$

Adding, we deduce that $1 + |\varkappa|^2 \leqq K\,(1 - |\varkappa|^2)$, and the required inequality (3.11) follows by a simple calculation.

The mapping w,

$$w(z) = z + i\sqrt{\frac{K - 1}{K + 1}}\,\bar{z}\,,$$

shows that the bound $K + \sqrt{K^2 - 1}$ is best possible.

3.7. Other special module conditions. In the proof of Theorem 3.4 we needed the module condition $M/K \leqq M(R') \leqq K\,M$ for large and small values of M. The validity of this condition for a fixed value of M is not enough to imply quasiconformality. For example it is easy to construct a homeomorphism of the plane which is not quasiconformal but leaves invariant the modules of all horizontal squares.

If we only know that the dilatation of *every* square R, $\bar{R} \subset G$, is at most K, the above methods cannot be applied. To our knowledge it is an open question whether this condition implies the quasiconformality of a homeomorphism $w: G \to G'$. If it does then w must in fact be K-quasiconformal as can be seen from the proof of Theorem 3.3 (cf. I.9.5).

By Theorem I.7.2 a sense-preserving homeomorphism $w: G \to G'$ is quasiconformal if it satisfies the module condition $M(w(B)) \leqq K\,M(B)$ for all *ring domains* B, $\bar{B} \subset G$. This theorem too can be sharpened to concern subclasses of the family of all ring domains. For results of this type we refer to Gehring-Väisälä [1].

Finally a result due to Renggli [1] should also be mentioned. It states that a sense-preserving homeomorphism is quasiconformal if it carries every curve family with module ∞ into a family with module ∞.

§ 4. Characterization of Quasiconformality with the Help of the Circular Dilatation

Besides using the geometric and analytic definitions we can characterize a quasiconformal mapping by means of the behaviour of the circular dilatation introduced in II.9.1. We then have the advantage that no assumptions about absolute continuity on lines are necessary.

4.1. Necessary conditions for the circular dilatation. Let $w : G \to G'$ be a homeomorphism and denote its circular dilatation by H as before. We shall first give necessary conditions on H for w to be K-quasiconformal.

We proved in II.9.1 that at every regular point $H(z)$ equals the dilatation quotient $D(z)$, which by I.9.6 nowhere exceeds the maximal dilatation $K(G)$. If $K(G)$ is finite, Theorem 1.4 states that almost all points of G are regular. Thus $H(z) \leqq K(G)$ almost everywhere in G.

On the other hand, Theorem II.9.2 shows that the circular dilatation of a K-quasiconformal mapping can exceed K at certain points. By Theorem II.9.1, however, it is always bounded by a number depending only on K. In view of the above remarks, this theorem can be extended as follows:

Theorem 4.1. *The circular dilatation of a K-quasiconformal mapping $w : G \to G'$ is bounded in G and at most equal to K almost everywhere in G.*

4.2. Sufficient conditions for the circular dilatation. The validity of the condition $H(z) \leqq K$ almost everywhere in G is not sufficient to imply that a sense-preserving homeomorphism $w : G \to G'$ is quasiconformal. This follows from the example constructed in 2.2, in which a non-quasiconformal homeomorphism has circular dilatation 1 almost everywhere in its domain. However, if we add the condition that the circular dilatation is bounded, then we can prove that w is K-quasiconformal. Thus Theorem 4.1 has a converse. The result can even be slightly sharpened; in fact H need not be bounded in G, but only everywhere finite there.[44]

Theorem 4.2. *A sense-preserving homeomorphism $w: G \to G'$ is K-quasiconformal if its circular dilatation is finite at every point of G and satisfies $H(z) \leqq K$ almost everywhere in G.*

Proof: The following proof is based on the Analytic definition of quasiconformality. In view of Theorem I.8.1 on removable singularities we can assume that neither G nor G' contains the point at infinity. First we show that a homeomorphism w whose circular dilatation satisfies the above conditions is absolutely continuous on lines in G.

The proof is similar to that of Lemma 1.1. We again take a horizontal rectangle $R = \{x + i y \mid a < x < b,\ c < y < d\}$, $\overline{R} \subset G$, and let I_y be the horizontal segment $a < x < b$ with ordinate y. Denote by R_y the part of R lying under I_y. Then the area $m(w(R_y)) = A(y)$ of the image of R_y is an increasing function of y. Hence, the finite derivative $A'(y)$ exists for almost all y in $c < y < d$.

Since $H(z) \leqq K$ almost everywhere in R, it follows from Fubini's theorem that the same inequality holds almost everywhere on I_y for almost all y, $c < y < d$. Hence it is enough to prove the absolute continuity of w on I_y under the hypotheses that $A'(y)$ exists and $H(x + i y) \leqq K$ holds for almost all x, $a < x < b$. We consider a fixed $I_{y_0} = I$ with these properties.

Let $F \subset I$ be a compact set. We suppose first that $H(z)$ does not exceed a finite bound N in F. As a first step in the proof we show that the linear measure of $w(F) = F'$ has an upper bound which tends to zero with $l(F)$. The result will then be extended to an arbitrary Borel subset of I, and it follows that w is absolutely continuous on I.

We denote by $M(z, r)$ and $m(z, r)$ the maximum and minimum of $|w(\zeta) - w(z)|$ on the circle $|\zeta - z| = r$. If the positive number ε is smaller than the distance between F and $-R$, we can construct the closed set

$$F_\varepsilon = \left\{ z \mid z \in F,\ (r \leqq \varepsilon) \Rightarrow \frac{M(z, r)}{m(z, r)} \leqq N + 1 \right\}.$$

In what follows we shall restrict ourselves to numbers $\varepsilon = 1/q$, where q is a natural number.

[44] In fact, it is enough to assume that H is finite outside a set of σ-finite linear measure (cf. V.3.4). The result was formulated by Yûjôbô [1] and proved by Gehring [1]. Our proof of Theorem 4.2 follows the method of Gehring.

As $\varepsilon \to 0$, the sets F_ε form a non-decreasing sequence which converges to F, i.e. $F = \bigcup F_\varepsilon$. The images F'_ε of F_ε then satisfy $\bigcup F'_\varepsilon = F'$, and consequently (cf. III.1.2)

$$\lim_{\varepsilon \to 0} l(F'_\varepsilon) = l(F') . \tag{4.1}$$

Since F_ε is compact it can be covered by a finite collection of discs $D_\varepsilon(z_h) = \{\zeta \mid |\zeta - z_h| < \varepsilon\}$, $z_h \in F_\varepsilon$, $h = 1, \ldots, n_\varepsilon$. We may assume that these discs cover their union $D_\varepsilon = \bigcup D_\varepsilon(z_h)$ at most twice, for if three discs $D_\varepsilon(z_h)$ have a non-empty intersection we can dispense with one of them without reducing the set $D_\varepsilon \cap I$. Since each of the segments $D_\varepsilon(z_h) \cap I$ has length 2ε, we have

$$l(D_\varepsilon \cap I) \geqq \varepsilon\, n_\varepsilon . \tag{4.2}$$

If now O, $I \supset O \supset F$, is open in I, then $D_\varepsilon \cap I \subset O$ as soon as ε is smaller than the distance between F and $I - O$. Since $l(O - F)$ can be chosen arbitrarily small, we therefore have $\limsup l(D_\varepsilon \cap I) \leqq l(F)$. On the other hand, $D_\varepsilon \cap I \supset F_\varepsilon$ for every ε and, since $\lim l(F_\varepsilon) = l(F)$, we obtain

$$\lim_{\varepsilon \to 0} l(D_\varepsilon \cap I) = l(F) . \tag{4.3}$$

As for the images $D'_\varepsilon(z_h)$ of $D_\varepsilon(z_h)$, $h = 1, \ldots, n_\varepsilon$, we see immediately that the area of their union is at least $\pi \sum (m(z_h, \varepsilon))^2/2$. Since all discs $D_\varepsilon(z_h)$ lie in the rectangle $\{x + iy \mid a < x < b,\ y_0 - \varepsilon < y < y_0 + \varepsilon\}$ we deduce that

$$2\left(A(y_0 + \varepsilon) - A(y_0 - \varepsilon)\right) \geqq \pi \sum_{h=1}^{n_\varepsilon} (m(z_h, \varepsilon))^2 .$$

The expression on the right hand side can be further estimated. Since the points z_h belong to F_ε, we have $M(z_h, \varepsilon)/m(z_h, \varepsilon) \leqq N + 1$, $h = 1, \ldots, n_\varepsilon$, and it follows that

$$2\left(A(y_0 + \varepsilon) - A(y_0 - \varepsilon)\right) \geqq \frac{\pi}{(N+1)^2} \sum_{h=1}^{n_\varepsilon} (M(z_h, \varepsilon))^2 .$$

By Schwarz's inequality,

$$2\left(A(y_0 + \varepsilon) - A(y_0 - \varepsilon)\right) \geqq \frac{\pi}{n_\varepsilon (N+1)^2} \left(\sum_{h=1}^{n_\varepsilon} M(z_h, \varepsilon)\right)^2 .$$

Thus by (4.2) we have

$$l(D_\varepsilon \cap I) \frac{A(y_0 + \varepsilon) - A(y_0 - \varepsilon)}{2\varepsilon} \geqq \frac{\pi}{4(N+1)^2} \left(\sum_{h=1}^{n_\varepsilon} M(z_h, \varepsilon)\right)^2 . \tag{4.4}$$

Letting $\varepsilon \to 0$ we obtain from (4.3)

$$\lim_{\varepsilon \to 0} l(D_\varepsilon \cap I) \frac{A(y_0 + \varepsilon) - A(y_0 - \varepsilon)}{2\varepsilon} = l(F)\, A'(y_0) . \tag{4.5}$$

To investigate the behaviour of the right hand side in (4.4) as $\varepsilon \to 0$ we first take a fixed ε_0 and note that the set F'_{ε_0} is contained in $\bigcup D'_\varepsilon(z_h)$ for $\varepsilon \leqq \varepsilon_0$. Since the set $D'_\varepsilon(z_h)$ lies inside a circle with radius $M(z_h, \varepsilon)$, there exists for every ε, $\varepsilon_0 \geqq \varepsilon > 0$, a covering of F'_{ε_0} by discs the sum of whose diameters is

$$2 \sum_{h=1}^{n_\varepsilon} M(z_h, \varepsilon) .$$

Since w is uniformly continuous in R, the greatest of the numbers $M(z_h, \varepsilon)$ tends to zero as $\varepsilon \to 0$. By the definition of the linear measure $l(F'_{\varepsilon_0})$ cannot therefore exceed

$$2 \liminf_{\varepsilon \to 0} \sum_{h=1}^{n_\varepsilon} M(z_h, \varepsilon) .$$

This holds for every ε_0, and by (4.1)

$$l(F') \leqq 2 \liminf_{\varepsilon \to 0} \sum_{h=1}^{n_\varepsilon} M(z_h, \varepsilon) .$$

Consequently, in view of (4.4) and (4.5), we obtain the estimate

$$(l(F'))^2 \leqq \frac{16}{\pi} (N + 1)^2 A'(y_0)\, l(F) . \tag{4.6}$$

This inequality holds for every compact set $F \subset I$ under the assumption that $H(z) \leqq N$ for $z \in F$.

The next step is to show that (4.6) also holds for all Borel subsets of I in which $H(z) \leqq N$. To do this we construct an approximation of a Borel set by closed sets, and we start by proving that the image $I' = w(I)$ of I is σ-finite with respect to the linear measure (cf. III.1.2).

Since $H(z)$ is finite for every $z \in I$, the same is true of the upper bound

$$\Phi(z) = \sup_r \frac{M(z, r)}{m(z, r)} ,$$

where r runs through the positive numbers for which both $z + r$ and $z - r$ belong to I. If E is a compact subinterval of I, then the set

$$E_N = E \cap I_N , \qquad I_N = \{z \mid \Phi(z) \leqq N\} ,$$

is closed. For $z \in E_N$ we have $H(z) \leqq N$, and by (4.6) the image E'_N of E_N has a finite linear measure. Since $E = \bigcup E_N$, where N runs through the positive integers, $E' = w(E)$ is σ-finite. The same is then also true of I', since I is the union of countably many closed segments.

Let $B \subset I$ be a Borel set in which $H(z) \leqq N$. The image B' of B is also a Borel set, and since I' is σ-finite, there exists by III.1.2 a sequence of closed sets $F_k' \subset B'$, $k = 1, 2, \ldots$, such that $\lim l(F_k') = l(B')$. The preimages $F_k = w^{-1}(F_k')$ are also closed and so (4.6) holds with $F = F_k$ for every k. By passing to the limit we deduce that (4.6) also holds if F and F' are replaced by B and B', respectively.

Finally we want to remove the restriction $H(z) \leqq N$. To do this we first consider a Borel set $B_0 \subset I$ of zero linear measure and its image B_0'. We have $H(z) \leqq N$ in the set $B_0 \cap I_N$. Since $l(B_0 \cap I_N) = 0$, the linear measure of the image $(B_0 \cap I_N)'$ also vanishes. Since $B_0' = \bigcup (B_0 \cap I_N)'$ we see that $l(B_0') = 0$.

Since I was chosen so that $H(z) \leqq K$ for almost all $z \in I$, there exists a Borel set $B_0 \subset I$ of zero linear measure such that $H(z)$ does not exceed K in $I - B_0$. Then $l(B_0') = 0$ as proved above. If now $B \subset I$ is an arbitrary Borel set, then its image satisfies $l(B') = l\left((B - B_0)'\right)$. However, $H(z) \leqq K$ in $B - B_0$, and (4.6) yields

$$\left(l(B')\right)^2 \leqq \frac{16}{\pi} (K + 1)^2 A'(y_0)\, l(B)\,.$$

Since this inequality holds for every Borel set $B \subset I$ and its image B', the homeomorphism w is absolutely continuous on I by the definition in III.2.2. In the same way we can prove that w is absolutely continuous on almost all vertical segments joining the horizontal sides of R.

Condition 1 of the Analytic definition is therefore satisfied. As regards condition 2, we remark that w is differentiable almost everywhere in G by Lemma III.3.1 and Theorem III.3.1. Let z be a point where this holds and where $H(z) \leqq K$. We can distinguish two cases according as z is a regular point of w or $\min |\partial_\alpha w(z)|$ vanishes. In the first case it follows from the discussion of 4.1 that $D(z) = H(z) \leqq K$. In the second case $m(z, r) = o(r)$. Since $H(z) \leqq K$ we also have $M(z, r) = o(r)$, which means that $\partial_\alpha w(z)$ vanishes for every α. Thus the dilatation condition $\max_\alpha |\partial_\alpha w(z)| \leqq K \min_\alpha |\partial_\alpha w(z)|$ is trivial in this case. Hence condition 2 of the Analytic definition is satisfied and w is quasiconformal.

Remark. Methods similar to the above can be applied to other problems in the theory of quasiconformal mappings. As an example we mention the question of the characterization of a quasiconformal mapping by means of its *distortion of angles*; we refer here to the work of Agard-Gehring [1] and Taari [1].

By slightly altering the above proof it can also be shown that the class of mappings which Volkovyskij [1] called K-quasiconformal with a pair of characteristics is a subclass of our K-quasiconformal mappings (Näätänen [1]).

§ 5. *Complex Dilatation*

5.1. Definition of the complex dilatation. This section will serve as a preparation for Chapter V, in which we deal with the problem of the existence of a quasiconformal mapping with prescribed dilatation. In order to formulate this problem it is convenient to introduce a new measure of dilatation, which will express not only the size of the dilatation quotient, but also the direction of the maximal distortion.

Let $w : G \to G'$ be a K-quasiconformal mapping and $z \in G$ a finite point at which w is differentiable. From the equation

$$\partial_\alpha w(z) = w_z(z) + w_{\bar{z}}(z)\, e^{-2i\alpha} \tag{5.1}$$

we deduce by a simple calculation that the dilatation condition (1.5) can also be written in the form

$$|w_{\bar{z}}(z)| \leqq \frac{K-1}{K+1} |w_z(z)| \,. \tag{5.2}$$

If z is a regular point, then $w_z(z) \neq 0$ (cf. 1.5). The quotient

$$\varkappa(z) = \frac{w_{\bar{z}}(z)}{w_z(z)} \tag{5.3}$$

thus exists at every finite regular point z where $w(z) \neq \infty$; by Theorem 1.4 it therefore exists almost everywhere in G. We call the function $\varkappa$ the *complex dilatation* of the mapping w. Being the quotient of two derivatives of a continuous function, $\varkappa$ is a Borel function.

The complex dilatation has a simple geometrical interpretation. We first see from equation (5.1) that $|\partial_\alpha w(z)|$ assumes its maximum $|w_z(z)|\, (1 + |\varkappa(z)|)$ when

$$\alpha = \alpha_M(z) = \frac{1}{2} \arg \varkappa(z) \,.$$

Next it follows from the definition of the dilatation quotient that

$$D(z) = \frac{1 + |\varkappa(z)|}{1 - |\varkappa(z)|} \,. \tag{5.4}$$

Thus $\varkappa(z)$ determines $D(z)$, and in the case $\varkappa(z) \neq 0$ also the direction $\alpha_M(z)$ of maximum distortion, the latter up to a multiple of π. Conversely, $D(z)$ and $\alpha_M(z)$ uniquely determine the complex dilatation $\varkappa(z)$.

By (5.4) (or (5.2)) we have

$$|\varkappa(z)| = \frac{D(z) - 1}{D(z) + 1} \leqq \frac{K - 1}{K + 1} \tag{5.5}$$

at every finite regular point z at which $w(z) \neq \infty$.

Since $\varkappa(z) = 0$ is equivalent to $D(z) = 1$, the complex dilatation of a conformal mapping vanishes identically. Conversely, if we suppose that the complex dilatation of a quasiconformal mapping $w: G \to G'$ vanishes almost everywhere in G, then w is 1-quasiconformal by the Analytic definition and so conformal by Theorem I.5.1.

5.2. Transformation formulae for the complex dilatation. Let $w: G \to G'$ be quasiconformal, f quasiconformal in G', and $f \circ w$ the composite mapping. Denote by $\varkappa_w$, $\varkappa_f$ and $\varkappa_{f \circ w}$ the corresponding complex dilatations. Suppose that z and $w(z)$ are finite regular points of w and f, respectively, and that $f(w(z))$ is finite. Then, by a formal calculation, we obtain

$$\varkappa_{f \circ w}(z) = \frac{\varkappa_w(z) + \varkappa_f(w(z))\, e^{-2i \arg w_z(z)}}{1 + \overline{\varkappa_w(z)}\, \varkappa_f(w(z))\, e^{-2i \arg w_{\bar{z}}(z)}} . \tag{5.6}$$

The set of non-regular points of f in G' is of zero area by Theorem 1.4. The same is true of the preimage in G of this set, because the quasiconformal mapping w^{-1} preserves null sets by Theorem 1.3. Since almost all points of G are also regular points of w, (5.6) holds almost everywhere in G.

By 5.1 the mapping f is conformal if and only if $\varkappa_f$ vanishes almost everywhere in G'. Since w and w^{-1} preserve null sets, this is equivalent to $\varkappa_f(w(z)) = 0$ for almost all $z \in G$. Thus by (5.6), f is conformal if and only if

$$\varkappa_{f \circ w}(z) = \varkappa_w(z) \tag{5.7}$$

almost everywhere in G.[45] For later references we state the sufficiency part of this result in the following form:

Uniqueness theorem. *Let $w: G \to G'$ be a quasiconformal mapping with the complex dilatation $\varkappa$. Then every quasiconformal mapping of G whose complex dilatation equals $\varkappa$ almost everywhere in G is of the form $f \circ w$, where f is a conformal mapping of G'.*

If the mapping w in (5.6) is conformal, then in general $\varkappa_f$ is not the same as $\varkappa_{f \circ w}$. In fact, in this case

$$\varkappa_{f \circ w}(z) = \varkappa_f(w(z))\, e^{-2i \arg w'(z)} . \tag{5.8}$$

---

[45] Note that the conformality of f follows from (5.7) only under the assumption that f is quasiconformal.

Finally, if f and w are quasiconformal mappings of the same domain, (5.6) yields

$$\varkappa_{f \circ w^{-1}}(\zeta) = \frac{\varkappa_f(z) - \varkappa_w(z)}{1 - \varkappa_f(z)\,\overline{\varkappa_w(z)}}\, e^{2\,i \arg w_z(z)}\,, \tag{5.9}$$

where $\zeta = w(z)$.

5.3. Beltrami's differential equation. The definition

$$w_{\bar z} = \varkappa\, w_z \tag{5.10}$$

of the complex dilatation of a quasiconformal mapping w can be interpreted as a differential equation. An equation of type (5.10) is called a *Beltrami* equation. If w is conformal, so that $\varkappa$ is identically zero, then (5.10) reduces to the Cauchy-Riemann equation $w_{\bar z} = 0$.

Now we suppose that $\varkappa$ is a measurable function defined in G with $\sup_z |\varkappa(z)| < 1$. In general there is no function w for which (5.10) holds at every point of G. We shall therefore explain what will be meant here by a solution of (5.10).

A function w is called a *generalized solution* of (5.10) in G if w is absolutely continuous on lines in G and the derivatives w_z, $w_{\bar z}$ (which by Lemma III.3.1 exist almost everywhere) satisfy (5.10) almost everywhere in G. Correspondingly, w will be called a *generalized L^p-solution* in G if it possesses L^p-derivatives and (5.10) holds almost everywhere in G.

Note that changing the values of $\varkappa$ in a null set does not affect the generalized solutions of (5.10). Thus $\varkappa$ need be defined only almost everywhere in G.

By 5.1 and Theorem 1.2, every K-quasiconformal mapping of G is a generalized L^2-solution of a Beltrami equation (5.10), where $\varkappa$ is the complex dilatation of the mapping. Then

$$|\varkappa(z)| \leqq \frac{K-1}{K+1} \tag{5.11}$$

holds at every point where $\varkappa(z)$ is defined.

Conversely, every homeomorphic generalized solution of (5.10) is K-quasiconformal, provided that $\varkappa$ satisfies (5.11). Since conditions 1 and 2 of the Analytic definition are then obviously satisfied, we need only show that w is sense-preserving. In any case w is either K-quasiconformal or the composition of a reflection and a K-quasiconformal mapping. By Theorem 1.4, the Jacobian J of w is therefore non-zero almost everywhere in G. On the other hand it follows from the equation

$$J(z) = |w_z(z)|^2 - |w_{\bar z}(z)|^2 = |w_z(z)|^2 \left(1 - |\varkappa(z)|^2\right),$$

together with (5.11), that $J(z) \geqq 0$ almost everywhere. Hence $J(z) > 0$ for almost all z, and it follows from I.1.6 that w is sense-preserving.

Theorem 5.1. *A K-quasiconformal mapping w of the domain G is a generalized L^2-solution in G of the Beltrami equation* (5.10) *where $\varkappa$ is the complex dilatation of w.*

If, conversely, $\varkappa$ is a measurable function in G which satisfies (5.11), *then every homeomorphic generalized solution of* (5.10) *in G is a K-quasiconformal mapping of G with complex dilatation $\varkappa(z)$ for almost all $z \in G$.*

This theorem is essentially contained in the Analytic definition of quasiconformality and in the theorems of § 1. Thus the characterization of quasiconformal mappings as homeomorphic solutions of Beltrami equations is here of a formal nature. These equations can however be transformed into integral equations and this leads to new methods of investigating quasiconformal mappings (see V.5).

5.4. Good approximation of a quasiconformal mapping. After introducing the complex dilatation we can complement the convergence theorems in II.5 by considering the convergence of the dilatations also. The rest of this section is devoted to problems of this kind.

Let w_n be a sequence of K-quasiconformal mappings of G with complex dilatations $\varkappa_n$, $n = 1, 2, \ldots$. We say that the sequence w_n is a *good approximation* of a quasiconformal mapping w of G with complex dilatation $\varkappa$ if the following two conditions are satisfied:

1. $\lim_{n \to \infty} w_n(z) = w(z)$, uniformly on compact subsets of G.
2. $\lim_{n \to \infty} \varkappa_n(z) = \varkappa(z)$ almost everywhere in G.

If G contains the point at infinity or the point $w^{-1}(\infty)$ then the convergence in condition 1 is to be with respect to the spherical metric.

We illustrate the meaning of condition 2 by a few remarks. From condition 1 alone it follows by Theorem I.5.2, that

$$\liminf_{n \to \infty} \left(\sup_{z \in G} |\varkappa_n(z)| \right) \geqq \sup_{z \in G} |\varkappa(z)| \, .$$

In the case when the mappings w_n are conformal and so each $\varkappa_n$ is identically zero we therefore have $\varkappa = 0$, and condition 2 follows from condition 1. However, in the general case the uniform convergence of w_n to w does not imply the convergence of the complex dilatations $\varkappa_n$ to $\varkappa$ almost everywhere. In fact, the following example shows that the local distortion properties of the mappings w_n can be entirely different from those of the limit mapping w:

For every ε, $0 < \varepsilon < 1$, there exists a sequence of quasiconformal mappings w_n of the square $R = \{x + iy \mid 0 < x < 1, 0 < y < 1\}$ with the following properties:

$$\lim_{n\to\infty} w_n(z) = z\,, \qquad \textit{uniformly in } R,$$

$$|\varkappa_n(z)| = 1 - \varepsilon \quad \textit{almost everywhere in } R \textit{ for every } n.$$

To prove this we divide R into n^2 squares $R_{hk} = \{x + iy \mid (h-1)/n < x < h/n, (k-1)/n < y < k/n\}$, $h, k = 1, \ldots, n$. We construct a quasiconformal mapping f_{hk} of each R_{hk} onto itself as follows; First map R_{hk} onto the unit disc $|\zeta| < 1$ by a conformal mapping φ such that the midpoint z_{hk} of R_{hk} goes into the origin. Then map the unit disc $(2/\varepsilon - 1)$-quasiconformally onto itself by the function t,

$$t(\zeta) = \zeta\,|\zeta|^{2/\varepsilon - 2}\,,$$

and finally map the unit disc back onto R_{hk} by means of φ^{-1}. The complex dilatation of the composite mapping $f_{hk} = \varphi^{-1} \circ t \circ \varphi$ has then the absolute value $1 - \varepsilon$ everywhere in R_{hk} except at the midpoint z_{hk}. This mapping can be extended to a homeomorphism of $\overline{R}_{hk}$ onto itself and the extended mapping leaves every boundary point of $\overline{R}_{hk}$ invariant.

Thus the formula $w_n(z) = f_{hk}(z)$ for $z \in \overline{R}_{hk}$ defines a homeomorphism w_n of R onto itself. By Theorem I.8.3 on the removability of an analytic arc it follows that w_n is quasiconformal with maximal dilatation $2/\varepsilon - 1$. Since w_n maps each R_{hk} onto itself, we have $|w_n(z) - z| \leqq \sqrt{2}/n$ at every point $z \in R$. If we perform the above construction for every $n = 1, 2, \ldots$, we obtain a sequence w_n with the required properties.

Since the limit mapping has complex dilatation zero everywhere in R, the sequence w_n is not a good approximation. We remark that this is so despite the fact that the absolute values $|\varkappa_n(z)|$ converge almost everywhere in R.

5.5. Weak convergence of the derivatives. In order to investigate the conditions under which a sequence of quasiconformal mappings is a good approximation we first make a few preliminary remarks.

Let f_n, $n = 1, 2, \ldots$, be a sequence of functions which are locally integrable in the domain G. We say that the sequence f_n converges *weakly* in G to a locally integrable function f if

$$\lim_{n\to\infty} \iint_R (f_n - f)\, d\sigma = 0 \tag{5.12}$$

for every horizontal rectangle R, $\overline{R} \subset G$.

We shall now prove that the uniform convergence of a sequence of quasiconformal mappings implies the weak convergence of the derivatives (Bers [1]). It follows from the example of 5.4 that the derivatives need not converge almost everywhere.

Lemma 5.1. *Let w_n, $n = 1, 2, \ldots$, be a sequence of K-quasiconformal mappings of G which converges uniformly to a finite function w on compact subsets of G. Then the derivatives $(w_n)_z$ and $(w_n)_{\bar z}$ converge weakly to w_z and $w_{\bar z}$, respectively.*

Proof: It follows from Theorem II.5.3 that the limit function w is either quasiconformal or constant. Therefore w and all the mappings w_n possess L^2-derivatives by Theorem 1.2.

Let R, $\overline{R} \subset G$, be an arbitrary horizontal rectangle. It follows from formulae (6.17) of III.6.5 that

$$\iint_R ((w_n)_z - w_z)\, d\sigma = \frac{i}{2} \int_{\partial R} (w_n - w)\, d\bar z\,,$$

$$\iint_R ((w_n)_{\bar z} - w_{\bar z})\, d\sigma = \frac{1}{2i} \int_{\partial R} (w_n - w)\, dz.$$

Since w_n tends uniformly to w on the boundary of R, the above integrals converge to zero, as was asserted.

As an extension of this result we shall show in V.5.6 that for a good approximation, the derivatives converge in the L^2 metric to the corresponding derivatives of the limit function.

5.6. A criterion for good approximation. We again consider the case when a sequence w_n of K-quasiconformal mappings of G converges to a quasiconformal mapping w uniformly on compact subsets of G. The example constructed in 5.4 shows that $|\varkappa_n(z)|$ can converge to a limit which is greater than $|\varkappa(z)|$ for almost every $z \in G$. If however the argument of $\varkappa_n(z)$ also converges almost everywhere in G, then we can show that w_n is a good approximation to w in the sense of the definition given in 5.4 (Bers [1]).

Theorem 5.2. *Let w_n, $n = 1, 2, \ldots$, be a sequence of K-quasiconformal mappings of G which converges to a quasiconformal mapping w with* *complex dilatation* $\varkappa$, *uniformly on compact subsets of* G. *If the complex dilatations* $\varkappa_n(z)$ *of* w_n *tend to a limit* $\varkappa_\infty(z)$ *almost everywhere in* G, *then* w_n *is a good approximation to* w, *i.e.* $\varkappa_\infty(z) = \varkappa(z)$ *almost everywhere in* G.

Proof: We may assume that G and its images $w_n(G)$ are finite domains. Since $w_{\bar z} - \varkappa\, w_z = 0$ and $w_z \neq 0$ almost everywhere in G, we have to prove that

$$\zeta(z) = w_{\bar z}(z) - \varkappa_\infty(z)\, w_z(z) = 0$$

for almost every $z \in G$. We write

$$\zeta = [w_{\bar z} - (w_n)_{\bar z}] + [(w_n)_{\bar z} - \varkappa_n (w_n)_z] + [\varkappa_n (w_n)_z - \varkappa_\infty (w_n)_z] + [\varkappa_\infty (w_n)_z - \varkappa_\infty\, w_z]\,,$$

and integrate ζ over an arbitrary horizontal rectangle R, $\overline{R} \subset G$. Since the term in the second bracket vanishes almost everywhere in G, we obtain

$$\iint_R \zeta\, d\sigma = I_{1,n} + I_{2,n} + I_{3,n}\,, \tag{5.13}$$

where

$$I_{1,n} = \iint_R \left(w_{\bar z} - (w_n)_{\bar z}\right) d\sigma\,,$$

$$I_{2,n} = \iint_R (\varkappa_n - \varkappa_\infty)\,(w_n)_z\, d\sigma\,,$$

$$I_{3,n} = \iint_R \varkappa_\infty \left((w_n)_z - w_z\right) d\sigma\,.$$

By Lemma 5.1 we have

$$\lim_{n\to\infty} I_{1,n} = 0\,. \tag{5.14}$$

Further, it follows from Schwarz's inequality and the dilatation condition (1.6) that

$$|I_{2,n}|^2 \leqq \iint_R |\varkappa_n - \varkappa_\infty|^2 d\sigma \iint_R |(w_n)_z|^2\, d\sigma \leqq K\, m(w_n(R)) \iint_R |\varkappa_n - \varkappa_\infty|^2\, d\sigma\,. \tag{5.15}$$

Since w is finite in G and w_n converges uniformly to w in R, $m(w_n(R))$ is uniformly bounded. The integral on the right hand side of (5.15) tends to zero as $n \to \infty$ by Lebesgue's convergence theorem (see III.1.5), and so

$$\lim_{n\to\infty} I_{2,n} = 0\,. \tag{5.16}$$

To estimate the third integral in (5.13) we refer to the result in III.5.3, by which the function $\varkappa_\infty$ can be approximated by uniformly bounded step functions φ_k with constant values in horizontal squares so that $\lim \varphi_k = \varkappa_\infty$ almost everywhere in R. We write

$$I_{3,n} = \iint_R (\varkappa_\infty - \varphi_k)\left((w_n)_z - w_z\right) d\sigma + \iint_R \varphi_k \left((w_n)_z - w_z\right) d\sigma\,. \tag{5.17}$$

A further application of Schwarz's inequality and Lebesgue's convergence theorem shows that to every $\varepsilon > 0$ there corresponds a $k_0 > 0$ such that for $k = k_0$ and every n the absolute value of the first integral on the right in (5.17) is less than ε. On the other hand, we deduce by applying Lemma 5.1 to the finitely many squares in which φ_{k_0} has a constant value that

$$\lim_{n \to \infty} \iint_R \varphi_{k_0} \left((w_n)_z - w_z\right) d\sigma = 0 .$$

Thus $I_{3,n}$ also tends to zero, and (5.13), (5.14) and (5.16) give

$$\iint_R \zeta \, d\sigma = 0 . \tag{5.18}$$

It follows that ζ vanishes almost everywhere in G. To prove this we first remark that every open set $O \subset G$ can be represented in the form $(\bigcup R_h) \cup E_0$ where the sets R_h, $h = 1, 2, \ldots$, are disjoint horizontal rectangles and E_0 is a set of zero area. By (5.18), the integral of ζ over O therefore vanishes. If A, $\bar{A} \subset G$, is an arbitrary measurable set, we can construct a decreasing sequence of open sets O_k, $A \subset O_k$, $\bar{O}_k \subset G$, $k = 1, 2, \ldots$, such that $\lim m(O_k - A) = 0$ (see III.1.2). Since the Lebesgue integral is locally absolutely continuous as a set function, we have

$$\iint_A \zeta \, d\sigma = 0 \tag{5.19}$$

for every measurable A, $\bar{A} \subset G$.

Let ξ be the real part of ζ. The set $E = \{z \mid z \in G, \xi(z) \neq 0\}$ is the union of the sets $E_k = \{z \mid z \in G, \xi(z) > 1/k\}$, $E_{-k} = \{z \mid z \in G, \xi(z) < -1/k\}$, $k = 1, 2, \ldots$. Since each of these sets has zero area by (5.19), we have $m(E) = 0$, and so ξ vanishes almost everywhere in G. In the same way we can show that the imaginary part of ζ vanishes almost everywhere, and the theorem is proved.

V. Quasiconformal Mappings with Prescribed Complex Dilatation

Introduction to Chapter V

In the previous chapter we introduced the complex dilatation of a quasiconformal mapping. It is a Borel function defined on the set of the regular points of the mapping i.e. at almost all points. Its absolute value is bounded by a number less than one.

We shall prove the following *Existence theorem*, which is fundamental in the theory of quasiconformal mappings: If $\varkappa$ is any measurable function in a domain G with $\sup_z |\varkappa(z)| < 1$, then there exists a quasiconformal mapping of G whose complex dilatation is equal to $\varkappa$ almost everywhere. By the Uniqueness theorem proved in IV.5 this mapping is uniquely determined up to a conformal mapping.

The Existence theorem will be proved in § 1 by a method which rests on the function theoretical sewing theorem of II.7. At the end of § 1 we briefly refer to other methods of proof.

In §§ 2—5 we make repeated use of the Existence theorem. First, in § 2 we examine systematically the mutual relationships between the dilatation measures introduced earlier: the local maximal dilatation, the dilatation quotient and the circular dilatation.

In § 3 three problems on the removability of sets are considered in detail. Such earlier results as the removability of a point (Theorem I.8.1) and the invariance of the maximal dilatation under removal of analytic arcs (Theorem I.8.3) turn out to be special cases of more general theorems.

In § 4 we show that every quasiconformal mapping admits a good approximation by regular and even real-analytic quasiconformal mappings. This leads us to a characterization of quasiconformal mappings as nonconstant limits of regular quasiconformal mappings.

In § 5 we apply the Hilbert transformation introduced in III.7 to the study of the derivatives of a quasiconformal mapping. Some earlier results of this type can be extended and sharpened in this way.

The last two sections, 6 and 7, are devoted to the study of the local behaviour of a mapping with a given complex dilatation. In particular, we give conditions under which a mapping is regular at a given point. Such questions can be attacked in various ways; our method is essentially based on module estimations with the help of extremal lengths.

§ 1. *Existence Theorem*

1.1. Formulation of the existence problem. We now pose the problem of constructing a quasiconformal mapping with given local dilatation properties. In view of the definition of quasiconformality, it would be natural to choose F or D as a measure of dilatation. Since, however, even the complex dilatation can be prescribed almost everywhere we shall prove the existence theorem in this formulation.

We draw attention to some restrictions which are necessary for the solvability of our problem. First of all we may not prescribe the value of the complex dilatation at every point where it is defined. This follows, for example, from the fact mentioned in IV.5.1 that the vanishing of the complex dilatation outside a null set already implies the conformality of a quasiconformal mapping. Secondly, we know that the complex dilatation is a Borel function. Since it coincides with the prescribed function only almost everywhere it is sufficient to assume that this function is measurable with respect to the Lebesgue area measure. Our problem can thus be formulated as follows:

Given an arbitrary measurable function $\varkappa$ in a domain G with

$$\sup_{z \in G} |\varkappa(z)| < 1 ,$$

construct a quasiconformal mapping of G with complex dilatation equal to $\varkappa(z)$ for almost every $z \in G$.

We remark that the problem can be transformed to the case when G is the whole plane. For an arbitrary domain G, we need only extend each measurable function $\varkappa$ by writing $\varkappa^*(z) = \varkappa(z)$ for $z \in G$, $\varkappa^*(z) = 0$ for $z \notin G$; the function $\varkappa^*$ is then measurable in the whole plane. If the problem can be solved for $\varkappa^*$, then the restriction of the solution to G solves the original problem.

1.2. Solution of a sewing problem. We first consider the special case when $\varkappa$ is a step function (see III.5.3). Let N be the net formed by the squares $Q_{hk} = \{x + i\,y \mid (h-1)\,\delta < x < h\,\delta,\ (k-1)\,\delta < y < k\,\delta\}$, $h, k = 0, \pm 1, \pm 2, \ldots,$ and $\varkappa$ a step function with $\sup_z |\varkappa(z)| < 1$ which has a constant value $\varkappa_{hk}$ in each square Q_{hk}. Since the complex dilatation of the required mapping is to coincide with $\varkappa$ only almost everywhere, the values of $\varkappa$ on the sides of the squares and at infinity are without significance; we may set $\varkappa(z) = 0$ at such points.

For each individual square Q_{hk} our problem can be solved immediately; for example the affine mapping η_{hk} defined by

$$\eta_{hk}(z) = z + \varkappa_{hk}\,\bar{z} \tag{1.1}$$

has the given complex dilatation $\varkappa_{hk}$ in Q_{hk}. We shall construct the required quasiconformal mapping of the plane starting with the functions (1.1), and joining the squares Q_{hk} together step by step. For this purpose we must first solve the following *sewing problem.*

Let R_1 and R_2 be two congruent non-intersecting rectangles having the open segment A as a common side. Let w_i, $i = 1, 2$, be a quasiconformal mapping of R_i with the (not necessarily constant) complex dilatation $\varkappa_i$. It is required to construct a quasiconformal mapping of the rectangle $R = R_1 \cup A \cup R_2$ having the complex dilatation $\varkappa_i(z)$ for almost all $z \in R_i$, $i = 1, 2$.

By IV.5.2 the mappings $f \circ w_i$ and w_i have the same complex dilatation if f is conformal. Hence there is no loss of generality in assuming that $w_1(R_1) = H_1$ and $w_2(R_2) = H_2$ are the upper and lower half-planes, respectively. The mappings w_1 and w_2 can then be extended topologically to $\bar{R}_1$ and $\bar{R}_2$. By means of suitable conformal mappings of H_1 and H_2 onto themselves we may ensure further that w_1 and w_2 map the segment A onto the same interval I of the real axis and that the points $w_1^{-1}(\infty)$ and $w_2^{-1}(\infty)$ are symmetrically placed with respect to A.

If the mappings w_1 and w_2 coincide on A then the sewing problem is solved. In fact the mapping w defined by $w(z) = w_1(z)$ for $z \in R_1 \cup A$, $w(z) = w_2(z)$ for $z \in R_2$ is then a homeomorphism of R onto the domain $H_1 \cup I \cup H_2$, and it follows from Theorem I.8.3 on the removability of an analytic arc that w is quasiconformal.

In the general case when w_1 and w_2 do not necessarily coincide on A we construct two conformal mappings f_1 and f_2 which map H_1 and H_2 onto disjoint Jordan domains such that $f_1 \circ w_1$ and $f_2 \circ w_2$ have the same boundary values on A. If we then write $w(z) = f_1(w_1(z))$ for $z \in R_1 \cup A$, $w(z) = f_2(w_2(z))$ for $z \in R_2$, then the mapping w is a homeomorphism. In the same way as above we deduce that w is quasiconformal. Since $f_i \circ w_i$, $i = 1, 2$, has the same complex dilatation as w_i, the sewing problem is solved.

We still have to prove the existence of the auxiliary conformal mappings f_1 and f_2. The above boundary condition $f_1 \circ w_1 = f_2 \circ w_2$ holds on A if, for every $x \in I$, f_1 and f_2 satisfy the equation

$$f_1(x) = f_2(w_2(w_1^{-1}(x)))\,.$$

In view of the sewing theorem in II.7.5 the existence of the required mappings f_1 and f_2 thus follows if we prove that the homeomorphism $\varphi : I \to I$, defined by $\varphi(x) = w_2(w_1^{-1}(x))$, is quasisymmetric. For this purpose we extend the quasiconformal mapping w_2 to R_1 by reflection in A. The composite mapping $w_2 \circ w_1^{-1}$ is then defined in the upper half-plane and maps it quasiconformally onto itself, keeping the point at infinity fixed. On I the boundary function of the mapping $w_2 \circ w_1^{-1}$ coincides with φ which is therefore quasisymmetric. The above sewing problem is thus solved.

1.3. Proof of the Existence theorem. Using the solution of the above sewing problem we now construct a quasiconformal mapping with prescribed complex dilatation, first in the case of a step function.

We again consider the net N introduced in 1.2 and the corresponding step function $\varkappa$. From the squares $Q_{hk} \in N$ and their sides we can obviously construct a sequence of rectangles R_n, $n = 1, 2, \ldots$, satisfying the following conditions:

1. R_1 is one of the squares Q_{hk}.
2. R_n, $n = 2, 3, \ldots$, consists of R_{n-1}, a rectangle congruent to R_{n-1}, and their common side.
3. $\bigcup_{n=1}^{\infty} R_n$ is the finite plane.

Starting with the affine mappings (1.1) we can then, by repeated solution of the sewing problem, construct a sequence of quasiconformal mappings $w_n : R_n \to R_n'$, such that the complex dilatation of w_n coincides with $\varkappa$ almost everywhere in R_n. The mappings w_n can be normalized by means of linear transformations in such a way that they do not assume the value ∞ and have 0 and 1 as fixed points from some n onwards.

By Theorem II.5.1 the family $\{w_n \mid n \geqq m\}$ is then normal in R_m. Consequently, for every R_m we can find a subsequence $w_{m,n}$, $n = 1, 2, \ldots$, of the sequence w_n, converging uniformly in compact subsets of R_m. These sequences can be chosen so that $w_{m+1,n}$ is a subsequence of $w_{m,n}$, $m = 1, 2, \ldots$. The diagonal sequence $w_{1,1}, w_{2,2}, \ldots$ then converges uniformly in every bounded domain. In view of the above normalization of the mappings it follows from the results of II.5 that $\lim w_{n,n} = w$ is a quasiconformal mapping of the finite plane onto itself. By Theorem IV.5.2 the complex dilatation of w coincides with $\varkappa$ almost everywhere in the plane. The existence problem of 1.1 is thus solved in the case when $\varkappa$ is a step function.

By a limiting process we can now easily treat the general case in which $\varkappa$ is an arbitrary measurable function with $\sup_z |\varkappa(z)| < 1$. By III.5.3,

$\varkappa$ can be approximated by a sequence of step functions $\varkappa_n$ such that $\sup_z |\varkappa_n(z)| \leqq \sup_z |\varkappa(z)|$, and $\varkappa_n(z) \to \varkappa(z)$ almost everywhere in the plane. For each $\varkappa_n$ we construct a quasiconformal mapping W_n of the finite plane onto itself, with complex dilatation coinciding with $\varkappa_n$ almost everywhere. If each W_n is normalized so that $W_n(0) = 0$, $W_n(1) = 1$, then the family $\{W_n\}$ is normal and we can choose a uniformly convergent subsequence of W_n which has as limit function W a quasiconformal mapping of the finite plane. This can be extended to the whole plane by writing $W(\infty) = \infty$. By Theorem IV.5.2 the complex dilatation of W coincides with $\varkappa$ almost everywhere.

We have thus obtained the desired fundamental result:

Existence theorem. *If G is an arbitrary domain and $\varkappa$ an arbitrary measurable function in G with*

$$\sup_{z \in G} |\varkappa(z)| < 1,$$

then there exists a quasiconformal mapping w of G whose complex dilatation coincides with $\varkappa$ almost everywhere in G.

We remark that the set of all quasiconformal mappings of G having complex dilatation $\varkappa$ almost everywhere coincides with the family $\{f \circ w\}$, where f runs through all conformal mappings of $w(G)$. This follows from the Uniqueness theorem proved in IV.5.2.

By the remark in I.8.1 the conformal and quasiconformal equivalence classes coincide for simply connected domains. In view of the Riemann mapping theorem we deduce from the Existence theorem the following

Mapping theorem. *Let G and G' be conformally equivalent simply connected domains and $\varkappa$ a measurable function in G with $\sup_z |\varkappa(z)| < 1$. Then there exists a quasiconformal mapping $w: G \to G'$ whose complex dilatation coincides with $\varkappa$ almost everywhere. This mapping is uniquely determined up to a conformal mapping of G' onto itself.*

1.4. Other proofs for the Existence theorem. The above method of proof resembles one used by Lavrentieff [1]. Here we indicate some others.

It should first be mentioned that one is allowed to make very restrictive assumptions about the given measurable function $\varkappa$. Indeed, by Theorem IV.5.2 it is enough to consider a class of functions which approximate every bounded measurable function almost everywhere (cf. III.5.2–3). In the above proof we have made use of this freedom by first choosing $\varkappa$ to be a step function.

Secondly, it should be emphasized that if $\varkappa$ is sufficiently regular, then it is relatively easy to find *locally* homeomorphic solutions of the equation $w_{\bar{z}} = \varkappa w_z$. This had already been proved by Gauss under the assumption that $\varkappa$ is real-analytic. In the special case when $\varkappa$ is a polynomial the existence of a locally homeomorphic solution will be shown by direct calculation in 4.2, in connection with another problem. We can also obtain such local solutions by applying the integral transformation introduced in III.7 (see § 5, where such questions are discussed in more detail).

The transition from local homeomorphisms to a global homeomorphic solution can be carried out in various ways. The classical method depends on the solvability of the boundary value problem for the real part of the solution (see Morrey [1], where references to the older literature are also to be found, and Martio [1]).

Another method is based on the application of uniformization theory. The plane is covered by discs U_i, $i = 1, 2, \ldots, n$, in such a way that in each U_i a homeomorphic solution of the equation $w_{\bar{z}} = \varkappa w_z$ can be constructed. By the Uniqueness theorem these solutions depend conformally on each other in common parts of the discs and define therefore a conformal structure for the plane. The Existence theorem then follows if we apply the uniformization theorem to the corresponding Riemann surface.

This proof is less direct than the one given in 1.1–1.3 in that the uniformization theory must be brought in. On the other hand, it is possible to give a proof of the uniformization theorem which is similar to that of the sewing theorem. We also refer to Kuusalo [1] who shows how the use of the uniformization theorem can be replaced by fairly simple normal family considerations.

Starting from the above-mentioned integral transformations we can also obtain solutions which are homeomorphic not only locally but globally. The proof was given in its final form by Bojarski [1] (cf. 5.7); see also Ahlfors-Bers [1]. If $\varkappa$ has bounded support, this method yields a representation for a normalized solution in terms of $\varkappa$. The formula is important as it can be used to study the dependence of the mapping on $\varkappa$.

§ 2. *Local Dilatation Measures*

2.1. Connections between D, F and H. The Existence theorem and the results of Chapter IV give us new information about the mutual dependence of the local dilatation measures D, F and H. Our earlier results are contained in I.9 and II.9; we have shown that for any quasiconformal mapping

$$F(z) \geqq D(z) = H(z) \tag{2.1}$$

at every regular point, while $H(z)$ can exceed $F(z)$ at a non-regular point.

Let w be a quasiconformal mapping of the domain G. Since almost all points of G are regular for w, (2.1) holds almost everywhere in G. We shall now show that these relations can be sharpened.

If U is a subdomain of G, then by Lemma I.9.1 the maximal dilatation $K(U)$ of w in U is equal to the least upper bound of $F(z)$ in U. Since $D(z) \leqq F(z)$ at every point z where $D(z)$ is defined, we have $K(U) \geqq \sup_z D(z)$ for $z \in U$.

On the other hand, it follows from the Analytic definition that the restriction of w to U is K-quasiconformal with $K = \sup_z D(z)$. Hence $K(U)$ cannot exceed this bound and so

$$K(U) = \sup_{z \in U} D(z) .$$

In view of the definition of $F(z)$ it follows that

$$F(z_0) = \inf_{U,\, z_0 \in U} \left(\sup_{z \in U} D(z) \right) \tag{2.2}$$

for every $z_0 \in G$.

Thus we obtain the following result:

F is the least upper semicontinuous majorant of D.

In (2.2) z only runs through the regular points of w lying in U. In addition to the set of non-regular points one may omit an arbitrary null set without diminishing the least upper bound of $D(z)$. This follows from the Analytic definition according to which the restriction of w to U is K-quasiconformal as soon as $D(z) \leqq K$ holds almost everywhere in U. Thus if we write as usual

$$\operatorname{sup\,ess}_{z \in U} D(z) = \inf_{E,\, m(E)=0} \left(\sup_{z \in U - E} D(z) \right),$$

$$\lim_{z \to z_0} \operatorname{sup\,ess} D(z) = \inf_{U,\, z_0 \in U} \left(\operatorname{sup\,ess}_{z \in U} D(z) \right),$$

then

$$K(U) = \operatorname{sup\,ess}_{z \in U} D(z)$$

and

$$F(z_0) = \lim_{z \to z_0} \operatorname{sup\,ess} D(z) \tag{2.3}$$

for every $z_0 \in G$. By (2.1), (2.3) and Theorem I.9.2 we also have

$$F(z_0) = \limsup_{z \to z_0} F(z) = \lim_{z \to z_0} \operatorname{sup\,ess} F(z) . \tag{2.4}$$

The circular dilatation $H(z)$ coincides with $D(z)$ almost everywhere in G (see (2.1)) and so

$$\limsup_{z \to z_0} \operatorname{ess} H(z) = \limsup_{z \to z_0} \operatorname{ess} D(z)$$

for every $z_0 \in G$. If we combine this with (2.2—4) we obtain the following result:

Theorem 2.1. *The dilatation measures D, F and H of a quasiconformal mapping $w : G \to G'$ satisfy the relations*

$$\begin{aligned} F(z_0) &= \limsup_{z \to z_0} F(z) = \limsup_{z \to z_0} \operatorname{ess} F(z) = \limsup_{z \to z_0} D(z) \\ &= \limsup_{z \to z_0} \operatorname{ess} D(z) = \limsup_{z \to z_0} \operatorname{ess} H(z) \end{aligned}$$

for every $z_0 \in G$.

2.2. Existence theorem for the local maximal dilatation. The Existence theorem of the previous section, in which the complex dilatation is prescribed, gives rise to the question of how far the real dilatation measures D, F and H can be prescribed. For the dilatation quotient $D = (1 + |\varkappa|)/(1 - |\varkappa|)$ this is already solved by the Existence theorem. Since $H(z) = D(z)$ at all regular points z, the circular dilatation H can also be prescribed almost everywhere. Using Theorem 2.1 we can prove a corresponding existence theorem for F.

Theorem 2.2. *Let Φ be a real-valued, upper semicontinuous, bounded function defined in G, with $\inf \Phi(z) \geqq 1$. Then there exists a quasiconformal mapping w of G whose local maximal dilatation $F(z)$ coincides with $\Phi(z)$ for almost all $z \in G$ and exceeds $\Phi(z)$ nowhere in G.*

Proof: By the Existence theorem of § 1, there is a quasiconformal mapping w of G with dilatation quotient $D(z) = \Phi(z)$ almost everywhere in G. For the local maximal dilatation of w we then have $F(z) \geqq \Phi(z)$ almost everywhere in G.

On the other hand, Theorem 2.1 shows that

$$F(z) = \limsup_{\zeta \to z} \operatorname{ess} D(\zeta) = \limsup_{\zeta \to z} \operatorname{ess} \Phi(\zeta) \leqq \limsup_{\zeta \to z} \Phi(\zeta) .$$

Since Φ is upper semicontinuous we thus have $F(z) \leqq \Phi(z)$ at every point of G. Thus the mapping w possesses the required properties.

Naturally there exist many different quasiconformal mappings with the same local maximal dilatation. We shall now show that not even the dilatation quotient D is determined by F.

2.3. An example where $F(z)$ is greater than $D(z)$ almost everywhere. Theorem 2.1 does not imply that $F(z)$ coincides with $D(z)$ almost everywhere. We can indeed construct an example in which $F(z)$ exceeds $D(z)$ almost everywhere and differs from it by an arbitrarily large amount in a set of positive measure.

Theorem 2.3. *Let G be a domain and ε and M two positive numbers. Then there exists a quasiconformal mapping w of G for which $F(z) > D(z)$ almost everywhere in G and $F(z) = D(z) + M$ except for a set of area $\leqq \varepsilon$.*

Proof: Let E be the set of points of G whose coordinates are rational numbers. Since E is countable it has zero area. Thus we can construct a decreasing sequence of open sets G_k, $E \subset G_k$, $k = 1, 2, \ldots$, such that $G_1 = G$, $m(G_2) \leqq \varepsilon$ and $\lim m(G_k) = 0$.

We set $D^*(z) = 1 + M\,(1 - 1/k)$ for $z \in G_k - G_{k+1}$, $k = 1, 2, \ldots$, which defines $D^*(z)$ for almost all $z \in G$. By the Existence theorem there exists a quasiconformal mapping w of G with dilatation quotient $D(z)$ equal to $D^*(z)$ for almost all $z \in G$. We shall determine the local maximal dilatation of w using Theorem 2.1.

Let U be an arbitrary subdomain of G. Since U contains points of E, $G_k \cap U$ has a positive area as a non-empty open set, for each $k = 1, 2, \ldots$. In G_k we have $D^*(z) \geqq 1 + M\,(1 - 1/k)$ almost everywhere, and so

$$\sup_{z \in U} \operatorname{ess} D(z) = M + 1 .$$

It follows from Theorem 2.1 that $F(z) = M + 1$ at every point z of G. Thus $F(z) > D(z)$ for almost every $z \in G$. In the set $G - G_2$ we have $D(z) = 1$ almost everywhere and consequently $F(z) - D(z) = M$. Since $m(G_2) \leqq \varepsilon$, the mapping w satisfies all the requirements of Theorem 2.3.

We see that two quasiconformal mappings with the same local maximal dilatation F can display quite different behaviour. The mapping w constructed above and an affine mapping with $F(z) = D(z) = M + 1$ will serve as examples.

Remark. Since $D(z) = H(z)$ almost everywhere in G, Theorem 2.3 will also hold if $D(z)$ is replaced by $H(z)$. However, a result of Gehring ([1], p. 25) shows that the inequality $F(z) > H(z)$ cannot hold at *every* point of G.

On the other hand, Gehring [1] has constructed *a quasiconformal mapping for which $H(z) > F(z)$ in a set of Hausdorff dimension* 2.

Since $H(z) \leqq F(z)$ at regular points we conclude: *The set of non-regular points of a quasiconformal mapping can possess Hausdorff dimension* 2. This set is, however, always of zero area, as has already been mentioned several times.

§ 3. Removable Point Sets

3.1. Three problems of removability. In I.8 and II.8 we investigated various continuation problems for quasiconformal mappings. These are related to a group of problems which will be treated here.

Let G be a domain and E a point set lying in G. For each fixed K, $1 \leqq K < \infty$, we consider the following three classes of mappings:

1. *The class* $\mathscr{W}_1$ consisting of all quasiconformal mappings of G whose local maximal dilatation satisfies the condition

$$\sup_{z \in G-E} F(z) \leqq K\,. \tag{3.1}$$

2. *The class* $\mathscr{W}_2$ consisting of all sense-preserving homeomorphisms of G satisfying (3.1).

3. *The class* $\mathscr{W}_3$, which is only defined when $G - E$ is a domain; it then consists of all K-quasiconformal mappings of $G - E$.

The set E is called *removable* with respect to the classes $\mathscr{W}_1$ and $\mathscr{W}_2$, respectively, if all mappings of the class concerned are K-quasiconformal in the whole domain G. Correspondingly, we say that E is removable with respect to the class $\mathscr{W}_3$ if every mapping of this class can be extended to a K-quasiconformal mapping of G.[46]

If $G - E$ is a domain, then in each of the above cases we have to continue a K-quasiconformal mapping of $G - E$ to a mapping of G. The removability of E in the case of the class $\mathscr{W}_1$ then means that every quasiconformal extension is K-quasiconformal; for $\mathscr{W}_2$ the set E is removable if every topological extension is K-quasiconformal and for $\mathscr{W}_3$ if a K-quasiconformal extension exists.

It should be noted that removability with respect to $\mathscr{W}_1$ or $\mathscr{W}_2$ is a monotonic property of E, i.e. all subsets of a removable set are removable. In 3.5 we prove that the same is true of $\mathscr{W}_3$; this is not quite evident from the above definition.

Of the results proved in I.8, Theorem I.8.1 states that a set consisting of a single point is removable with respect to $\mathscr{W}_3$, and Theorem I.8.3

---

[46] For a fourth related problem see Renggli [1].

expresses the removability of an analytic arc with respect to $\mathfrak{W}_2$. Our subsequent results will contain these theorems as special cases.

Remark. To obtain an analogy between the cases 1, 2 and 3, the removability of E with respect to $\mathfrak{W}_i$, $i = 1, 2$, should be defined as follows: E is removable with respect to $\mathfrak{W}_i$ if there corresponds to every mapping $w \in \mathfrak{W}_i$ a K-quasiconformal mapping of G whose restriction to $G - E$ coincides with w.

If E has no interior points, then the two definitions of removability coincide. In fact, in this case a homeomorphism of G is uniquely determined by its restriction to $G - E$. If E has interior points, then we can only deduce that a set which is removable in the sense of the first definition also satisfies the requirements of the second. The converse of this is false; for example, a set E containing the whole of G apart from a single point is removable according to the second definition but not to the first.

It also follows from this example that removability in the sense of the second definition is not a monotonic property of the set. One can however avoid this disadvantage by excluding certain sets E. Since such restrictions carry with them additional difficulties we have preferred the first definition.

3.2. Independence of the removability on the maximal dilatation. We have not discussed so far the dependence of the classes $\mathfrak{W}_1$, $\mathfrak{W}_2$, $\mathfrak{W}_3$ on K. Actually, the value of K has no influence on our removability problems: if a set is removable with respect to $\mathfrak{W}_i$ for some value of K, then the same is true for every other K. This assertion, which we shall return to later, is proved in part in the following lemma.

Lemma 3.1. *Let $E \subset G$ be a set which is removable with respect to one of the classes $\mathfrak{W}_1$, $\mathfrak{W}_2$, $\mathfrak{W}_3$ for $K = 1$. Then E has no interior points and is removable relative to the class concerned for every K.*

Proof: If E has interior points we can choose a disc D, $\overline{D} \subset E$, and map the complement of $\overline{D}$ conformally onto a Jordan domain with a non-analytic quasiconformal boundary curve (see II.8). By Theorem II.8.3 this conformal mapping can be extended to a quasiconformal mapping w of the whole plane. However, w cannot be conformal on the boundary of D. Hence E is not removable with respect to $\mathfrak{W}_1$ and $\mathfrak{W}_2$ for $K = 1$. In the case of $\mathfrak{W}_3$, $G - E$ is a domain by definition. It then follows by the uniqueness of analytic continuation that a conformal mapping of G cannot coincide with w in $G - E$, since both mappings would assume the same values on the boundary of D, and w (Fr D) is a non-analytic

Jordan curve. Hence for $K = 1$ the set E is not removable with respect to $\mathscr{W}_3$ either, and the first part of the lemma is proved.

In order to prove the second part we assume that E is removable with respect to one of the classes $\mathscr{W}_i$, $i = 1, 2, 3$, for $K = 1$. Let w be a mapping which belongs to this class for some $K > 1$. We have to prove either that w is a K-quasiconformal mapping of G or, in the case of $\mathscr{W}_3$, that it can be extended to such a mapping.

In the case of $\mathscr{W}_3$ the set E is closed in G by hypothesis. This assumption may also be made in the other two cases without loss of generality. In fact, it follows from the semicontinuity of the local maximal dilatation that the set $E_\varepsilon = \{z \mid F(z) \geqq K + \varepsilon\}$ is closed in G for every $\varepsilon > 0$. As a subset of E, E_ε is also removable for $K = 1$. If the assertion holds with E_ε and $K + \varepsilon$ instead of E and K, then $F(z) \leqq K + \varepsilon$ at every point of G, and so it is enough to prove this for every $\varepsilon > 0$.

The set E may thus be assumed to be closed in G. The mapping w is then quasiconformal in every component of the open set $G - E$ and consequently preserves null sets (see IV.1.4). If $\varkappa$ denotes the complex dilatation of w, then we can define a measurable function $\varkappa^*$ almost everywhere in the plane by writing

$$\begin{aligned} \varkappa^*(w(z)) &= -\varkappa(z)\, e^{2i \arg w_z(z)} \qquad &&\text{for} \quad z \in G - E\,, \\ \varkappa^*(\zeta) &= 0 &&\text{for} \quad \zeta \notin w(G - E)\,. \end{aligned}$$

By the Existence theorem, there is a K-quasiconformal mapping w^* of the whole plane with complex dilatation $\varkappa^*$ almost everywhere.

The composite mapping $f = w^* \circ w$ is defined in the same domain as w and has complex dilatation zero almost everywhere in $G - E$ by equation (5.6) of IV.5.2. Since $G - E$ is open it follows from Theorem 2.1 that the local maximal dilatation of f is equal to 1 everywhere in $G - E$. Since E is removable for $K = 1$ by hypothesis, there exists a conformal mapping $\tilde{f}$ of G whose restriction to $G - E$ coincides with f.

The mapping $(w^*)^{-1} \circ \tilde{f}$ is then K-quasiconformal in G and coincides with w in $G - E$. In the cases $w \in \mathscr{W}_i$, $i = 1, 2$, we have $w = (w^*)^{-1} \circ \tilde{f}$ everywhere in G, since E has no interior points. The lemma is thus proved.

3.3. Removability of a set with respect to the class $\mathscr{W}_1$.

For the class $\mathscr{W}_1$ the removability problem is solved in the following theorem.

Theorem 3.1. *Let G be a domain and E a subset of G. Then*

$$\sup_{z \in G} F(z) = \sup_{z \in G - E} F(z)$$

for the local maximal dilatation of every quasiconformal mapping of G if and only if every compact subset of E has zero area.

Proof: The condition is necessary. For let $E_0 \subset E$ be a compact set with $m(E_0) > 0$. Then for every $K \geqq 1$ the function F defined by $F(z) = K + 1$ for $z \in E_0$, $F(z) = K$ for $z \in G - E_0$ is upper semicontinuous. Hence by Theorem 2.2 there exists a quasiconformal mapping of G whose local maximal dilatation has the value $K + 1$ almost everywhere in E_0 and does not exceed K anywhere in $G - E$.

To prove that the condition is sufficient suppose next that every compact subset of E is a null set. Then all subsets of E which are closed in G have zero area, since G can be exhausted by a sequence of compact sets.

Let w be a quasiconformal mapping of G whose local maximal dilatation has the upper bound K in $G - E$. In order to prove that $F(z) \leqq K$ everywhere in G we again denote by E_ε, $\varepsilon > 0$, the set of points z where $F(z) \geqq K + \varepsilon$. This set is closed in G and since $E_\varepsilon \subset E$ we have $m(E_\varepsilon) = 0$. Hence $F(z) < K + \varepsilon$ almost everywhere in G. By Theorem 2.1 we have therefore $F(z) \leqq K + \varepsilon$ everywhere in G, and since $\varepsilon > 0$ was arbitrary, it follows that $F(z) \leqq K$ in G as required.

From Theorem 3.1 we see that the removability of a set with respect to the class $\mathcal{W}_1$ is independent of K (cf. 3.2).

3.4. Removability of a set with respect to $\mathcal{W}_2$. We now investigate the class $\mathcal{W}_2$. In this case it is not possible to give a necessary and sufficient condition for the removability of E in terms of Hausdorff measures (cf. 3.7).

Since the class $\mathcal{W}_1$ is a subclass of $\mathcal{W}_2$, the condition of Theorem 3.1 is *necessary* for the removability of a set with respect to $\mathcal{W}_2$. The following theorem, which contains Theorem I.8.3 as a special case, gives a *sufficient* condition (Strebel [1]; this work also contains results on the classes $\mathcal{W}_1$ and $\mathcal{W}_3$.)

Theorem 3.2. *If every compact subset of the set $E \subset G$ has σ-finite linear measure, then*

$$\sup_{z \in G} F(z) = \sup_{z \in G - E} F(z)$$

for every sense-preserving homeomorphism of the domain G.

Proof: We may assume that w is a sense-preserving homeomorphism of G, whose local maximal dilatation has a finite upper bound K in $G - E$. To prove that w is K-quasiconformal in G we make use of the Analytic definition. We show first that w is absolutely continuous on lines in G.

Let us consider a rectangle $R = \{x + i\,y \mid a < x < b,\ c < y < d\}$, $\overline{R} \subset G - w^{-1}(\infty)$. We choose an arbitrary $\varepsilon > 0$ and let E_ε have the same meaning as before. Since this set is contained in E, the compact set $E_\varepsilon \cap \overline{R}$ has σ-finite linear measure by hypothesis.

Let N_k be the net (see III.5.3) whose squares have sides of length 2^{-k}. For every $k = 1, 2, \ldots$, choose those squares of N_k whose closures lie in $R - E_\varepsilon$ and which are not contained in any square chosen earlier. The union of the squares Q_h, $h = 1, 2, \ldots$, so chosen covers the rectangle R with the exception of a set E_0 consisting of $R \cap E_\varepsilon$ and the sides of Q_h. Thus E_0 also has σ-finite linear measure and its area is zero.

Next, we consider an arbitrary Q_h and denote by $I_h(y)$ the open horizontal segment with ordinate y which joins the vertical sides of Q_h. Since w is quasiconformal in every component of the open set $R - E_\varepsilon$ which contains $\overline{Q}_h$, we see by Lemma IV.1.1 that w is absolutely continuous on all segments $I_h(y)$, except for a set A_h of values of y with zero linear measure.

We next consider horizontal segments $I(y) = \{x + i\,y \mid a < x < b\}$, $c < y < d$. To prove the absolute continuity of w on almost every $I(y)$ we need three preparatory results. First it follows from the above that for $y \notin \bigcup A_h$, thus for almost all y, $c < y < d$, w is absolutely continuous on all (non-empty) segments $I(y) \cap Q_h = I_h(y)$.

By Lemma III.1.3 a set of finite linear measure has at most finitely many points on almost all horizontal segments $I(y)$. Since the linear measure of $E_0 = R - \bigcup Q_h$ is σ-finite, we conclude secondly that the set $E_0 \cap I(y) = I(y) - \bigcup I_h(y)$ is countable for almost all y, $c < y < d$.

To obtain the third preliminary result we consider the restriction of w to an arbitrary square Q_h. Since it is $(K + \varepsilon)$-quasiconformal, we have $(|w_z(z)| + |w_{\bar z}(z)|)^2 \leqq (K + \varepsilon)\, J(z)$ for almost all $z \in Q_h$. Thus by Lemma III.3.3,

$$\iint_{Q_h} (|w_z| + |w_{\bar z}|)^2\, d\sigma \leqq (K + \varepsilon)\, m(w(Q_h))$$

for each $h = 1, 2, \ldots$. Summing over h and noting that $m\,(R - \bigcup Q_h) = m(E_0) = 0$ we deduce that

$$\iint_{R} (|w_z| + |w_{\bar z}|)^2\, d\sigma \leqq (K + \varepsilon)\, m(w(R))\,.$$

Since $w(R)$ is a bounded domain, Fubini's theorem (III.1.5) states that $(|w_z| + |w_{\bar z}|)^2$, and so also $|w_x| = |w_z + w_{\bar z}|$, is integrable with respect to linear measure on almost all segments $I(y)$, $c < y < d$.

Combining the above results we deduce that for almost all y, $c < y < d$, the segment $I(y) = I$ has the following three properties: 1°. I consists of the countable set $E_0 \cap I = I_0$ and a sequence of segments I_h, $h = 1, 2, \ldots$. 2°. w is absolutely continuous on every I_h, $h = 1, 2, \ldots$. 3°. The integral $\int_I |w_x|\, dx$ is finite.

From these conditions we shall deduce that w is absolutely continuous on I. By III.2.1—2 it is enough to show that $l(w(I))$ is finite and that w maps every set $A \subset I$ of zero linear measure onto a set of zero linear measure.

It follows from 2° that the length of the image of I_h is

$$l(w(I_h)) = \int_{I_h} |w_x|\, |dx|$$

for each $h = 1, 2, \ldots$ (see III.2.6—7). Since I_0 and $w(I_0)$ are countable and consequently have zero linear measure, it follows from 3° that

$$l(w(I)) = \sum_h \int_{I_h} |w_x|\, |dx| = \int_I |w_x|\, |dx| < \infty .$$

If $A \subset I$ is a set with $l(A) = 0$ then it follows from 2° that $l(w\,(I_h \cap A)) = 0$ for $h = 1, 2, \ldots$. Since $w\,(I_0 \cap A)$ is countable, $l(w(A))$ vanishes, and the absolute continuity of w on I is proved.

In the same way we can prove that w is absolutely continuous on almost all vertical segments lying in R. The first requirement of the Analytic definition is thus satisfied.

It is now easy to verify that w satisfies the second requirement of the Analytic definition. At every point z of the open set $G - E_\varepsilon$ we have $F(z) < K + \varepsilon$, i.e. w is $(K + \varepsilon)$-quasiconformal in every component of $G - E_\varepsilon$. By Theorem IV.1.1,

$$\max_\alpha |\partial_\alpha w(z)| \leqq (K + \varepsilon) \min_\alpha |\partial_\alpha w(z)|$$

for almost all $z \in G - E_\varepsilon$. Since every compact subset of E has σ-finite linear measure, we have $m(E_\varepsilon) = 0$ and the inequality therefore holds almost everywhere in G. Hence w is $(K + \varepsilon)$-quasiconformal in G by the Analytic definition. Since $\varepsilon > 0$ was arbitrary, the theorem is proved.

As in the case of $\mathscr{W}_1$, we remark that the removability of a set $E \subset G$ with respect to $\mathscr{W}_2$ is independent of K. To see this we consider a set E which is removable with respect to $\mathscr{W}_2$ for some $K > 1$. Then every sense-preserving homeomorphism w of G with $F(z) = 1$ in $G - E$ is a K-quasiconformal mapping of G. Since E is removable with respect to $\mathscr{W}_1$, and this is true of every $K \geqq 1$, it follows that w is conformal in G. Thus E is removable with respect to $\mathscr{W}_2$ for $K = 1$ and so for every K by Lemma 3.1.

3.5. Removability of a set with respect to $\mathscr{W}_3$. To deal with the removability problem for the class $\mathscr{W}_3$ we need some results from the theory of analytic functions. Using such tools, we first show that every set removable with respect to $\mathscr{W}_3$ has area zero. This follows from the following, somewhat sharper, result.

Lemma 3.2. *Let E be a subset of the domain G such that $G - E$ is a domain. If the area of E is positive, then there exists a conformal mapping of $G - E$ which cannot be quasiconformally extended to G.*

Proof: By a theorem of Koebe [1] every domain can be mapped onto a slit domain whose complement has zero area. Let f be such a conformal mapping of $G - E$. If f could be extended to a quasiconformal mapping of G then this would carry E onto a set of zero area. However, this contradicts the fact that a quasiconformal mapping preserves null sets (see IV.1.4), and the lemma is proved.

If $m(E) = 0$, then E cannot have interior points and a quasiconformal mapping of $G - E$ can be extended to E in at most one way. By Lemma 3.2, this holds for every set E removable with respect to $\mathscr{W}_3$, and so such a set is removable also with respect to $\mathscr{W}_1$ and $\mathscr{W}_2$. For the same reason every subset of a set removable with respect to $\mathscr{W}_3$ is removable.

Every conformal mapping of $G - E$ can thus be extended to a quasiconformal mapping of G only if $m(E) = 0$. If such an extension exists it is conformal by Theorem 3.1, and so E is removable with respect to $\mathscr{W}_3$ for $K = 1$. By Lemma 3.1 the same holds for every K, and we have thus proved the following result (cf. Pesin [1]):

Theorem 3.3. *Let G be a domain and E a subset of G such that $G - E$ is also a domain. If every conformal mapping of $G - E$ has a quasiconformal extension to G then every K-quasiconformal mapping of $G - E$ can be extended to a K-quasiconformal mapping of G.*

It follows from this theorem that the third type of removability is also independent of K. In fact, if E is removable for some $K = K_0$, then in particular all conformal mappings of $G - E$ have a K_0-quasiconformal (in fact conformal) extension to G. By Theorem 3.3, E is then removable for every K.

3.6. A function-theoretical removability problem. Since the removability problem for $\mathscr{W}_3$ is independent of K, it is equivalent to the following purely function-theoretical problem: Under what conditions can every conformal mapping of $G - E$ be extended to a conformal mapping of G?

We limit ourselves to the case when E is compact. Since $G - E$ is a domain, it lies in a component of $-E$. The boundary of each component of $-E$ is contained in E, and so $-E$ is connected and hence a domain.[47]

We say that E belongs to the null class O_{AD} if there exists no non-constant analytic function f defined in $-E$ such that

$$\iint_{-E} |f'|^2 \, d\sigma < \infty .$$

The significance of the class O_{AD} for our removability problem can be seen from the following function-theoretical equivalence theorem (Sario [1], Ahlfors-Beurling [1]):

The set E belongs to the null class O_{AD} if and only if every conformal mapping of $G - E$ can be extended to a conformal mapping of G.

By this theorem, the removability of E with respect to $\mathscr{W}_3$ is equivalent to $E \in O_{AD}$, provided that E and G satisfy the above conditions.

The null class O_{AD} has been exhaustively investigated in function theory; see for example the work of Ahlfors and Beurling [1] mentioned above, where further references are given. Of the results proved we mention the following: *Every set E of zero linear measure belongs to O_{AD}.*

3.7. Examples of removable and non-removable sets. By Theorem 3.1, sets which are removable with respect to the class $\mathscr{W}_1$ can be characterized in terms of Hausdorff measures. On the other hand, this is not possible for $\mathscr{W}_2$ and $\mathscr{W}_3$. By means of some examples we shall now test the sharpness of the necessary or sufficient conditions obtained above for these classes.

By Theorem 3.2, a set is removable with respect to $\mathscr{W}_2$ if all compact subsets of E have σ-finite linear measure. This sufficient condition cannot be improved very much. In fact, in IV.2.2 we constructed a non-quasiconformal homeomorphism of a square with the property that $F(z) = 1$ outside a set of dimension 1. It follows from this example that *there exist sets of dimension 1 which are not removable with respect to $\mathscr{W}_2$.*

By 3.6 a compact set of zero linear measure[48] is removable with respect to $\mathscr{W}_3$. This sufficient condition cannot be sharpened very much either:

[47] Conversely it follows directly from Lemma I.1.3 that if E is a compact subset of G and $-E$ is a domain, then $G - E$ is a domain.

[48] Note that the complement of a compact set of zero linear measure is connected.

There exist sets of finite linear measure which are not removable with respect to $\mathcal{W}_3$. A simple example is a segment.

On the other hand, we saw in 3.4 and 3.5 that the inner area of a set which is removable with respect to $\mathcal{W}_2$ or $\mathcal{W}_3$ vanishes. As concerns the sharpness of this necessary condition we refer to a result of Sario [1]. He has proved that the Cartesian product E of two congruent Cantor sets $E_x = E_x(p_1, p_2, \ldots)$ and $E_y = E_y(p_1, p_2, \ldots)$, lying on the x- and y-axes, respectively, belongs to O_{AD} (and so is removable for $\mathcal{W}_3$ by 3.6) provided that $l(E_x) = 0$. If we choose $p_n = (n+1)^{-1}$, then $l(E_x) = 0$, while it follows from Lemma III.1.2 that $\dim E = 2$. Since sets removable for $\mathcal{W}_3$ are also removable for $\mathcal{W}_2$ we deduce that *there exist sets of dimension 2 which are removable with respect to* $\mathcal{W}_2$ *and* $\mathcal{W}_3$.

§ 4. Approximation of a Quasiconformal Mapping

4.1. Approximation by mappings with prescribed complex dilatation. The proof of the Existence theorem depends essentially on good approximation (see IV.5.4) of the required mapping by mappings whose complex dilatations are step functions. With the help of the Existence theorem this approximation process can be generalized as we shall now show (cf. Bers [1]).

Theorem 4.1. *Let* $w: G \to G'$ *be a quasiconformal mapping with complex dilatation* $\varkappa$. *Let* $\varkappa_n$, $n = 1, 2, \ldots$, *be a sequence of measurable functions in* G *with* $\sup_z |\varkappa_n(z)| \leqq k < 1$, *which converges to* $\varkappa$ *almost everywhere in* G. *Then given a domain* D, $\overline{D} \subset G$, *there exists a sequence* n_ν *of positive integers and a sequence* w_{n_ν} *of quasiconformal mappings of* D, $\nu = 1, 2, \ldots$, *with the following two properties*:

1. *The mappings* w_{n_ν} *give a good approximation to* w *in* D.
2. *The complex dilatation of* w_{n_ν} *coincides with* $\varkappa_{n_\nu}$ *almost everywhere in* D.

Proof: By applying Theorem II.8.1 we first construct a quasiconformal mapping w^* of the plane which coincides with w in D. Let $\varkappa^*$ be the complex dilatation of w^*.

By the Existence theorem, for every $n = 1, 2, \ldots$, there exists a quasiconformal mapping w_n^* of the plane whose complex dilatation coincides almost everywhere in D with $\varkappa_n$ and in $-D$ with $\varkappa^*$. Since the mappings w_n^* are determined up to a linear transformation we may set $w_n^*(z_h) = w^*(z_h)$ for three fixed points z_h, $h = 1, 2, 3$. It then follows from Theorem II.5.1 that the family of mappings w_n^* is normal. Hence we can find a subsequence $w^*_{n_\nu}$, $\nu = 1, 2, \ldots$, which converges uniformly in the plane (with respect to the spherical metric) to a limit function w^*_0. By Theorem II.5.3, w^*_0 is a quasiconformal mapping of the plane.

The complex dilatations of the mappings $w^*_{n_\nu}$ converge almost everywhere in the plane to the complex dilatation $\varkappa^*$ of w^*. By Theorem IV.5.2 the complex dilatation of w^*_0 therefore coincides with $\varkappa^*$ almost everywhere. It follows from the Uniqueness theorem that $w^* = w^*_0$, since w^* and w^*_0 have the same values at the points z_1, z_2, z_3. If we set w_{n_ν} equal to the restriction of $w^*_{n_\nu}$ to D, then we obtain a sequence which satisfies conditions 1 and 2 of our theorem.

Remark. If G is simply connected, we may replace D by G. In fact, by the Mapping theorem in 1.3 there is for every n a quasiconformal mapping $w_n : G \to G'$ whose complex dilatation coincides with $\varkappa_n$ almost everywhere in G. If these mappings are suitably normalized, then the results of II.5 show that we can choose a subsequence which converges to w uniformly in every compact subset of G. The sequence thus constructed is then a good approximation to w in G.

If G is not simply connected, there may not exist a quasiconformal mapping of G onto G' with complex dilatation $\varkappa_n$ almost everywhere in G, and the above method of proof is not applicable if $D = G$. In any case, given a sequence of domains $D_1 \subset D_2 \subset \cdots$, $\overline{D}_i \subset G$, $\lim D_i = G$, by Theorem 4.1 we can construct a sequence of quasiconformal mappings w_{n_i}, $i = 1, 2, \ldots$, of the plane which converges uniformly to w on compact subsets of G and has the property that w_{n_i} has complex dilatation $\varkappa_{n_i}$ almost everywhere in D_i. However, this sequence is not a good approximation of w if the maximal dilatations of w_{n_i} are not uniformly bounded in G.

4.2. Mappings whose complex dilatation is a polynomial. The significance of Theorem 4.1 lies in the possibility of choosing the functions $\varkappa_n$ so that the approximating mappings w_n are not only regular but even *real-analytic*. This can be achieved if we approximate the complex dilatation $\varkappa$ by polynomials (cf. III.5.3).

By definition a function is real-analytic in a domain G of the finite z-plane if it can be represented by an absolutely convergent power series in z and $\bar{z}$ in a neighbourhood of every point of G. A real-analytic function in G belongs to the class C^∞ there.

Lemma 4.1. *Let G and G' be finite domains and $w : G \to G'$ a quasiconformal mapping whose complex dilatation coincides almost everywhere in G with a polynomial P in z and $\bar{z}$. Then w is regular and real-analytic.*

Proof: We have to prove that w has a power series expansion in a neighbourhood of an arbitrary point $z_0 \in G$, and that the Jacobian J of w does not vanish at z_0. It follows from the Uniqueness theorem that instead of w we can consider an arbitrary quasiconformal mapping $\hat{w}$ defined in a neighbourhood U of z_0 and having complex dilatation P almost everywhere in U.

Such a mapping $\hat{w}$ can be constructed directly by means of a power series. Let

$$P(z) = \sum_{m,n=0}^{N} a_{mn} (z - z_0)^m (\bar{z} - \bar{z}_0)^n \tag{4.1}$$

and write

$$\hat{w}(z) = \sum_{h,k=0}^{\infty} c_{hk} (z - z_0)^h (\bar{z} - \bar{z}_0)^k . \tag{4.2}$$

The coefficients c_{hk} are to be so chosen that $\hat{w}$ will satisfy the Beltrami differential equation

$$\hat{w}_{\bar{z}}(z) = P(z)\, \hat{w}_z(z) \tag{4.3}$$

at every point z with a neighbourhood where the series (4.2) converges absolutely.

This will be done by a direct calculation. In what follows we shall set $c_{hk} = 0$ if h or k is negative, and $a_{mn} = 0$ if a_{mn} does not appear in (4.1).

Differentiating (4.2) termwise with respect to z and $\bar{z}$ and substituting in (4.3) we obtain the equations

$$k\, c_{hk} = \sum_{m,n=0}^{N} (h + 1 - m)\, a_{mn}\, c_{h+1-m,\, k-1-n} \, . \tag{4.4}$$

For $k = 0$ these are of the form $0 = 0$. Hence the coefficients c_{h0} may be chosen arbitrarily. We set $c_{h0} = 0$ for $h \neq 1$ and $c_{10} = 1$. Starting from these initial conditions we then determine all the other coefficients c_{hk} recursively by (4.4).

To investigate the convergence of (4.2) we show first that

$$c_{hk} = 0 \qquad \text{for } h > k\,N + 1 \, . \tag{4.5}$$

For $k = 0$ this follows from the initial condition. For $k = 1$, $c_{h1} = a_{h0}$ by (4.4), and so $c_{h1} = 0$ for $h > N$. The general case can be proved by induction: Suppose that (4.5) is true for $k < k_0$ and set $k = k_0$, $h = h_0 > k_0 N + 1$ in (4.4). It then follows from $0 \leqq m, n \leqq N$ that

$$h_0 + 1 - m > k_0 N + 2 - m > (k_0 - 1 - n)\, N + 1 \, ,$$

and all the coefficients $c_{h_0+1-m,\, k_0-1-n}$ vanish. Thus (4.5) holds for $k = k_0$ too.

In order to derive an upper bound for $|c_{hk}|$ we write $A = \max |a_{mn}|$. We first have $|c_{h0}| \leqq 1$ because of the initial condition, and $|c_{h1}| \leqq A$ since $c_{h1} = a_{h0}$. For $k > 1$ we obtain from (4.4)

$$|c_{hk}| \leqq A \, \frac{h+1}{k} \sum_{m,n=0}^{N} |c_{h+1-m,\,k-1-n}| \, .$$

In view of (4.5) we may assume that $h \leqq k\,N + 1$. Then $(h+1)/k \leqq N + 2/k \leqq N + 1$, and consequently

$$|c_{hk}| \leqq A\,(N+1) \sum_{m,n=0}^{N} |c_{h+1-m,\,k-1-n}| \, . \tag{4.6}$$

From this we deduce that

$$\begin{aligned} |c_{h2}| &\leqq A\,(N+1) \sum_{m=0}^{N} (|c_{h+1-m,0}| + |c_{h+1-m,1}|) \\ &\leqq A\,(A+1)\,(N+1)^2 \leqq (A\,(N+1)^2 + 1)^2 \, . \end{aligned}$$

If we write $\alpha = A\,(N+1)^2 + 1$, we have

$$|c_{hk}| \leqq \alpha^k \tag{4.7}$$

for $k = 0, 1, 2$. It is easily seen by induction that this inequality holds for every k.

We have therefore

$$|c_{hk}\,(z - z_0)^h\,(\bar z - \bar z_0)^k| \leqq \alpha^k\,|z - z_0|^{h+k} \, ,$$

and the series (4.2) is absolutely convergent in the disc $|z - z_0| < 1/\alpha$. Furthermore it follows from (4.4) that $\hat w$ satisfies the equation (4.3) there.

For the Jacobian of $\hat w$ at the point z_0 we have

$$J(z_0) = |\hat w_z(z_0)|^2 - |\hat w_{\bar z}(z_0)|^2 = |c_{10}|^2 - |c_{01}|^2 = 1 - |a_{00}|^2 = 1 - |P(z_0)|^2 \, ,$$

so that $J(z_0) > 0$, since $|P(z_0)| < 1$. Hence $\hat w$ is a regular quasiconformal mapping in a neighbourhood of z_0.

4.3. Quasiconformal mappings as limits of regular mappings. Let G and G' be finite domains and $w : G \to G'$ a K-quasiconformal mapping with complex dilatation $\varkappa$. Then $|\varkappa| \leqq (K-1)/(K+1) = k < 1$. By III.5.3, we can construct in every domain D, $\bar D \subset G$, a sequence of polynomials P_n in z and $\bar z$, $n = 1, 2, \ldots$, with the properties that $|P_n(z)| \leqq k$ for $z \in D$ and

$$\lim_{n \to \infty} P_n(z) = \varkappa(z)$$

almost everywhere in D.

By Theorem 4.1, there exists in D a good approximation w_{n_ν}, $\nu = 1, 2, \ldots$, of w such that the complex dilatation of w_{n_ν} coincides with the polynomial P_{n_ν} almost everywhere in D. It follows from Lemma 4.1 that every w_{n_ν} is a regular real-analytic quasiconformal mapping of D which thus has complex dilatation $P_{n_\nu}(z)$ for every $z \in D$. Since $|P_{n_\nu}| \leqq k$, w_{n_ν} is K-quasiconformal. We thus conclude:

Every finite K-quasiconformal mapping is locally the limit of a sequence of regular real-analytic K-quasiconformal mappings.

Conversely, it follows from Theorem II.5.3 that a function defined in G which can be uniformly approximated by a sequence of K-quasiconformal mappings in every domain D, $\bar{D} \subset G$, is either a K-quasiconformal mapping of G or a constant. We have thus obtained the following new characterization of quasiconformality, emphasizing once more the connection between regular and general quasiconformal mappings.

Theorem 4.2. *A non-constant finite function w in a finite domain G is a K-quasiconformal mapping of G if and only if in every domain D, $\bar{D} \subset G$, there exists a good approximation of w by regular K-quasiconformal mappings.*

§ 5. Application of the Hilbert Transformation to Quasiconformal Mappings

5.1. Transformation of the Beltrami equation into an integral equation. We shall now examine the local behaviour of the derivatives of a finite K-quasiconformal mapping $w: G \to G'$ with the aid of the Hilbert transformation introduced in III.7. In order to be able to apply Lemma III.7.2 we first replace w by a K-quasiconformal mapping $\tilde{w}$ of the plane, which in a prescribed bounded domain $D \subset G$ depends conformally on w.

Using the Existence theorem we define the mapping $\tilde{w}$ as follows: The complex dilatation $\varkappa$ of $\tilde{w}$ is defined to be that of w in D and zero elsewhere. By the Uniqueness theorem, $\tilde{w}$ is uniquely determined up to a linear transformation of the plane and is conformal outside $\bar{D}$. Thus by a suitable normalization we can ensure that $\tilde{w}$ has the expansion

$$\tilde{w}(z) = z + \sum_{n=1}^{\infty} a_n z^{-n} \tag{5.1}$$

in a neighbourhood of infinity.

In D we have

$$\tilde{w} = \varphi \circ w , \tag{5.2}$$

where φ is a conformal mapping of $w(D)$. Thus from the local behaviour of the derivatives of $\tilde{w}$ in D we can draw conclusions about those of w.

As a consequence of the normalization (5.1) the function f, defined by $f(z) = \tilde{w}(z) - z$, satisfies the hypothesis of Lemma III.7.2: f is absolutely continuous on lines in the plane and has L^2-integrable derivatives f_z and $f_{\bar{z}}$. Since $f_z = \tilde{w}_z - 1$, $f_{\bar{z}} = \tilde{w}_{\bar{z}}$, we deduce from Lemma III.7.2 that

$$\tilde{w}_z = 1 + S\, \tilde{w}_{\bar{z}} \tag{5.3}$$

almost everywhere in the plane. On the other hand $\tilde{w}$, being a quasiconformal mapping with complex dilatation $\varkappa$, satisfies the Beltrami equation $\tilde{w}_{\bar{z}} = \varkappa\, \tilde{w}_z$ almost everywhere. If we combine this with (5.3), we conclude that $\tilde{w}_{\bar{z}}$ satisfies the relation

$$\tilde{w}_{\bar{z}} = \varkappa + \varkappa\, S\, \tilde{w}_{\bar{z}}$$

almost everywhere.

5.2. Solution of the integral equation. We are thus led to the singular integral equation

$$\omega = \varkappa + \varkappa\, S\, \omega , \tag{5.4}$$

where $\varkappa$ is a measurable function in the finite plane Ω with compact support and with the property that

$$\sup_{z \in \Omega} |\varkappa(z)| = k < 1 .$$

By a solution of this equation we shall mean a function ω in the class $L^2(\Omega)$, which satisfies (5.4) almost everywhere in Ω.

The solution of the equation (5.4) *is uniquely determined up to its values in a set of area zero.* In fact, if ω_1 and ω_2 are two solutions, then the difference $\psi = \omega_1 - \omega_2$ satisfies the equation $\psi = \varkappa\, S\, \psi$ almost everywhere, and so

$$\|\psi\|_2 \leqq k\, \|S\, \psi\|_2 .$$

By III.7.4 and 6, however, $\|\psi\|_2 = \|S\, \psi\|_2$ for every function $\psi \in L^2(\Omega)$. Since $k < 1$, it follows that $\|\psi\|_2 = 0$, and consequently $\omega_1 = \omega_2$ almost everywhere in Ω.

The integral equation (5.4) can be solved in the following way by iteration. We set

$$S_0\, \varkappa = 1 , \quad S_i\, \varkappa = S\, (\varkappa\, S_{i-1}\, \varkappa) , \quad i = 1, 2, \ldots ,$$

and

$$\omega_n = \sum_{i=0}^{n} \varkappa\, S_i\, \varkappa\,, \qquad n = 0, 1, 2, \ldots. \tag{5.5}$$

Then

$$S\,\omega_{n-1} = \sum_{i=1}^{n} S_i\, \varkappa\,,$$

and consequently

$$\omega_n = \varkappa + \varkappa\, S\, \omega_{n-1}\,, \qquad n = 1, 2, \ldots. \tag{5.6}$$

Using the relation $||\omega||_2 = ||S\,\omega||_2$, we prove that the sequence ω_n converges. First we obtain

$$||S_i\,\varkappa||_2 = ||\varkappa\, S_{i-1}\,\varkappa||_2 \leqq k||S_{i-1}\,\varkappa||_2\,.$$

Since this inequality holds for every $i = 2, 3, \ldots$, we deduce that

$$||S_i\,\varkappa||_2 \leqq k^{i-1}\,||\varkappa||_2 \leqq k^i\, C\,, \tag{5.7}$$

where C^2 equals the area of the support of $\varkappa$.

From (5.5) and (5.7) it follows that, for $n > m$,

$$||\omega_n - \omega_m||_2 \leqq k \sum_{i=m+1}^{n} ||S_i\,\varkappa||_2 \leqq C \sum_{i=m+1}^{n} k^{i+1} \leqq C\,\frac{k^{m+2}}{1-k}\,.$$

Since the space $L^2(\Omega)$ is complete (see III.7.5), this implies the existence of a function $\omega \in L^2(\Omega)$ such that

$$\lim_{n\to\infty} ||\omega - \omega_n||_2 = 0\,.$$

Since $||S\,(\omega - \omega_n)||_2 = ||\omega - \omega_n||_2$ we also have $\lim ||S\,\omega - S\,\omega_n||_2 = 0$. Hence by III.7.5 we can find a subsequence ω_{n_i} such that

$$\lim_{i\to\infty} \omega_{n_i}(z) = \omega(z)\,, \qquad \lim_{i\to\infty} S\,\omega_{n_i}(z) = S\,\omega(z)$$

for almost all z. It follows from (5.6) that ω is a solution of the integral equation (5.4).

5.3. L^p-integrability of the solution of the integral equation. In the above iteration method of solving equation (5.4) we made use of the fact that the Hilbert transformation leaves the L^2-norm invariant. It is however only necessary that the transformation does not increase the norm too much. In fact, there is a more general result on the norms of the Hilbert transformation, which we state here without proof (Calderón-Zygmund [1]; a simplified proof is due to Vekua [1], pp. 55—61, and also given by Ahlfors [4], pp. 106—112):

The Hilbert transformation is bounded in every space $L^p(\Omega)$, $p > 1$. In other words, the number

$$||S||_p = \sup_{||f||_p = 1} ||S f||_p ,$$

the *L^p-norm of the Hilbert transformation*, is finite for every $p > 1$.

By Riesz's convexity theorem (see for example Dunford-Schwartz [1], p. 525, Zygmund [1], p. 95, or Ahlfors [4], pp. 113–115), log $(||S_p||)$ is a convex function of $1/p$ which increases with p for $p > 2$ and tends to infinity as $p \to \infty$. From the convexity it follows that $||S_p||$ is a continuous function of p. By III.7.6 we have $||S||_2 = 1$. Furthermore Calderón and Zygmund [1] have shown that

$$||S||_p/p < M < \infty \tag{5.8}$$

as $p \to \infty$.

From 5.2 we obtain the following result (Bojarski [1]):

Lemma 5.1. *The solution of the integral equation* (5.4) *belongs to the class* $L^p(\Omega)$ *for every* p *for which* $||S||_p < 1/k$.

Proof: Using the notation of 5.2 we have

$$||S_i \varkappa||_p = ||S (\varkappa S_{i-1} \varkappa)||_p \leqq k ||S||_p \, ||S_{i-1} \varkappa||_p , \qquad i = 2, 3, \ldots,$$

and consequently

$$||S_i \varkappa||_p \leqq (k ||S||_p)^i \, C^{2/p} .$$

It follows that for $n > m$

$$||\omega_n - \omega_m||_p \leqq k \, C^{2/p} \sum_{i=m+1}^{n} (k ||S||_p)^i .$$

For $k||S||_p < 1$, ω_n is thus a Cauchy sequence in L^p. Since L^p is complete (see III.7.5), the solution of (5.4), being the L^p-limit of ω_n, belongs to L^p.

5.4. Integrability of the derivatives of a quasiconformal mapping. We now return to the given finite-valued K-quasiconformal mapping w of G and the associated K-quasiconformal mapping $\tilde{w}$ of the plane. In 5.1 we showed that $\tilde{w}_{\bar{z}}$ satisfies the integral equation (5.4) with $\sup |\varkappa(z)| = k \leqq (K-1)/(K+1)$. Thus it follows from Lemma 5.1 that $\tilde{w}_{\bar{z}} \in L^p$ for every p for which $||S||_p < (K+1)/(K-1)$. In view of (5.3) $\tilde{w}_z$ belongs locally to the same classes L^p. From (5.2) we see finally that the derivatives of w also belong to these L^p-classes locally in G.

Let $p(K)$ be the supremum of all numbers p for which all K-quasiconformal mappings have L^p-derivatives. By Theorem IV.1.2 we have $p(K) \geqq 2$. From the above we obtain for $p(K)$ the lower estimation

$$||S||_{p(K)} \geqq \frac{K+1}{K-1} .$$

Since $||S||_2 = 1$, we thus have $p(K) > 2$ for every K.

On the other hand, the K-quasiconformal mapping w defined by

$$w(z) = z|z|^{1/K-1} \tag{5.9}$$

shows that $p(K)$ is at most $2\,K/(K-1)$. This can be seen by considering the derivatives of w in the neighbourhood of $z = 0$.

Summarizing these results we obtain the following generalization of Theorem IV.1.2:

Theorem 5.1. *A K-quasiconformal mapping has L^p-derivatives for every $p < p(K)$, where $p(K)$ satisfies the inequalities*

$$||S||_{p(K)} \geqq \frac{K+1}{K-1}, \qquad 2 < p(K) \leqq \frac{2\,K}{K-1} . \tag{5.10}$$

Although the value of $||S||_p$ is not known for $p > 2$ and the lower bound for $p(K)$ in (5.10) is therefore implicit, (5.10) does give rise to some further conclusions on $p(K)$. The double inequality on the right in (5.10) shows that

$$\lim_{K \to \infty} p(K) = 2 .$$

This result shows that in Theorem IV.1.2, which refers to an arbitrary quasiconformal mapping, the exponent $p = 2$ cannot be replaced by any larger number.

From the first inequality (5.10) and the estimate (5.8) we deduce further that

$$\lim_{K \to 1} p(K) = \infty .$$

In particular, $p(K) \to \infty$ in such a way that $(K-1)\,p(K)$ remains between two finite positive bounds.

An interesting open problem is to determine the exact value of $p(K)$. One might conjecture that the upper bound in (5.10) is the solution to this problem, i.e. that $p(K) = 2\,K/(K-1)$.

5.5. Change of area under quasiconformal mappings. By using Theorem 5.1 we can obtain a sharpened version of Theorem IV.1.3 on absolute continuity with respect to the area measure (Bojarski [2]).

Theorem 5.2. *Let w be a K-quasiconformal mapping of a domain G and F a compact subset of $G - \{\infty\} - \{w^{-1}(\infty)\}$. Then for every $\delta < 1 - 2/p(K)$ there exists a finite number C such that*

$$m(w(E)) \leqq C(m(E))^{\delta} \tag{5.11}$$

for every measurable set $E \subset F$.

Proof: By IV.1.4 we have

$$m(w(E)) = \iint_E J\, d\sigma .$$

From $w_z, w_{\bar z} \in L^p$ it follows that $J = |w_z|^2 - |w_{\bar z}|^2 \in L^{p/2}$. Applying Hölder's inequality for an arbitrary p, $2 < p < p(K)$, we deduce that

$$m(w(E)) \leqq \left(\iint_E J^{p/2}\, d\sigma\right)^{2/p} (m(E))^{\delta} ,$$

where $\delta = 1 - 2/p$; (5.11) follows from this.

The least upper bound $\delta(K)$ of values δ for which Theorem 5.2 holds is unknown. Example (5.9) shows that $\delta(K) \leqq 1/K$.[49]

5.6. L^p-convergence of the derivatives. If a sequence of K-quasiconformal mappings w_n of a domain G of the finite plane converges uniformly in every compact subset of G to a finite-valued function w, then the derivatives of w_n converge weakly to the corresponding derivatives of w, by Lemma IV.5.1. We can now extend this result by showing that in the case of good approximation the convergence takes place in the following stronger sense:

Theorem 5.3. *Let w_n be a sequence of finite-valued K-quasiconformal mappings of a domain G of the finite plane. If w_n is a good approximation of the quasiconformal mapping w in G, then for some $p > 2$ the derivatives $(w_n)_z$ and $(w_n)_{\bar z}$ converge in the L^p metric of any compact subset F of G to w_z and $w_{\bar z}$, respectively.*

Proof: We may assume, without loss of generality, that F is a closed disc. Let D be an open disc such that $F \subset D$, $\bar D \subset G$. As in 5.1 we find mappings $\tilde w_n$ and $\tilde w$ of the plane which have the same complex dilatations as w_n and w in D, and are conformal outside $\bar D$. At infinity they have an expansion of the form (5.1).

By Lemma III.7.2, we have for the function $\psi_n = \tilde w - \tilde w_n$

$$(\psi_n)_z = S(\psi_n)_{\bar z} \tag{5.12}$$

[49] On the other hand, $\delta(K) \geqq 1/K^{2/\alpha}$, where $\alpha = \liminf_{K \to 1} (K - 1)\, p(K)$ (Lehto [2]); by 5.4 we have $0 < \alpha \leqq 2$. See also Gehring-Reich [1].

almost everywhere in D. Denote by $\varkappa_n$ and $\varkappa$ the complex dilatations of $\tilde{w}_n$ and $\tilde{w}$, respectively. Then

$$\varkappa_n(\psi_n)_z = \varkappa\,\tilde{w}_z - \varkappa_n(\tilde{w}_n)_z + \varkappa_n\,\tilde{w}_z - \varkappa\,\tilde{w}_z = (\psi_n)_{\bar{z}} + (\varkappa_n - \varkappa)\,\tilde{w}_z$$

almost everywhere. Thus up to a null set

$$(\psi_n)_{\bar{z}} = \varkappa_n\, S(\psi_n)_{\bar{z}} + (\varkappa - \varkappa_n)\,\tilde{w}_z$$

and consequently

$$\|(\psi_n)_{\bar{z}}\|_p \leqq k\,\|S\|_p\,\|(\psi_n)_{\bar{z}}\|_p + \|(\varkappa - \varkappa_n)\,\tilde{w}_z\|_p \tag{5.13}$$

with $k = (K-1)/(K+1)$.

We now fix p so that simultaneously $p > 2$ and $k\,\|S\|_p < 1$; this is possible by 5.3. Since $|\varkappa - \varkappa_n|$ vanishes in $-\overline{D}$ and is less than 2 in D, we conclude from (5.13) that

$$\|(\psi_n)_{\bar{z}}\|_p \leqq \frac{\|(\varkappa - \varkappa_n)\,\tilde{w}_z\|_p}{1 - k\,\|S\|_p} < \frac{2}{1 - k\,\|S\|_p}\left(\iint_D |\tilde{w}_z|^p\, d\sigma\right)^{1/p}. \tag{5.14}$$

The right hand side is bounded, since $\tilde{w}_z$ belongs locally to L^p by Theorem 5.1. As $n \to \infty$, $\varkappa_n$ tends to $\varkappa$ almost everywhere and by Lebesgue's convergence theorem (III.1.5) we have $\lim \|(\varkappa - \varkappa_n)\,\tilde{w}_z\|_p = 0$. It then follows from (5.14) that

$$\lim_{n\to\infty} \|(\psi_n)_{\bar{z}}\|_p = \lim_{n\to\infty} \|\tilde{w}_{\bar{z}} - (\tilde{w}_n)_{\bar{z}}\|_p = 0\,. \tag{5.15}$$

By (5.12) we also have $\lim \|\tilde{w}_z - (\tilde{w}_n)_z\|_p = 0$.

The derivative $(\psi_n)_{\bar{z}}$ vanishes outside $\overline{D}$, and owing to the normalization (5.1) we obtain from Lemma III.7.1 by proceeding to the limit

$$\psi_n(z) = T_D(\psi_n)_{\bar{z}}\,(z)$$

for every finite z. Hence, by Hölder's inequality,

$$|\psi_n(z)| \leqq \frac{1}{\pi}\left(\iint_D \frac{d\sigma}{|\zeta - z|^q}\right)^{1/q} \|(\psi_n)_{\bar{z}}\|_p\,,$$

where $q = p/(p-1) < 2$. In view of (5.15) we deduce that $\psi_n = \tilde{w} - \tilde{w}_n$ tends uniformly to zero in the plane as $n \to \infty$. It follows that $\tilde{w}_n^{-1}$ tends to $\tilde{w}^{-1}$ uniformly in $\tilde{w}(D)$.

In D we have $w_n = f_n \circ \tilde{w}_n$, $w = f \circ \tilde{w}$, where f_n and f are conformal mappings of $\tilde{w}_n(D)$ and $\tilde{w}(D)$. Since $\tilde{w}_n$ tends to $\tilde{w}$ uniformly in D we see that $\tilde{w}(F) \subset \tilde{w}_n(D)$ from some n onwards. For these values of n the inverses $\tilde{w}_n^{-1}$ map a neighbourhood U of $\tilde{w}(F)$, $U \subset \tilde{w}(D)$, into D. Since by hypothesis w_n tends to w uniformly in D it follows that $f_n \to f$ uniformly in U. The derivatives f_n' therefore also converge uniformly in a neighbourhood $V \subset U$ of $\tilde{w}(F)$, and since $\tilde{w}_n(F)$ lies in V from some n onwards, we have

$$\lim f'_n(\tilde{w}_n(z)) = f'(\tilde{w}(z)) \tag{5.16}$$

uniformly in F.

Next, from

$$w_{\bar{z}}(z) - (w_n)_{\bar{z}}(z) = f'_n(\tilde{w}_n(z))\,(\tilde{w}_{\bar{z}}(z) - (\tilde{w}_n)_{\bar{z}}(z)) + (f'(\tilde{w}(z)) - f'_n(\tilde{w}_n(z)))\,\tilde{w}_{\bar{z}}(z)$$

it follows that

$$\left(\iint_F |w_{\bar{z}} - (w_n)_{\bar{z}}|^p\, d\sigma\right)^{1/p} \leqq \max_{z \in F} |f'_n(\tilde{w}_n(z))|\; \|\tilde{w}_{\bar{z}} - (\tilde{w}_n)_{\bar{z}}\|_p$$
$$+ \left(\iint_F |(f'(\tilde{w}(z)) - f'_n(\tilde{w}_n(z)))\,\tilde{w}_{\bar{z}}(z)|^p\, d\sigma\right)^{1/p}.$$

From (5.16) and (5.15) we see that the first term on the right tends to zero as $n \to \infty$. By Lebesgue's convergence theorem the same is true for the second term.

It can be shown in the same way that $(w_n)_z$ tends to w_z in the L^p metric of F, and the theorem is proved.

As a consequence of the theorem we conclude that for a good approximation $w_n \to w$ in G there exists a subsequence w_{n_ν}, for which the derivatives tend to the corresponding derivatives of the limit mapping w almost everywhere in G (see III.7.5).

5.7. Representation of a mapping with prescribed complex dilatation. Let $\varkappa$ be a measurable function in the plane, having bounded support and satisfying $\sup_z |\varkappa(z)| < 1$. By the Existence theorem there is a quasiconformal mapping w of the plane with complex dilatation equal to $\varkappa$ almost everywhere. Since w is conformal in a neighbourhood of infinity it can be normalized as in (5.1).

Application of Lemma III.7.1 to the function ψ, $\psi(z) = w(z) - z$, now gives $\psi = T\,\psi_{\bar{z}}$.[50] In other words we have

$$w(z) = z + T\,\omega(z) \tag{5.17}$$

for every z, where $\omega = w_{\bar{z}}$ can be constructed as a solution of the integral equation (5.4) as we showed in 5.1 and 5.2. The formula (5.17) is then a *representation* of a quasiconformal mapping with complex dilatation $\varkappa$ almost everywhere.

On the other hand, starting with the integral equation (5.4), we can construct the right hand side of (5.17) without using the Existence

[50] Here $T = T_D$, where D contains the support of $\varkappa$ (cf. the notation in III.7.2).

theorem. It is then possible to show directly that the function defined by (5.17) is a quasiconformal homeomorphism of the plane whose complex dilatation is equal to $\varkappa(z)$ for almost all z (see Bojarski [1], or, for a detailed proof, Vekua [1], pp. 68—85; see also Ahlfors-Bers [1]). In this way a new proof of the Existence theorem can be obtained (cf. 1.4).

§ 6. *Conformality at a Point*

6.1. Presentation of the problem. In 4.2 we have already discussed the problem of deducing the regularity of a quasiconformal mapping from the properties of its complex dilatation. This section and the next are devoted to the investigation of this problem. To start with we consider the following situation:

Let $\varkappa$ be a measurable function with $\sup_z |\varkappa(z)| < 1$ in a domain G. By the Existence theorem there is a quasiconformal mapping w of G with complex dilatation equal to $\varkappa$ almost everywhere. Let z_0 be a fixed point of G. We seek sufficient conditions on the function $\varkappa$ such that w will be regular at z_0.

It follows from the Uniqueness theorem that the regularity of w at z_0 depends on the function $\varkappa$ and indeed only on its values in an arbitrarily small neighbourhood U of z_0: two quasiconformal mappings whose complex dilatations coincide almost everywhere in U are simultaneously regular or not at z_0. Our problem is therefore meaningful.

Without loss of generality we may assume that z_0 is the origin. Furthermore, by means of an affine mapping we can normalize our problem in such a way that w, if it is regular at the origin, has an expansion of the form

$$w(z) = w_z(0)\, z + o(z)\,, \qquad w_z(0) \neq 0\,. \tag{6.1}$$

If (6.1) holds, then $w(z)/z$ tends to the finite, non-zero limit $w_z(0)$ as $z \to 0$. In this case the mapping w is said to be *conformal at the origin.* In general, a homeomorphism w of a neighbourhood of a point $z_0 \neq \infty$ is called conformal at z_0 if $\big(w(z) - w(z_0)\big)/(z - z_0)$ tends to a finite non-zero limit as $z \to z_0$.

In this section we shall study conditions on $\varkappa$ under which the mapping w is conformal at the origin.[51]

---

[51] The somewhat related problem on the effect of the dilatation quotient on the boundary behaviour of quasiconformal mappings has been studied by Carleson [1].

6.2. Examples. By Lemma 4.1 w is conformal at $z = 0$ when $\varkappa$ is a polynomial with $\varkappa(0) = 0$. It is not difficult to see that this condition can be replaced by essentially weaker ones. For example, it would suffice that $\varkappa$ be differentiable and $\varkappa(0) = 0$. To obtain a still weaker condition we consider the following example.

Let w,

$$w(z) = f(|z|)\, e^{i \arg z} = f\left(\sqrt{z\bar{z}}\right) \sqrt{\frac{z}{\bar{z}}}\,, \tag{6.2}$$

be a mapping of the unit disc onto itself, where f is a non-negative function which increases strictly with $|z|$, vanishes for $|z| = 0$ and takes the value 1 for $|z| = 1$. If f has a positive continuous derivative in the open interval $(0, 1)$, then w is a regular C^1-mapping of the punctured unit disc $0 < |z| < 1$. For the complex dilatation $\varkappa$ of w we obtain

$$\varkappa(z) = \frac{z^2}{|z|^2}\, \frac{f'(|z|)\,|z| - f(|z|)}{f'(|z|)\,|z| + f(|z|)}\,.$$

In order to obtain a simpler expression we consider the dilatation quotient $D = (1 + |\varkappa|)/(1 - |\varkappa|)$. Assuming that $f(|z|)/|z|$ does not decrease with $|z|$ we have

$$D(z) = |z|\, \frac{f'(|z|)}{f(|z|)}\,.$$

It follows that

$$\frac{w(z)}{z} = \frac{f(|z|)}{|z|} = e^{-\int\limits_{|z|}^{1} \frac{D(r) - 1}{r}\, dr} \tag{6.3}$$

The mapping w is therefore conformal at $z = 0$ if and only if the integral

$$\int\limits_0^1 \frac{D(r) - 1}{r}\, dr \tag{6.4}$$

has a finite value.

The integral (6.4) can of course be infinite even if $D(z) - 1$, and hence also $\varkappa(z)$, tends to zero as $z \to 0$. As an example we have the mapping w defined by

$$w(z) = \frac{z}{1 - \log|z|}\,,$$

with dilatation quotient

$$D(z) = 1 + \frac{1}{1 - \log|z|}\,.$$

This mapping is not conformal at $z = 0$ since

$$\lim_{z \to 0} \frac{w(z)}{z} = 0 .$$

Conversely it is easy to see that there exist quasiconformal mappings which are conformal at the origin even though their maximal dilatations differ from 1 by an arbitrary amount in every neighbourhood of this point. In fact we can obviously find a function Φ, non-negative and continuous for $0 < r < 1$, such that $\limsup_{r \to 0} \Phi(r)$ is arbitrarily large while the integral

$$\int_0^1 \frac{\Phi(r)}{r} \, dr$$

is finite. If we then set $D(r) - 1 = \Phi(r)$ in (6.3), we obtain a mapping w with the required properties.

For a general quasiconformal mapping w the dilatation quotient D is not a function of $|z|$ alone, and we have to replace the integral (6.4) by

$$\int_0^R \int_0^{2\pi} \frac{D(r\, e^{i\varphi}) - 1}{r} \, dr \, d\varphi = \iint_U \frac{D(z) - 1}{|z|^2} \, d\sigma , \qquad (6.5)$$

$U = \{z \mid |z| < R\}$. We shall prove that the convergence of this integral is sufficient to ensure the conformality of w at the origin.

Since $D(z) - 1 = 2\,|\varkappa(z)|/(1 - |\varkappa(z)|)$ the convergence of the integral (6.5) is equivalent to

$$\iint_U \frac{|\varkappa(z)|}{|z|^2} \, d\sigma < \infty .$$

6.3. Module estimations in terms of the means of *D*. In order to draw conclusions from the convergence of the integral (6.5), we first derive some estimations for the dilatations of quadrilaterals and ring domains using certain mean values of D.

Let $w : G \to G'$ be a quasiconformal mapping with dilatation quotient D. We consider a family $\mathscr{C}$ of arcs or curves $C \subset G$ and the family $\mathscr{C}'$ of their images $w(C)$. By applying formula (3.8) in IV.3.3 to the inverse of w we obtain

$$M(\mathscr{C}') \leqq \iint_G D\, \varrho^2 \, d\sigma \qquad (6.6)$$

for every ϱ admissible for $\mathscr{C}$.

As a first application of (6.6) we consider a horizontal rectangle $Q = \{x + iy \mid a < x < b,\ c < y < d\}$, $\overline{Q} \subset G$. If we set $\varrho(z) = 1/(d - c)$

for $z \in Q$, $\varrho(z) = 0$ for $z \notin Q$, then ϱ is admissible with respect to the family of arcs C connecting the a-sides of Q within Q. In view of IV.3.4 we deduce therefore from (6.6) that

$$M(w(Q)) \leqq \frac{1}{(d-c)^2} \iint_Q D \, d\sigma .$$

Since $M(Q) = (b-a)/(d-c)$ we can write this estimate in the form

$$\frac{M(w(Q))}{M(Q)} \leqq \frac{1}{m(Q)} \iint_Q D \, d\sigma , \tag{6.7}$$

where $m(Q)$ denotes the area of Q. If we exchange the roles of the a- and b-sides, we obtain the same upper bound for $M(Q)/M(w(Q))$.

In order to find a corresponding inequality for an annulus $B = \{z \mid r < |z| < R\}$, $\bar{B} \subset G$, we set $\varrho(z) = 1/(2\pi|z|)$ for $z \in B$, $\varrho(z) = 0$ for $z \notin B$. Then ϱ is admissible with respect to the family of curves $C \subset B$ which separate the boundary components of B (cf. I.6.3). In view of IV.3.4 it follows therefore from (6.6) that

$$M(w(B)) \leqq \frac{1}{2\pi} \iint_B \frac{D(z)}{|z|^2} \, d\sigma .$$

Since

$$M(B) = \frac{1}{2\pi} \iint_B \frac{d\sigma}{|z|^2}$$

we have

$$M(w(B)) - M(B) \leqq \frac{1}{2\pi} \iint_B \frac{D(z)-1}{|z|^2} \, d\sigma . \tag{6.8}$$

Another estimate will be obtained in the case $M(B) < \infty$ if we set $\varrho(z) = 1/(|z| \log (R/r)) = 1/(|z| \, M(B))$ for $z \in B$. Then ϱ is admissible for the family of arcs which join the boundary components of B. By (6.6) and IV.3.4 we have

$$\frac{1}{M(w(B))} \leqq \frac{1}{2\pi \, (M(B))^2} \iint_B \frac{D(z)}{|z|^2} \, d\sigma ,$$

and so

$$\frac{M(B)}{M(w(B))} \leqq \frac{1}{2\pi \, M(B)} \iint_B \frac{D(z)}{|z|^2} \, d\sigma = 1 + \frac{1}{2\pi \, M(B)} \iint_B \frac{D(z)-1}{|z|^2} \, d\sigma .$$

From this and (6.8) we obtain

$$|M(w(B)) - M(B)| \leqq \frac{1}{2\pi} \iint_B \frac{D(z)-1}{|z|^2} \, d\sigma . \tag{6.9}$$

6.4. Transition to the logarithmic plane. The proof that the convergence of the integral (6.5) implies the conformality of the quasiconformal mapping w at the origin will be divided into two parts: We first deduce that $|w(z)/z|$ tends to a non-zero finite limit as $z \to 0$, and secondly, that $\arg(w(z)/z)$ also converges. In order to simplify certain calculations it is advisable to consider $\log w$ as a function of $\log z$. In performing this transformation of coordinates we shall simultaneously determine the branches of $\arg z$ and $\arg w$ appearing in the second limit problem.

Let w be a K-quasiconformal mapping of the plane with $w(0) = 0$, $w(\infty) = \infty$. We map the slit plane $G_0 = \{z \mid 0 < \arg z < 2\pi\}$ conformally onto the strip $S = \{\zeta = \xi + i\eta \mid 0 < \eta < 2\pi\}$ using a branch of the logarithm. Similarly the image $w(G_0)$ of G_0 is mapped by the logarithm onto a domain S'. The composite mapping $f: S \to S'$, $f(\zeta) = \log w(e^\zeta)$, is then K-quasiconformal and has a continuous extension to all finite boundary points of S.

The exponential function maps a segment $\{\zeta = \xi_0 + i\eta \mid 0 \leqq \eta \leqq 2\pi\}$ onto the circle $|z| = e^{\xi_0}$. Since w is sense-preserving, we see that $f(\zeta) = \log|w(e^\zeta)| + i \arg w(e^\zeta)$ increases by $2\pi i$ as η goes from 0 to 2π. Hence we can extend $f = u + iv$ throughout the finite plane as a continuous function by the formula

$$f(\zeta + 2n\pi i) = f(\zeta) + 2n\pi i\,, \qquad n = 0, \pm 1, \ldots, \tag{6.10}$$

so that $v(\zeta)$ coincides with a branch of $\arg w(e^\zeta)$. The function f is then a homeomorphism of the plane and since it is locally the composition of a K-quasiconformal mapping and two conformal mappings, f is K-quasiconformal. At the points $\log z + 2n\pi i$, $n = 0, \pm 1, \ldots$, the dilatation quotient D of f has the same value as the dilatation quotient of w at the point z, provided that w is regular at z.

By this transition to the logarithmic plane we have determined a branch of $\arg z$ and $\arg w$ as a function of ζ. In what follows the notation $\eta = \arg z$, $v = \arg w$ will always refer to these branches. We note that $\arg(w/z)$ is then a single-valued function of z for $z \neq 0, \infty$.

It is important to remark on the uniform Hölder continuity of the family $\mathcal{F}$ of all K-quasiconformal mappings of the plane satisfying (6.10). This follows from the fact that the image of the strip $T = \{\xi + i\eta \mid 0 < \eta < 4\pi\}$ under every $f \in \mathcal{F}$ contains no closed vertical segment of length 4π. By Theorems II.4.1 and II.4.3, $\mathcal{F}$ is therefore equicontinuous in T and Hölder continuous at every point ζ_0 of T. Since f and $f - f(\zeta_0)$ are simultaneously members of $\mathcal{F}$, the family $\mathcal{F}$ is Hölder continuous with respect to the euclidean metric too. This holds not only in T but in the whole finite plane, since every point is equivalent modulo $2\pi i$ to a point in T.

Thus for every finite point ζ_0 there exist two positive numbers C and d such that

$$|f(\zeta) - f(\zeta_0)| \leqq C\,|\zeta - \zeta_0|^{1/K} \tag{6.11}$$

for $f \in \mathcal{F}$ and $|\zeta - \zeta_0| \leqq d$. Since the family $\mathcal{F}$ is invariant under a translation of the ζ-plane, (6.11) holds for fixed C and d depending only on K and not on the choice of ζ_0.

6.5. Convergence of the absolute value. After these preparations we shall now prove that w is conformal at $z = 0$ if the integral (6.5) is finite, by showing first the existence of the non-zero limit of $|w(z)/z|$. This result is due to Teichmüller [1] and Wittich [1] with the slight restriction that the mapping is regular for $z \neq 0$.

Lemma 6.1. *Let w be a K-quasiconformal mapping of the plane with $w(0) = 0$, $w(\infty) = \infty$, and*

$$I(r) = \frac{1}{2\pi} \iint\limits_{|z|<r} \frac{D(z) - 1}{|z|^2}\, d\sigma < \infty \quad \textit{for } r < \infty\,. \tag{6.12}$$

Then there exists a constant c,

$$\min_{|z|=1} |w(z)|\, e^{-I(1)} \leqq c \leqq \max_{|z|=1} |w(z)|\, e^{I(1)}\,, \tag{6.13}$$

such that

$$\left|\left|\frac{w(z)}{z}\right| - c\right| \leqq c\,\varepsilon(|z|)\,, \tag{6.14}$$

where $\varepsilon(|z|) \to 0$ as $z \to 0$ and the function ε depends only on K and I.

Proof: We write

$$\min_{|z|=e^{\xi}} |w(z)| = e^{h(\xi)}\,, \qquad \max_{|z|=e^{\xi}} |w(z)| = e^{h(\xi)+\psi(\xi)}\,, \tag{6.15}$$

and show first that $\psi(\xi) \to 0$ as $\xi \to -\infty$.

Consider an annulus $B = \{z \mid e^s < |z| < e^t\}$ and its image $w(B)$. The module of B is $t - s$ and for the module of $w(B)$ formula (6.9) yields the estimations

$$t - s - I(e^t) \leqq M(w(B)) \leqq t - s + I(e^t)\,. \tag{6.16}$$

The ring domain $w(B)$ lies in the annulus $B' = \{w \mid e^{h(s)} < |w| < e^{h(t)+\psi(t)}\}$ and, if $h(t) > h(s) + \psi(s)$, contains the annulus $B'' = \{w \mid e^{h(s)+\psi(s)} < |w| < e^{h(t)}\}$. From the monotonicity of the module it thus follows that

$$h(t) - h(s) - \psi(s) \leqq M(w(B)) \leqq h(t) - h(s) + \psi(t)\,.$$

Consequently, by (6.16),

$$t - s - I(e^t) - \psi(t) \leqq h(t) - h(s) \leqq t - s + I(e^t) + \psi(s) . \quad (6.17)$$

As in 6.4 we introduce the variables $\zeta = \xi + i\,\eta = \log z$, $f = u + i\,v = \log w$. The annulus B, slit along the positive real axis, corresponds to the horizontal rectangle $Q = \{\xi + i\,\eta \mid s < \xi < t,\ 0 < \eta < 2\,\pi\}$ of the ζ-plane and to the Jordan domain $f(Q)$ of the f-plane. Since $w(B)$ is contained in the annulus B' the area of $f(Q)$ satisfies the inequality

$$m(f(Q)) \leqq 2\,\pi\,(h(t) - h(s) + \psi(t)) . \quad (6.18)$$

The integral in (6.12) transforms to

$$I(e^{\xi}) = \frac{1}{2\pi} \int_{-\infty}^{\xi} \int_{0}^{2\pi} (D(\zeta) - 1)\, d\xi\, d\eta ,$$

where D is the dilatation quotient of f.

The function f maps the vertical segment $C_\xi = \{\xi + i\,\eta \mid 0 \leqq \eta \leqq 2\,\pi\}$ onto a Jordan arc C'_ξ with the endpoints $f(\xi)$ and $f(\xi + 2\,\pi\,i) = f(\xi) + 2\,\pi\,i$ (cf. (6.10)). Since C'_ξ connects the vertical lines $u = h(\xi)$ and $u = h(\xi) + \psi(\xi)$, its length has the lower bound $2\sqrt{\pi^2 + (\psi(\xi))^2}$.

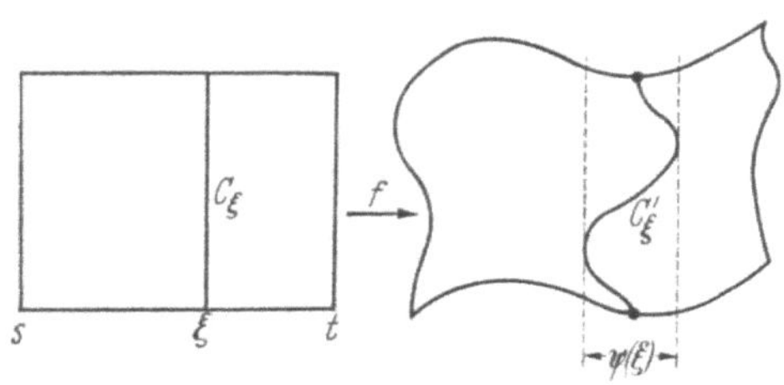

Fig. 12

At regular points $\max_\alpha |\partial_\alpha f(\zeta)|^2 = D(\zeta)\,J(\zeta)$, where J is the Jacobian of f. If f is absolutely continuous on C_ξ and regular at almost all points of C_ξ, we therefore have

$$2\sqrt{\pi^2 + (\psi(\xi))^2} \leqq \int_0^{2\pi} \sqrt{D\,(\xi + i\,\eta)\; J\,(\xi + i\,\eta)}\; d\eta . \quad (6.19)$$

By Lemma IV.1.1 and Theorem IV.1.4 this inequality holds for almost all ξ. Integrating with respect to ξ and using Schwarz's inequality we thus obtain

$$\left(2 \int_s^t \sqrt{\pi^2 + (\psi(\xi))^2}\; d\xi\right)^2 \leqq \iint_Q D\, d\sigma \iint_Q J\, d\sigma . \quad (6.20)$$

Here

$$\iint_Q D\,d\sigma = m(Q) + \iint_Q (D-1)\,d\sigma \leqq 2\pi\,(t - s + I(e^t)),$$

and by (6.18) and (6.17)

$$\iint_Q J\,d\sigma = m(f(Q)) \leqq 2\pi\,(t - s + I(e^t) + \psi(s) + \psi(t)).$$

In order to estimate the left-hand integral in (6.20) we refer to the inequality (9.1) in II.9.2, which yields the upper bound πK for $\psi(\xi)$. A simple calculation then shows that

$$\sqrt{\pi^2 + (\psi(\xi))^2} \geqq \pi + \frac{(\psi(\xi))^2}{4\pi K}.$$

Combining the above estimations with (6.20) we obtain

$$2\pi\,(t-s) + \frac{1}{2\pi K}\int_s^t (\psi(\xi))^2\,d\xi$$

$$\leqq 2\pi\,(t-s)\left(1 + \frac{2\,I(e^t) + \psi(s) + \psi(t)}{t-s} + \frac{I(e^t)\,(I(e^t) + 2\pi K)}{(t-s)^2}\right)^{1/2}$$

$$\leqq 2\pi\,(t-s) + \pi\left[2\,I(e^t) + \psi(s) + \psi(t) + I(e^t)\,\big(I(e^t) + 2\pi K\big)/(t-s)\right].$$

Letting $s \to -\infty$ we conclude that

$$\int_{-\infty}^t (\psi(\xi))^2\,d\xi \leqq 2\pi^2 K\,\big(2\,I(e^t) + \psi(t) + \limsup_{s\to-\infty} \psi(s)\big)$$

$$\leqq 4\pi^2 K\,\big(I(e^t) + \pi K\big). \tag{6.21}$$

From this inequality we derive an upper bound for ψ, making use of the fact that by (6.11) the function ψ fulfills the Hölder condition

$$|\psi(\xi + \Delta\xi) - \psi(\xi)| \leqq C|\Delta\xi|^{1/K}$$

for $|\Delta\xi| \leqq d$, where C and d depend only on K. If $M = \sup \psi(\xi)$ for $\xi \leqq t$, it follows that $\psi(\xi) \geqq M/2$ on an interval of length $\min\,(d, (M/(2\,C))^K)$. Hence

$$\int_{-\infty}^t (\psi(\xi))^2\,d\xi \geqq \frac{M^2}{4}\min\left(d, \left(\frac{M}{2C}\right)^K\right) = \frac{M^{2+K}}{4}\min\left(\frac{d}{M^K}, \frac{1}{(2\,C)^K}\right)$$

$$\geqq \frac{M^{2+K}}{4}\min\left(\frac{d}{(\pi K)^K}, \frac{1}{(2\,C)^K}\right).$$

In other words,

$$\int_{-\infty}^t (\psi(\xi))^2\,d\xi \geqq C' \sup_{\xi\leqq t}\,(\psi(\xi))^{K+2}, \tag{6.22}$$

where C' depends only on K. Since the left-hand integral converges by (6.21), we see that $\psi(\xi) \to 0$ as $\xi \to -\infty$.

Next we show that $\psi(\xi)$ has a majorant which is independent of the mapping f and tends to zero as $\xi \to -\infty$. Setting $t = 0$ in (6.21) we first see that for every $x < 0$

$$\min_{x \leqq \xi \leqq x/2} (\psi(\xi))^2 \leqq \frac{C''}{|x|},$$

where $C'' = 8\pi^2 K \big(I(1) + \pi K\big)$. After this we choose t in (6.21) to be the point on the interval $x \leqq \xi \leqq x/2$ where ψ takes its minimum. Then (6.21) yields

$$\int_{-\infty}^{t} (\psi(\xi))^2 d\xi \leqq 2\pi^2 K \big(2\,I(e^t) + (C''/|x|)^{1/2}\big) \leqq 2\pi^2 K \big(2\,I(e^{x/2}) + (C''/|x|)^{1/2}\big),$$

and we obtain from (6.22) the desired estimation

$$C'\big(\psi(x)\big)^{K+2} \leqq 2\,\pi^2 K \,\big(2\,I(e^{x/2}) + (C''/|x|)^{1/2}\big). \qquad (6.23)$$

In order to complete the proof we first notice that $h(\xi) - \xi$ tends to a limit as $\xi \to -\infty$. This now follows from the double inequality (6.17), written in the form

$$-\,I(e^t) - \psi(t) \leqq \big(h(t) - t\big) - \big(h(s) - s\big) \leqq I(e^t) + \psi(s)\,, \qquad (6.24)$$

since $\psi(t)$ and $I(e^t)$ tend to 0 as $t \to -\infty$. Denoting

$$\lim_{\xi \to -\infty} \big(h(\xi) - \xi\big) = \log c\,,$$

and using (6.23) we see from $e^{h(\xi)-\xi} \leqq |w(z)/z| \leqq e^{h(\xi)-\xi+\psi(\xi)}$ that c has the property (6.14).

Finally, (6.24) yields for $t = 0$, $s \to -\infty$,

$$h(0) - I(1) \leqq \log c \leqq h(0) + \psi(0) + I(1)\,.$$

This agrees with (6.13), and the lemma is proved.

6.6. Convergence of the argument. The lemma which we have to prove is as follows (cf. Belinskij [1]):

Lemma 6.2. *Under the hypothesis of Lemma* 6.1, $\arg w(z) - \arg z$ *tends to a limit* α *as* $z \to 0$. *The rate of convergence depends on* K *and* I *but not otherwise on the mapping* w.

Proof: We go over to the logarithmic plane as in 6.4. Hypothesis (6.12) then becomes

$$I(e^x) = \frac{1}{2\pi} \int_{-\infty}^{x} \int_{0}^{2\pi} \big(D(\zeta) - 1\big)\, d\xi\, d\eta < \infty \qquad (6.25)$$

for $x < \infty$, where D is the dilatation quotient of the mapping $f = u + i v$. We have to prove that $v(\xi + i\eta) - \eta$ tends to a limit α as $\xi \to -\infty$ and in such a way that $|v(\xi + i\eta) - \eta - \alpha|$ is bounded by a number $\varepsilon(\xi)$ depending only on K, I and ξ and tending to zero as $\xi \to -\infty$.

By Lemma 6.1 we know that

$$|u(\xi + i\eta) - u(\xi' + i\eta') - (\xi - \xi')| < \varepsilon_1(x) \tag{6.26}$$

for every pair of points $\xi + i\eta$, $\xi' + i\eta'$ with $\xi, \xi' \leqq x$, where ε_1 is an increasing function of x depending only on K and I and tending to zero as $x \to -\infty$. We shall consider only values of x so small that both $\varepsilon_1(x)$ and $2\pi I(e^x)$ are less than min $(1, d^2)$, where d is the constant mentioned in connection with the distortion formula (6.11).

We fix two values η_1 and η_2, write

$$\eta_1 - \eta_2 = \beta\,, \qquad v(\xi + i\eta_1) - v(\xi + i\eta_2) = b(\xi)\,,$$

and show that $b(\xi)$ tends to β as $\xi \to -\infty$. First we suppose that $\eta_2 < \eta_1$.

Consider a horizontal rectangle $Q_x = \{\xi + i\eta \mid x - \varepsilon_2(x) < \xi < x, \eta_2 < \eta < \eta_1\}$, where $\varepsilon_2(x) = \max\left(\sqrt{\varepsilon_1(x)}, \sqrt{2\pi I(e^x)}\right)$. The module of Q_x is $\varepsilon_2(x)/\beta$. To estimate the module of the image $Q_x' = f(Q_x)$, we note that by (6.26) the b-sides of Q_x' are separated by a vertical strip S_x' of width $\varepsilon_2(x) - \varepsilon_1(x)$. If ϱ denotes the function having the constant value $1/(\varepsilon_2(x) - \varepsilon_1(x))$ in $Q_x' \cap S_x'$ and vanishing elsewhere, then $l_\varrho(C) \geqq 1$ for every rectifiable arc $C \subset Q_x'$ connecting the b-sides of Q_x'. Hence IV.3.4 gives

$$\frac{1}{M(Q_x')} \leqq \iint\limits_{Q_x'} \varrho^2\, d\sigma = \frac{m(Q_x' \cap S_x')}{(\varepsilon_2(x) - \varepsilon_1(x))^2}\,. \tag{6.27}$$

From (6.11) it follows that each a-side of Q_x' lies in a horizontal strip of width $C(\varepsilon_2(x))^{1/K}$ (Fig. 13). The maximal height of the set $Q_x' \cap S_x'$ is therefore not greater than $|b(x)| + 2C(\varepsilon_2(x))^{1/K}$, and its area is at most $(\varepsilon_2(x) - \varepsilon_1(x))\left(|b(x)| + 2C(\varepsilon_2(x))^{1/K}\right)$. In view of (6.27) it follows that

$$M(Q_x') \geqq \frac{\varepsilon_2(x) - \varepsilon_1(x)}{|b(x)| + 2C(\varepsilon_2(x))^{1/K}}\,. \tag{6.28}$$

On the other hand, for $\beta \leqq 4\pi$ we conclude from (6.7) and (6.10) that

$$M(Q_x') \leqq M(Q_x) + \frac{4\pi M(Q_x)}{m(Q_x)} I(e^x) = \frac{\varepsilon_2(x)}{\beta} + \frac{4\pi I(e^x)}{\beta^2}\,.$$

Consequently,

$$|b(x)| + 2\,C\big(\varepsilon_2(x)\big)^{1/K} \geqq \frac{\varepsilon_2(x) - \varepsilon_1(x)}{\dfrac{\varepsilon_2(x)}{\beta} + \dfrac{4\pi I(e^x)}{\beta^2}} \geqq \beta - \frac{\varepsilon_1(x)\,\beta}{\varepsilon_2(x)} - \frac{4\pi I(e^x)}{\varepsilon_2(x)}\,.$$

Since $\varepsilon_2(x) = \max\left(\sqrt{\varepsilon_1(x)}, \sqrt{2\pi I(e^x)}\right)$ and $\beta \leqq 4\pi$, we further obtain

$$|b(x)| \geqq \beta - 2\,C\big(\varepsilon_2(x)\big)^{1/K} - 4\pi\sqrt{\varepsilon_1(x)} - 2\sqrt{2\pi I(e^x)} = \beta - \varepsilon_3(x)\,, \tag{6.29}$$

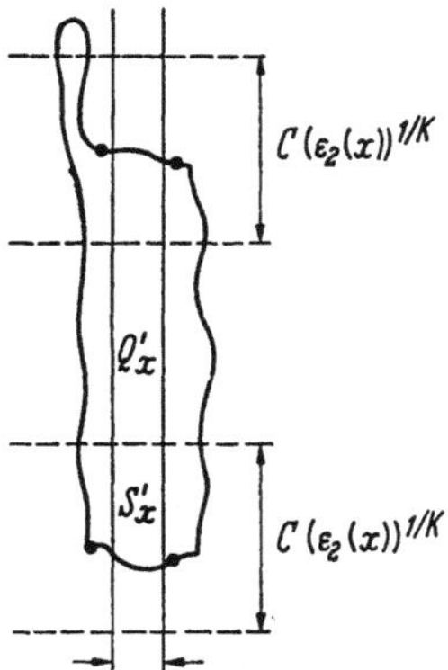

Fig. 13

where $\varepsilon_3(x)$ depends only on K and I and tends to zero monotonically as $x \to -\infty$. Henceforth we shall assume that x is so small that $\varepsilon_3(x) < 2\pi$.

Inequality (6.29) holds for all pairs η_1 and η_2 such that $\eta_1 - \eta_2 = \beta$ lies between 0 and 4π. If x is fixed, then $b(x)$ is a continuous function of η_1 and η_2 and therefore by (6.29) has the same sign for all pairs η_1, η_2 such that $\varepsilon_3(x) < \eta_1 - \eta_2 \leqq 4\pi$. However, by (6.10), $b(x) = 2\pi$ when $\beta = 2\pi$, and so (6.29) becomes

$$b(x) \geqq \beta - \varepsilon_3(x)\,. \tag{6.30}$$

Furthermore it follows from (6.10) that (6.30) holds not only for $\varepsilon_3(x) < \beta \leqq 4\pi$, but for every β.

To obtain an upper bound for $b(x)$ we need only interchange η_1 and η_2. Then $b(x)$ and β become $-b(x)$ and $-\beta$ and we obtain finally

$$|b(x) - \beta| = |v(x + i\eta_1) - v(x + i\eta_2) - (\eta_1 - \eta_2)| \leqq \varepsilon_3(x)\,. \tag{6.31}$$

In order to complete the proof we fix two numbers x and x_1, $x_1 < x$, and consider the expression

$$v(x + i\eta) - \eta - (v(x_1 + i\eta_1) - \eta_1) = \delta(\eta, \eta_1) = \delta \tag{6.32}$$

as a function of η and η_1. First it follows from (6.31) that

$$\sup \delta(\eta, \eta_1) - \inf \delta(\eta, \eta_1) \leqq 2\,\varepsilon_3(x)\,. \tag{6.33}$$

If δ vanishes for some pair η, η_1, then $|\delta| \leqq 2\,\varepsilon_3(x)$. If not, then δ assumes only positive or only negative values. We suppose first that $\delta > 0$.

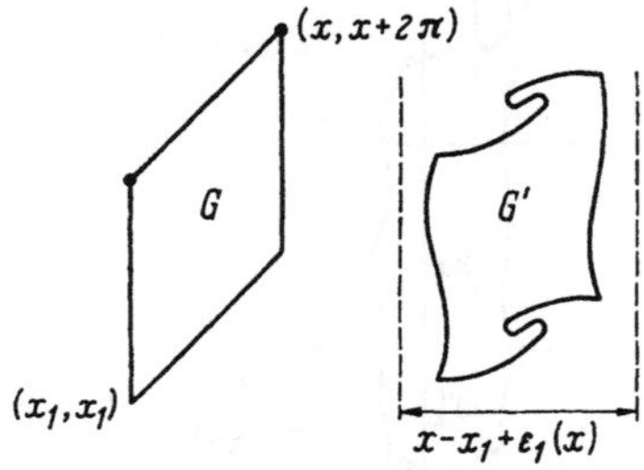

Fig. 14

Consider the parallelogram $G = \{\xi + i\eta \mid x_1 < \xi < x, \xi < \eta < \xi + 2\pi\}$ and its image $G' = f(G)$ (Fig. 14). The area of G' can be estimated from (6.10) and (6.26) with the help of Fubini's theorem; this gives

$$m(G') \leqq 2\pi\,(x - x_1 + \varepsilon_1(x))\,. \tag{6.34}$$

To obtain an estimate for δ we cover the parallelogram G with parallel segments $L_y = \{\xi + i\eta \mid x_1 < \xi < x, \eta = \xi + y\}$, $0 < y < 2\pi$. By Lemma IV.1.1 and Theorem IV.1.4, f is absolutely continuous on L_y and regular at almost all points of L_y, up to a null set of y-values. For y not belonging to this null set, let $\Lambda(y)$ denote the length of the image arc $L_y' = f(L_y)$. Then (cf. (6.19))

$$\int_0^{2\pi} \Lambda(y)\,dy \leqq \sqrt{2}\int_0^{2\pi}\left(\int_{L_y}\sqrt{D(\zeta)\,J(\zeta)}\,d\xi\right)dy\,.$$

By Fubini's theorem,

$$\int_0^{2\pi} \Lambda(y)\,dy \leqq \sqrt{2}\iint_G \sqrt{D(\zeta)\,J(\zeta)}\,d\sigma\,,$$

and it follows from Schwarz's inequality and (6.10) that

$$\int_0^{2\pi} \Lambda(y)\,dy \leqq \left(2 \iint_G D(\zeta)\,d\sigma \iint_G J(\zeta)\,d\sigma\right)^{1/2}$$
$$\leqq \left(2\left(2\pi\, I(e^x) \iint_G J(\zeta)\,d\sigma + \iint_G d\sigma \iint_G J(\zeta)\,d\sigma\right)\right)^{1/2}.$$

Here

$$\iint_G J(\zeta)\,d\sigma = m(G'), \qquad \iint_G d\sigma = m(G) = 2\pi\,(x - x_1).$$

Hence, by (6.34), we have

$$\int_0^{2\pi} \Lambda(y)\,dy \leqq 2\pi\left(2\,(x - x_1 + \varepsilon_1(x))\,(x - x_1 + I(e^x))\right)^{1/2}$$
$$= 2\sqrt{2}\,\pi\,(x - x_1) + \varepsilon_4(x), \tag{6.35}$$

where $\varepsilon_4(x) \to 0$ as $x \to -\infty$.

To find a lower bound for $\Lambda(y)$ we recall that $\Lambda(y)$ is the length of the arc L'_y which joins the points $f(x_1 + i\,(x_1 + y))$ and $f(x + i\,(x + y))$. Hence, by (6.26) and (6.32), we have for $x - x_1 > \varepsilon_1(x)$,

$$\Lambda(y) \geqq \left((x - x_1 - \varepsilon_1(x))^2 + (x - x_1 + \inf\delta)^2\right)^{1/2}$$
$$\geqq \sqrt{2}\,(x - x_1 - \varepsilon_1(x)) + \frac{\inf\delta}{\sqrt{2}}.$$

For $x - x_1 \leqq \varepsilon_1(x)$ we have

$$\Lambda(y) \geqq \inf\delta.$$

Inserting these estimates into (6.35) we see that

$$\inf\delta \leqq 2\,\varepsilon_1(x) + \frac{\varepsilon_4(x)}{\sqrt{2}\,\pi}$$

in both cases. Thus in view of (6.33) we have

$$\delta \leqq 2\,\varepsilon_1(x) + 2\,\varepsilon_3(x) + \frac{\varepsilon_4(x)}{\sqrt{2}\,\pi} = \varepsilon_5(x). \tag{6.36}$$

We have proved this relation under the assumption that $\delta > 0$. If $\delta < 0$ we replace the parallelogram G by $\{\xi + i\eta \mid x_1 < \xi < x, -\xi < \eta < -\xi + 2\pi\}$ and use the above argument.

We then obtain the upper bound (6.36) for $-\delta$. In the case when δ assumes both positive and negative values, we have already stated that $|\delta| \leqq 2\,\varepsilon_3(x)$. Hence in all cases we have the inequality

$$|\delta| = |(v(x + i\eta) - \eta) - (v(x_1 + i\eta_1) - \eta_1)| \leqq \varepsilon_5(x)$$

for $x_1 \leqq x$.

It follows that $v(x + i\eta) - \eta$ tends to a limit α as $x \to -\infty$, and in such a way that

$$|v(x + i\eta) - \eta - \alpha| \leqq \varepsilon_5(x)$$

for every finite η. Lemma 6.2 is now proved.

6.7. Summary of the results. The desired result on conformality at a point follows easily from Lemmas 6.1 and 6.2. In fact, for any two complex numbers $z_1 = r_1 e^{i\varphi_1}$, $z_2 = r_2 e^{i\varphi_2}$ the inequality $|z_1 - z_2| \leqq |r_1 - r_2| + r_2 |\varphi_1 - \varphi_2|$ holds and it follows that $w(z)/z - c\, e^{i\alpha}$ tends to zero if $|w(z)|/|z|$ and $\arg(w(z)/z)$ tend to $c \neq 0, \infty$ and α, respectively, as $z \to 0$. By definition w is then conformal at $z = 0$, and from the expansion

$$w(z) = c\, e^{i\alpha} z + o(z)$$

we see that $w_z(0) = c\, e^{i\alpha}$ and $w_{\bar z}(0) = 0$.

Theorem 6.1. *Let w be a K-quasiconformal mapping of the finite plane onto itself with $w(0) = 0$ and*

$$I(r) = \frac{1}{2\pi} \iint\limits_{|z|<r} \frac{D(z) - 1}{|z|^2}\, d\sigma < \infty \qquad \textit{for} \quad r < \infty .$$

Then w is conformal at $z = 0$ and

$$\left| \frac{w(z)}{z} - w_z(0) \right| < |w_z(0)|\, \varepsilon(|z|)\,, \qquad \lim_{|z|\to 0} \varepsilon(|z|) = 0\,,$$

where the function ε depends only on K and I and not otherwise on the mapping w. The derivative $w_z(0)$ satisfies the inequalities

$$\min_{|z|=1} |w(z)|\, e^{-I(1)} \leqq |w_z(0)| \leqq \max_{|z|=1} |w(z)|\, e^{I(1)}\,. \tag{6.37}$$

Remarks. From the proofs of Lemmas 1 and 2 we see that our result can be expressed in a slightly more general form: If $I(r) \leqq \Delta(r)$, where $\Delta(r) \to 0$ as $r \to 0$, then the conclusions of the theorem hold for a function ε depending only on K and Δ. This monotonic dependence of ε on I will be needed in § 7.

Secondly we remark that the inequality on the right hand side of (6.37) can be sharpened if $\max |w(z)|$ is much larger than $\min |w(z)|$ on $|z| = 1$. The restriction of w to the unit disc D can in fact be written in the form $w = f \circ \hat{w}$, where $\hat{w}$ is a quasiconformal mapping of D onto itself with $\hat{w}(0) = 0$, and f maps the unit disc conformally onto $w(D)$. By Koebe's one-quarter theorem (cf. II.1.3) we then have

$$|f'(0)| \leqq 4 \min_{|z|=1} |w(z)|\,.$$

By (6.37) $|\hat{w}_z(0)| \leqq e^{I(1)}$, since $\hat{w}$ has the same dilatation quotient as w. For $|w_z(0)| = |f'(0)|\,|\hat{w}_z(0)|$ we thus obtain the estimate

$$|w_z(0)| \leqq 4 \min_{|z|=1} |w(z)|\, e^{I(1)} .$$

For further generalizations of Koebe's one-quarter theorem for quasiconformal mappings we refer to Pfluger [1] and Juve [1].

§ 7. Regularity of a Mapping with Prescribed Complex Dilatation

7.1. Complex dilatation and regular points. Having dealt with the problem of finding a sufficient condition for local conformality, we can treat without difficulty the more general problem of the regularity of a quasiconformal mapping at a point. In what follows we consider only finite points; of course we can get rid of this restriction by means of a linear mapping.

Contrary to the local conformality, the regularity of a quasiconformal mapping at a point z_0 cannot be deduced from the behaviour of its dilatation quotient for $z \neq z_0$. We must therefore consider the complex dilatation of the mapping. In 6.2 we remarked that the conformality condition in Theorem 6.1 is equivalent to the convergence of the integral

$$\iint\limits_{|z|<r} \frac{|\varkappa(z)|}{|z|^2}\, d\sigma .$$

This gives rise to the following generalization of Theorem 6.1:

Theorem 7.1. *Let G and G' be domains of the finite plane, z_0 a point of G, and $w: G \to G'$ a quasiconformal mapping with complex dilatation $\varkappa$, where $|\varkappa(z)| \leqq k < 1$ almost everywhere in G. If there exists a number $\varkappa_0$ such that*

$$\iint\limits_{|z-z_0|<r} \frac{|\varkappa(z)-\varkappa_0|}{|z-z_0|^2}\, d\sigma < \infty \tag{7.1}$$

for some $r > 0$, then w is regular at z_0 and $\varkappa(z_0) = \varkappa_0$.

Proof: By Theorem II.8.1 the restriction of w to a neighbourhood of z_0 can be extended to a quasiconformal mapping of the plane. By means of a linear transformation we ensure that the point at infinity is mapped onto itself. Hence we may assume, without loss of generality, that G and G' are planes.

In order to be able to apply Theorem 6.1 we construct the auxiliary mapping f,

$$f(\zeta) = w(z_0 + \zeta - \varkappa_0 \bar{\zeta}) - w(z_0), \tag{7.2}$$

where $\varkappa_0$ is the number appearing in (7.1). We show that f is conformal at $\zeta = 0$.

From (7.1) it follows that $|\varkappa_0| \leqq k < 1$ and so f is quasiconformal. Therefore, by Theorem 6.1, it is enough to show that

$$\iint\limits_{|\zeta| < r} \frac{|\varkappa_f(\zeta)|}{|\zeta|^2} \, d\sigma < \infty \tag{7.3}$$

holds for some $r > 0$, where $\varkappa_f$ denotes the complex dilatation of f.

To prove (7.3) we apply the transformation formula (5.6) of IV.5.2 to calculate $\varkappa_f$. Setting

$$z = z_0 + \zeta - \varkappa_0 \bar{\zeta} \tag{7.4}$$

we obtain

$$|\varkappa_f(\zeta)| = \left| \frac{\varkappa(z) - \varkappa_0}{1 - \bar{\varkappa}_0 \varkappa(z)} \right| \leqq \frac{|\varkappa(z) - \varkappa_0|}{1 - k^2}$$

almost everywhere. Since the Jacobian of the affine mapping defined by (7.4) is $1 - |\varkappa_0|^2$, it follows that

$$\iint\limits_{|\zeta| < r} \frac{|\varkappa_f(\zeta)|}{|\zeta|^2} \, d\sigma \leqq \frac{1}{(1 - k^2)^2} \iint\limits_{|\zeta(z)| < r} \frac{|\varkappa(z) - \varkappa_0|}{|\zeta(z)|^2} \, d\sigma ,$$

where by (7.4)

$$\zeta(z) = \frac{z - z_0 + \varkappa_0(\bar{z} - \bar{z}_0)}{1 - |\varkappa_0|^2} . \tag{7.5}$$

Because

$$|\zeta(z)| \geqq \frac{|z - z_0|}{1 + |\varkappa_0|} \geqq \frac{|z - z_0|}{1 + k} ,$$

we obtain finally

$$\iint\limits_{|\zeta| < r} \frac{|\varkappa_f(\zeta)|}{|\zeta|^2} \, d\sigma \leqq \frac{1}{(1 - k)^2} \iint\limits_{|z - z_0| < (1 + k) r} \frac{|\varkappa(z) - \varkappa_0|}{|z - z_0|^2} \, d\sigma . \tag{7.6}$$

Thus (7.3) follows from (7.1). Hence by Theorem 6.1 we have

$$f(\zeta) = f_\zeta(0) \, \zeta + o(\zeta) . \tag{7.7}$$

In view of (7.2), (7.4) and (7.5) this is equivalent to

$$w(z) = w(z_0) + \frac{f_\zeta(0)}{1 - |\varkappa_0|^2} (z - z_0) + \frac{\varkappa_0 f_\zeta(0)}{1 - |\varkappa_0|^2} (\bar{z} - \bar{z}_0) + o(z - z_0) . \tag{7.8}$$

From this we see that w has the required properties: it is regular at z_0 and $\varkappa(z_0) = \varkappa_0$.

7.2. Continuous differentiability. We shall now investigate the question of when a quasiconformal mapping w with prescribed complex dilatation is regular in a domain G. By definition this means that w is to be regular at every point of G and have continuous partial derivatives in G. The latter property does not follow from the first and in fact we need a stronger condition than the validity of (7.1) for every $z_0 \in G$ (cf. 6.2).

The continuous differentiability of w in G does, however, follow if we impose upon the integral of (7.1) the additional requirement that it converge uniformly in every compact subset F of G. This means that for every $\varepsilon > 0$ there exists a $\delta > 0$ such that

$$\iint\limits_{|z-z_0|<\delta} \frac{|\varkappa(z) - \varkappa(z_0)|}{|z - z_0|^2}\, d\sigma < \varepsilon$$

for every $z_0 \in F$, where by Theorem 7.1, $\varkappa(z_0)$ must be the complex dilatation of w at z_0.

As previously we restrict ourselves to domains of the finite plane.

Theorem 7.2. *Let $\varkappa$, $|\varkappa| \leqq k < 1$, be a measurable function in a domain G such that the integral*

$$\iint \frac{|\varkappa(z) - \varkappa(z_0)|}{|z - z_0|^2}\, d\sigma_z$$

converges uniformly in every compact subset of G. Then every quasiconformal mapping $w: G \to G'$ whose complex dilatation coincides with $\varkappa$ almost everywhere in G is a regular quasiconformal mapping of G and has the complex dilatation $\varkappa(z)$ for every $z \in G$.

Proof: As in the proof of Theorem 7.1 we can extend w to a quasiconformal mapping of the plane and normalize this so that $w(\infty) = \infty$. We fix a domain U, $\overline{U} \subset G$, and show that the derivatives w_z and $w_{\bar{z}}$ are continuous in U.

For an arbitrary point $z_0 \in U$ the equation (7.8) holds with $\varkappa_0 = \varkappa(z_0)$. Comparing this with the equivalent relation (7.7) and applying Theorem 6.1 we obtain for the remainder term in (7.8) the estimate

$$|o(z - z_0)| \leqq |f_\zeta(0)\, \zeta|\, \varepsilon(|\zeta|) \leqq |\zeta|\, \varepsilon(|\zeta|) \max_{|\zeta|=1} |f(\zeta)|\, e^{I(1)}$$

$$\leqq \frac{|z - z_0|}{1 - k}\, \varepsilon\left(\frac{|z - z_0|}{1 - k}\right) \max_{|z-z_0|=1+k} |w(z) - w(z_0)|\, e^{I(1)}. \quad (7.9)$$

Here ε and I are the functions appearing in Theorem 6.1, the function I being defined with respect to f. We can choose ε to be decreasing with $|\zeta|$.

Since the maximal dilatation of the mapping f is at most $(1+k)^2/(1-k)^2$ we have

$$D_f(\zeta) - 1 = 2\frac{|\varkappa_f(\zeta)|}{1-|\varkappa_f(\zeta)|} \leqq 2\frac{1+k^2}{(1-k)^2}|\varkappa_f(\zeta)| .$$

In view of (7.6) it follows that

$$I(r) \leqq \frac{1+k^2}{\pi(1-k)^4} \iint\limits_{|z-z_0|<(1+k)r} \frac{|\varkappa(z)-\varkappa(z_0)|}{|z-z_0|^2}\, d\sigma = \Delta(r) .$$

Here $\Delta(r)$ depends on z_0, but tends by hypothesis uniformly to zero as $r \to 0$. Hence $\Delta(1)$ and thus also $e^{I(1)}$ in (7.9) are uniformly bounded for $z_0 \in U$. By the first remark at the end of 6.7, the function ε depends only on k and Δ. Since $|w(z)-w(z_0)|$ is uniformly bounded for $z_0 \in U$, $|z-z_0| \leqq 1+k$, we can thus write (7.9) in the form $|o(z-z_0)| \leqq |z-z_0|\,\varepsilon_1(|z-z_0|)$, where $\varepsilon_1(|z-z_0|) \to 0$ as $|z-z_0| \to 0$ and ε_1 is the same function of $|z-z_0|$ for every point $z_0 \in U$. From (7.8) we then obtain

$$|w(z) - w(z_0) - w_z(z_0)(z-z_0) - w_{\bar z}(z_0)(\bar z - \bar z_0)| < |z-z_0|\,\varepsilon_1(|z-z_0|) . \tag{7.10}$$

For real $z - z_0 = h$ it follows from (7.10) that

$$\left|\frac{w(z_0+h) - w(z_0)}{h} - w_x(z_0)\right| < \varepsilon_1(|h|) .$$

Hence w_x is continuous in U as a uniform limit of continuous functions. In the same way we can show that w_y is continuous, and the theorem is proved.

Besides the investigations of Teichmüller, Wittich and Belinskij mentioned in § 6, results related to Theorems 7.1 and 7.2 have been obtained by Šabat [1]; see also Volkovyskij [1]. In the above formulation these theorems can be found in Lehto [1]. By applying the representation formula (5.17) in 5.7, Bojarski [2] has demonstrated regularity at a point under the condition

$$\iint\limits_{|z-z_0|<r} \left|\frac{\varkappa(z)-\varkappa_0}{z-z_0}\right|^p d\sigma < \infty , \qquad p > 2 . \tag{7.11}$$

This result is contained in Theorem 7.1: By Hölder's inequality (7.1) follows from (7.11), while the example $|\varkappa(z)| = |z|^{1-2/p}$, $z_0 = \varkappa_0 = 0$, shows that the converse is false.

7.3. Sharpness of the conditions. Simple examples suffice to show that the criteria given in Theorems 7.1 and 7.2 are not best possible.

In fact, the integral in (7.1) need not be finite even for a mapping w which is regular throughout G and not only at z_0.

To see this we choose as our domain G the square $\{x + i\,y \mid -1/2 < x < 1/2,\ -1/2 < y < 1/2\}$ and define a mapping w by the formula

$$w\,(x + i\,y) = x + i\left(y - \int_0^y \frac{dt}{\log |t|}\right).$$

This mapping is regular in G and has complex dilatation

$$\varkappa\,(x + i\,y) = \frac{1}{2\log |y| - 1}.$$

An easy calculation shows that the integral

$$\int_{-1/2}^{1/2} \int_{-1/2}^{1/2} \frac{|\varkappa\,(x + i\,y)|}{x^2 + y^2}\,dx\,dy$$

is infinite.

It is thus to be expected that a non-trivial necessary regularity condition must depend not only on $|\varkappa(z) - \varkappa(z_0)|$ but also on the argument of $\varkappa(z) - \varkappa(z_0)$. In fact, Reich and Walczak [1] have shown that for every measurable function g of the real variable r, $0 < r < 1$, such that $0 \leqq g(r) \leqq k < 1$, there exists a mapping of the unit disc onto itself which is conformal at the origin and whose complex dilatation has absolute value $g(|z|)$ for almost all z, $|z| < 1$.[52] An example can be constructed as follows:

We go over to the $\zeta = \xi + i\,\eta = \log z$-plane and write

$$h(\xi) = \frac{2\,g(e^{\xi})}{\sqrt{1 - (g(e^{\xi}))^2}}. \tag{7.12}$$

First we define a function h^* of ξ in the following way: If the integral

$$\int_{-\infty}^{0} h\,d\xi$$

is finite, then we write $h^* = h$. If not, then there exists a decreasing sequence of numbers $\xi_0 = 0, \xi_1, \xi_2, \ldots$, such that

$$\int_{\xi_n}^{\xi_{n-1}} h\,d\xi = \frac{1}{n}, \qquad n = 1, 2, \ldots.$$

[52] Reich and Walczak [1] have also given a refined version of the sufficient condition (7.1), in which the argument of $\varkappa$ is considered.

Then we set $h^*(\xi) = h(\xi)$ for $\xi_{2n+1} < \xi \leqq \xi_{2n}$ and $h^*(\xi) = -h(\xi)$ for $\xi_{2n+2} < \xi \leqq \xi_{2n+1}$, $n = 0, 1, \ldots$.

The function ψ,

$$\psi(\xi) = \int_0^\xi h^* \, d\xi ,$$

tends to a finite limit α as $\xi \to -\infty$ in both cases. Being an integral, ψ is absolutely continuous on every compact interval of the half-line $\xi < 0$ and has the derivative

$$\psi'(\xi) = h^*(\xi)$$

for almost all ξ.

We now define a mapping f of the left half-plane onto itself by the formula

$$f(\xi + i\,\eta) = \xi + i(\eta + \psi(\xi)) .$$

This mapping is absolutely continuous on lines and its derivatives satisfy the equations

$$f_\zeta(\zeta) = 1 + \frac{i}{2} h^*(\xi) , \qquad f_{\bar\zeta}(\zeta) = \frac{i}{2} h^*(\xi)$$

for almost all $\zeta = \xi + i\,\eta$. Since $|h^*(\xi)| = h(\xi)$ is bounded, f is quasiconformal. By (7.12) the modulus of the complex dilatation of f satisfies

$$\left|\frac{f_{\bar\zeta}(\zeta)}{f_\zeta(\zeta)}\right| = \frac{h(\xi)}{\sqrt{4 + (h(\xi))^2}} = g(e^\xi)$$

for almost all ζ.

Finally we return to the $z = e^\zeta$-plane. The mapping f corresponds to the mapping w,

$$w(z) = e^{f(\log z)} = z\, e^{i\,\psi(\log|z|)} ,$$

of the unit disc onto itself. This has the required properties: its complex dilatation has absolute value $g(|z|)$ for almost all z, and $w(z)/z$ tends to the limit $e^{i\alpha}$ as $z \to 0$.

VI. Quasiconformal Functions

Introduction to Chapter VI

In previous chapters we considered quasiconformal homeomorphisms between plane domains. We now extend the concept of quasiconformality to mappings which are not necessarily one-to-one.

The simplest generalization retaining the essential features of the theory of quasiconformal homeomorphisms is the class of functions which are composed of a quasiconformal mapping and an analytic function. Since we shall consider mappings into the whole plane we allow the analytic functions to have poles.

Definition. *A function f is called K-quasiconformal in a plane domain G if it permits a representation*

$$f = \varphi \circ w ,$$

where $w : G \to G'$ is a K-quasiconformal homeomorphism and φ is a non-constant analytic function in G'. A function is quasiconformal if it is K-quasiconformal for some K.

If $f = \varphi \circ w$ is quasiconformal in a domain G, then f is locally homeomorphic in G up to points which w maps onto zeros of the derivative φ' or onto multiple poles of φ, i.e. f is locally homeomorphic with the possible exception of isolated points. In addition, f satisfies the module condition $M(f(Q)) \leqq K\, M(Q)$ for every quadrilateral Q such that $\overline{Q} \subset G$ and $\overline{Q}$ is mapped topologically by f. These properties of f are called *geometric* in analogy with our previous terminology.

The following *analytic* properties of a K-quasiconformal function f are likewise immediate consequences of the definition: f has L^2-derivatives (or even L^p-derivatives for every $p < p(K)$; cf. V.5.4), and the dilatation condition $\max_\alpha |\partial_\alpha f(z)| \leqq K \min_\alpha |\partial_\alpha f(z)|$ holds almost everywhere. In other words, f is a generalized L^p-solution of a Beltrami equation $f_{\bar z} = \varkappa f_z$, where $|\varkappa| \leqq (K-1)/(K+1)$, $p < p(K)$.

In the present chapter we shall establish geometric and analytic properties which are sufficient to characterize a quasiconformal function.

Geometric conditions will be given in § 1 and analytic conditions in § 2.

In view of the definition of a quasiconformal function it is obvious that many theorems of Chapters I, II, IV and V can be carried over to quasiconformal functions, either as such or with suitable modifications. Here such generalizations will not be dealt with.

§ 1. *Geometric Characterization of a Quasiconformal Function*

1.1. Quasiconformality and module conditions. According to the definition given in the Introduction, a quasiconformal function is continuous and, with the possible exception of isolated points, locally homeomorphic. Conversely, let us consider a continuous function f which is locally homeomorphic except at isolated points in a domain G. If the restriction of f to a domain $U \subset G$ is a sense-preserving homeomorphism, then a simple connectivity argument shows that f preserves the orientation in every subdomain of G where it is homeomorphic.

We now prove that the geometric properties of K-quasiconformal functions mentioned in the Introduction do in fact characterize these functions.

Theorem 1.1. *Let f be a continuous*[53] *mapping of a domain G which is locally homeomorphic and sense-preserving up to a set $E \subset G$ of isolated points. If f satisfies the module condition*

$$M(f(Q)) \leqq K\, M(Q) \tag{1.1}$$

for every quadrilateral Q such that $\overline{Q} \subset G$ and f maps $\overline{Q}$ topologically, then f is K-quasiconformal in G.

Proof: By hypothesis, every point $z \in G - E$ possesses a neighbourhood U which is mapped topologically, and by (1.1) K-quasiconformally, onto $f(U)$. The complex dilatation of f is therefore defined almost everywhere in U.

Let $\varkappa$ be a function which coincides with the complex dilatation of f at every point where the latter is defined and vanishes elsewhere in G. Then $\varkappa$ is measurable and satisfies

$$|\varkappa(z)| \leqq \frac{K-1}{K+1}$$

everywhere in G.

By the Existence theorem there is a K-quasiconformal homeomorphism w of G whose complex dilatation is equal to $\varkappa(z)$ for almost all $z \in G$.

---

[53] Note that continuity here is in the sense of the spherical metric.

The composite mapping $\varphi = f \circ w^{-1}$ is locally homeomorphic in $w(G - E)$. It follows from the Uniqueness theorem (IV.5.2) that every point of $w(G - E)$ has a neighbourhood which is mapped conformally by φ. Since φ is continuous throughout $w(G)$ and $w(E)$ consists of isolated points, φ is analytic in $w(G)$. The function $f = \varphi \circ w$ is hence K-quasiconformal in G.

1.2. Interior mappings. The topological condition in Theorem 1.1 that f be locally homeomorphic except at isolated points can be expressed in another non-trivially equivalent form. For this purpose we introduce the concept of an interior mapping.

A mapping of a domain G into a domain G' is called *open* if the image of every open subset of G is open. If f is continuous and open, then for every set A we have

$$f^{-1}(\bar{A}) = \overline{f^{-1}(A)} \cap G. \tag{1.2}$$

The mapping f is called *light* if every component of the preimage $f^{-1}(\zeta)$ of any point $\zeta \in f(G)$ consists of a single point. If f is continuous, open and light, then it is called an *interior* mapping of G.

Non-constant analytic functions provide examples of interior mappings. On the other hand, by a theorem of Stoïlow [1] an interior mapping can always be represented in the form $f = \varphi \circ w$, where w is a homeomorphism of G and φ is analytic in $w(G)$.

From this it follows that an interior mapping of a domain G is locally homeomorphic in G with the possible exception of isolated points. Thus in Theorem 1.1 we can replace the words "continuous and locally homeomorphic up to a set $E \subset G$ of isolated points" by the word "interior".

We shall establish this equivalent version of Theorem 1.1 completely without reference to Stoïlow's theorem. For this it is necessary to prove directly the part of Stoïlow's theorem which states that an interior mapping is locally homeomorphic up to isolated points. For the proof we first give three lemmas on open, light and interior mappings, respectively.

1.3. A lemma on open mappings. The following property of an open mapping will be of frequent application in our proof.

Lemma 1.1. *Let f be a continuous open mapping of a domain G, D a subdomain of $f(G)$, and A, $\bar{A} \subset G$, a component of $f^{-1}(D)$. Then $f(A) = D$.*

Proof: In view of the continuity of f, no boundary point of A is mapped onto a point of D. Consequently

$$f(A) = f(\bar{A}) \cap D .$$

Since f is an open mapping and A is an open set, $f(A)$ is open. On the other hand, $f(\bar{A})$ is closed, being the continuous image of the compact set $\bar{A}$. The set $f(A)$ is hence both open and closed in D. Since D is connected our assertion follows.

1.4. A lemma on light mappings. Of light mappings we need the following property:

Lemma 1.2. *Let f be a continuous light mapping of a domain G, z_0 a point of G, and U a neighbourhood of z_0 with $\bar{U} \subset G$. Then there exists a neighbourhood V of $w_0 = f(z_0)$ such that the component of $f^{-1}(\bar{V})$ which contains z_0 lies in U.*

Proof: We may assume that U is a disc. Let V_n, $n = 1, 2, \ldots$, be discs with the same centre w_0 and with radii r_n, $r_1 > r_2 > \cdots$, $\lim r_n = 0$. We show that the z_0-component A_n of the set $f^{-1}(V_n)$ is contained in U from some n onwards. Our assertion follows from this, since the z_0-component of $f^{-1}(\bar{V}_{n+1})$ is a closed subset of A_n.

We suppose that on the contrary A_n does not lie in U for any value of n and derive a contradiction. First we have for the z_0-component B_n of $A_n \cap U$ the relation

$$\bar{B}_n \cap \operatorname{Fr} U \neq \emptyset . \tag{1.3}$$

For otherwise $\bar{B}_n \subset U$ and so $\bar{B}_n \cap A_n = \bar{B}_n \cap (A_n \cap U)$. Since B_n, as a component of $A_n \cap U$, is closed in this set, it follows that $B_n = \bar{B}_n \cap A_n$. Consequently B_n is not only open but also closed in A_n. Since A_n is connected this is impossible and (1.3) follows.

We now make use of the following well-known result on compact sets (Newman [1], pp. 47 and 81): *The intersection of a decreasing sequence of non-empty (connected) compact sets is non-empty (connected).*

From this we first deduce that $B = \bigcap \bar{B}_n$ is connected, since $\bar{B}_n$ is connected as the closure of a connected set. Further, in view of (1.3), it follows from $B \cap \operatorname{Fr} U = \bigcap (\bar{B}_n \cap \operatorname{Fr} U)$ that $B \cap \operatorname{Fr} U$ is non-empty. On the other hand, $f(B) \subset \bar{V}_n$ for every n so that $f(B)$ consists of w_0 only. Hence B lies in the z_0-component of $f^{-1}(w_0)$, which therefore contains points of $\operatorname{Fr} U$. This contradicts the assumption that f is light, i.e. that the z_0-component of $f^{-1}(w_0)$ consists of the point z_0 alone.

1.5. A lemma on interior mappings. Using the two Lemmas above we prove the following result on interior mappings:

Lemma 1.3. *Let f be an interior mapping of a simply connected domain G such that $f(G)$ is not the whole plane, and V, $\overline{V} \subset f(G)$, a disc. Let D_1 and D_2 be discs, $\overline{D}_1 \cup \overline{D}_2 \subset V$, $\overline{D}_1 \cap \overline{D}_2 = \emptyset$, and $w \in V$ a point outside $\overline{D}_1 \cup \overline{D}_2$. If A, $\overline{A} \subset G$, is a component of $f^{-1}(V)$, then the sets $f^{-1}(w)$, $f^{-1}(D_1)$, $f^{-1}(D_2)$ each have a finite positive number n, n_1, n_2 of components in A, and these numbers satisfy $n \leqq n_1 n_2$.*

Proof: We first remark that every component of the preimage of a set $E \subset V$ either lies completely in A or completely outside A. We also note that for a compact $E \subset V$ the set $f^{-1}(E) \cap A$ is compact.

Let w_1 be an arbitrary point of D_1. As a compact set $A \cap f^{-1}(w_1)$ is covered by a finite number of components of $f^{-1}(D_1)$. In fact, this finite covering consists of all components of $f^{-1}(D_1)$ lying in A, since every such component is mapped onto the whole of D_1 by Lemma 1.1. Consequently the number n_1 is finite and the same reasoning shows that n_2 is finite.

From Lemma 1.1 it follows that n, n_1 and n_2 are positive. We suppose now that $n > n_1 n_2$, contrary to our assertion. In order to derive a contradiction we choose $n_1 n_2 + 1$ points z_i in the set $f^{-1}(w) \cap A$ and cover them with discs U_i whose closures are disjoint and lie in A. By Lemma 1.2 there exists a disc D_3, $\overline{D}_3 \subset V$, with midpoint w, such that the z_i-component of $f^{-1}(\overline{D}_3)$ is contained in U_i for every i. The number n_3 of the components A_{3k} of $f^{-1}(\overline{D}_3) \cap A$ is thus at least $n_1 n_2 + 1$. On the other hand we deduce as above that $f^{-1}(D_3) \cap A$ and by (1.2) also $f^{-1}(\overline{D}_3) \cap A$, has only finitely many components.

If necessary we extend the domains D_1 and D_2 so that both meet the disc D_3, and carry out the extension in such a way that $\overline{D}_1$ and $\overline{D}_2$ remain disjoint subsets of V and the complement of $\overline{D}_1 \cup \overline{D}_2 \cup \overline{D}_3$ is connected. We thus remove the restriction that D_1 and D_2 are discs; it was made only to ensure that the above extension would be possible. It follows from Lemma 1.1 that the number of components of $f^{-1}(D_h) \cap A$, $h = 1, 2$, is at most n_h after the extension. By (1.2) the same is true for the components A_{hk}, $h = 1, 2$, of the sets $f^{-1}(\overline{D}_h) \cap A$.

Since $D_1 \cap D_3$ and $D_2 \cap D_3$ are non-empty, every A_{3k} has points in common with at least one A_{1i} and one A_{2j} by Lemma 1.1. Since the number of pairs (A_{1i}, A_{2j}) is at most $n_1 n_2$ there exists a pair A_{1p}, A_{2q} which meets two components A_{3r} and A_{3s} (Fig. 15). We can now apply

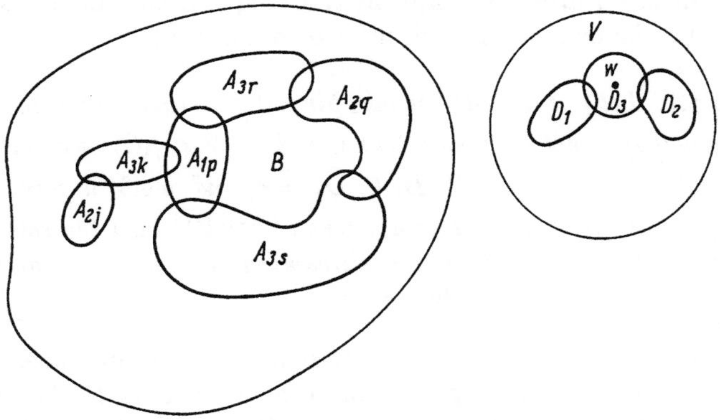

Fig. 15

Lemma I.1.2′ in the following way: We set $F_1 = A_{1p} \cup A_{3r}$, $F_2 = A_{2q} \cup A_{3s}$, and choose as the sets F_ν, $\nu = 3, 4, \ldots$, the remaining A_{ik}. Then the conditions of Lemma I.1.2′ are satisfied and it follows that the complement of the set $F = \bigcup A_{ik} = f^{-1}(\bigcup \overline{D}_i) \cap A$ is disconnected.

Since G is simply connected, $-G$ is contained in a component of $-F$. Hence one component B of $-F$ lies in G. If z^* is a boundary point of the domain B, then $z^* \in F$, since z^* would otherwise have a neighbourhood lying in the complement of F. Hence $f(z^*) \in \bigcup \overline{D}_i$, and therefore $f(B) - \bigcup \overline{D}_i = f(\overline{B}) - \bigcup \overline{D}_i$. The set $f(B) - \bigcup \overline{D}_i$ is consequently both open and closed in the complement of $\bigcup \overline{D}_i$. Since this complement is connected and $f(G)$ is not the whole plane it follows that $f(B) - \bigcup \overline{D}_i = \emptyset$, and so $f(B) \subset \bigcup \overline{D}_i$. From $f(A - F) \cap (\bigcup \overline{D}_i) = \emptyset$ we thus conclude that $B \cap (A - F) = B \cap A$ must be empty.

On the other hand $B \cap A \neq \emptyset$, since the set F lies in A and the boundary of B lies in F. Hence we are led to a contradiction and the lemma is proved.

1.6. Local behaviour of interior mappings. The required result on interior mappings can now easily be proved.

Theorem 1.2. *An interior mapping f of a domain G is locally homeomorphic in G with the possible exception of isolated points.*

Proof: Let z_0 be an arbitrary point of G and $w_0 = f(z_0)$. It follows from Lemmas 1.2 and 1.3 that the points of $f^{-1}(w_0)$ have no limit points in G. The point z_0 lies therefore in a disc U, $\overline{U} \subset G$, which contains no other points of $f^{-1}(w_0)$. Let U be so small that $f(U)$ is not the whole plane. By Lemma 1.2 there exists a disc $V \subset f(U)$ with midpoint w_0 such that the closure of the z_0-component A of $f^{-1}(V)$ lies in U.

Let w_1 and w_2 be two other points in V and n_i the number of the points of $f^{-1}(w_i) \cap A$, $i = 1, 2$. We cover the points w_0 and w_1 by discs D_0 and D_1 such that $\overline{D}_0 \cup \overline{D}_1 \subset V$, $\overline{D}_0 \cap \overline{D}_1 = \emptyset$, and $w_2 \notin \overline{D}_0 \cup \overline{D}_1$. According to Lemma 1.1 $f^{-1}(D_0)$ has precisely one component in A, and the number of components of $f^{-1}(D_1)$ in A is at most n_1.

Since U is simply connected, Lemma 1.3 can be applied to the restriction of f to U, and so $n_2 \leqq 1 \cdot n_1 = n_1$. By interchanging the roles of w_1 and w_2 we also deduce that $n_1 \leqq n_2$. With the possible exception of the point w_0, which has precisely one preimage point in A, all points of V have therefore the same finite number n of preimage points in A.

If $n = 1$, then the mapping $f: A \to V$ is homeomorphic. In the case $n > 1$ the mapping cannot be one-to-one in any neighbourhood of z_0. We show that every $z_1 \neq z_0$ of A then has a neighbourhood in which f is homeomorphic. The point z_0 is therefore the only exceptional point in A.

We consider the set $f^{-1}(f(z_1)) \cap A = \{z_1, z_2, \ldots, z_n\}$ and cover the points z_i with disjoint discs $U_i \subset A$, $i = 1, 2, \ldots, n$. By Lemma 1.2 there exists a disc V_1 with midpoint $f(z_1)$ such that the z_i-component of $f^{-1}(V_1)$ lies in U_i for every $i = 1, 2, \ldots, n$. Thus by Lemma 1.1, $f^{-1}(V_1)$ has exactly n components in A, which are all mapped onto V_1. If now some point of V_1 had more than one preimage point in any of these n components, then it would have at least $n + 1$ preimage points in A. This shows that the mapping of the z_1-component of $f^{-1}(V_1)$ onto V_1 is homeomorphic, and the theorem is proved.

1.7. Other forms of the geometric characterization. Once in possession of Theorem 1.2 we can define when an interior mapping is sense-preserving (cf. the remark at the beginning of 1.1). We then obtain from Theorem 1.2 the following modification of Theorem 1.1.

Theorem 1.1′. *Let f be a sense-preserving interior mapping of the domain G. If every quadrilateral Q, $\overline{Q} \subset G$, which is mapped by f topologically onto a quadrilateral Q' satisfies the module condition $M(Q') \leqq K\, M(Q)$, then f is K-quasiconformal in G.*

By the hypothesis of Theorem 1.1 or Theorem 1.1′ the domain G can be represented in the form

$$G = E \cup \left(\bigcup_{i=1}^{\infty} U_i \right),$$

where E consists of isolated points, the sets U_i are domains, and the restriction of f to U_i, $i = 1, 2, \ldots$, is a sense-preserving homeomorphism of U_i. It follows from the proof of Theorem 1.1 that Theorems 1.1 and 1.1′ remain true if the module condition (1.1) is replaced by any such condition which ensures the K-quasiconformality of f in every U_i. For example we can replace (1.1) by a corresponding condition for analytic quadrilaterals, ring domains or rectangles, as follows from Theorems I.5.3, I.7.2 and IV.3.3. Furthermore it is possible to assume instead of (1.1) that

$$\sup_{z \in G-E} F(z) \leqq K$$

(Theorem I.9.1), or that $H(z)$ is finite in $G - E$ and $\leqq K$ almost everywhere in $G - E$ (Theorem IV.4.2).

§ 2. *Analytic Characterization of a Quasiconformal Function*

2.1. L^1-solutions of the Cauchy-Riemann equation. We shall set out analytic conditions implying the quasiconformality of a function in two steps. In the simplest special case, that of a 1-quasiconformal, i.e. analytic function, we obtain the conditions from a previous result. The general case can then be reduced to this with the aid of the Existence theorem.

Since the results of Chapter III which find application here were stated only for domains of the finite plane and finite-valued functions, we shall restrict ourselves to this case. In particular, this means that analytic functions will have no poles. As a special case of Lemma III.7.1 we then have

Theorem 2.1. *Let f be a generalized L^1-solution of the Cauchy-Riemann equation in a domain G. Then f is analytic in G.*

This theorem is closely related to the Uniqueness theorem according to which a homeomorphic generalized solution of the Cauchy-Riemann equation is a conformal mapping. Since we have removed the assumption the the mapping be one-to-one here, we need the additional condition that the derivatives be locally integrable.

2.2. L^p-solutions of Beltrami equations. Let $p(K)$ be the extremal exponent for K-quasiconformal mappings as defined in V.5.4, and $1 - 1/p(K) = 1/q(K)$. We remind the reader that $p(1) = \infty$ and $p(K)$ decreases monotonically to 2, as $K \to \infty$. Hence $q(K)$ increases monotonically from 1 to 2 as K increases from 1 to ∞.

For K-quasiconformal functions we can generalize Theorem 2.1 as follows:

Theorem 2.2. *In a domain G let f be a non-constant generalized L^q-solution of a Beltrami equation*

$$f_{\bar{z}} = \varkappa f_z \tag{2.1}$$

where $\varkappa$ is a measurable function in G with $|\varkappa(z)| \leqq (K-1)/(K+1)$ and $q > q(K)$. Then f is a K-quasiconformal function in G.

Proof: By the Existence theorem there is a K-quasiconformal homeomorphism w of G with complex dilatation $\varkappa(z)$ for almost all $z \in G$. We shall show that the function $\varphi = f \circ w^{-1}$ is analytic in the domain $w(G)$, which will prove the theorem.

By hypothesis f has L^q-derivatives for a $q > q(K)$. On the other hand w and w^{-1}, being K-quasiconformal mappings, have L^p-derivatives for every $p < p(K)$. We fix p such that $1/p + 1/q = 1$ and apply Lemma III.6.4 to the composite function $\varphi = f \circ w^{-1}$. It follows that φ has L^1-derivatives in $w(G)$.

In view of Theorem 2.1 the analyticity of φ thus follows if we show that $\varphi_{\bar{\zeta}}(\zeta) = 0$ for almost all $\zeta \in w(G)$. For this we need only repeat the calculations in IV.5.2 without the assumption that f be a homeomorphism. Formula (5.9) in IV.5.2 then yields the desired result.

Hence φ is analytic in $w(G)$. The function $f = \varphi \circ w$ is then K-quasiconformal as was asserted.

Remark. As was remarked above, $q(K) < 2$ for every K. Hence Theorem 2.2 holds under the assumption that f is a generalized L^2-solution of the Beltrami equation (2.1). The proof now becomes simpler: Without drawing on the results of V.5, we deduce the existence of L^1-derivatives of $f \circ w^{-1}$ from Lemma III.6.4 using the fact that a quasiconformal mapping has L^2-derivatives.

2.3. Other forms of the analytic characterization. In view of the connection between the Beltrami equation and the dilatation condition (1.6) in IV.1.3 we see immediately that Theorem 2.2 can also be expressed in the following form:

Let f be a non-constant function with L^q-derivatives for a $q > q(K)$ in a domain G, and suppose that the inequality

$$\max_{\alpha} |\partial_\alpha f(z)|^2 \leqq K\, J(z)$$

holds almost everywhere in G. Then f is a K-quasiconformal function in G.

A K-quasiconformal function $f = \varphi \circ w$ naturally has L^p-derivatives for every $p < p(K)$. Hence we conclude from Theorem 2.2:

If the function f is a generalized L^p-solution of the Beltrami equation $f_{\bar z} = \varkappa f_z$, with $|\varkappa(z)| \leqq (K-1)/(K+1)$ for some $p > q(K)$, then f is a generalized L^p-solution for every $p < p(K)$.

This theorem naturally yields the more information the closer K lies to 1, since $q(1) = 1$, $p(1) = \infty$, $q(K) \to 2$, $p(K) \to 2$ as $K \to \infty$.

The analytic characterizations of a K-quasiconformal function given here are implicit in the sense that the precise value of $p(K)$ is unknown. Apart from this we do not know how high a degree of integrability of the derivatives must be assumed in Theorems 2.1 and 2.2.

Bibliography

Agard, S. B., and F. W. Gehring: [1] Angles and quasiconformal mappings. Proc. London Math. Soc. **3** 14 A, 1—21 (1965).

Ahlfors, L.: [1] On quasiconformal mappings. J. Analyse Math. **3**, 1—58 and 207—208 (1954). — [2] Conformality with respect to Riemannian metrics. Ann. Acad. Sci. Fenn. A I **206** (1955). — [3] Quasiconformal reflections. Acta Math. **109**, 291—301 (1963). — [4] Lectures on quasiconformal mappings. Princeton, N. J.: Van Nostrand 1966.

Ahlfors, L., and L. Bers: [1] Riemann's mapping theorem for variable metrics. Ann. Math. **72**, 385—404 (1960).

Ahlfors, L., and A. Beurling: [1] Conformal invariants and function-theoretic null-sets. Acta Math. **83**, 101—129 (1950).

Apostol, T. M.: [1] Mathematical Analysis. Reading, Mass.: Addison-Wesley 1957.

Behnke, H., und F. Sommer: [1] Theorie der analytischen Funktionen einer komplexen Veränderlichen. Berlin-Göttingen-Heidelberg: Springer 1962.

Belinskij, P. P. (П. П. Белинский): [1] Поведение квазиконформного отображения в изолированной точке. Докл. Акад. Наук СССР **91**, 709—710 (1953).

Bers, L.: [1] On a theorem of Mori and the definition of quasiconformality. Trans. Amer. Math. Soc. **84**, 78—84 (1957). — [2] The equivalence of two definitions of quasiconformal mappings. Comment. Math. Helv. **37**, 148—154 (1962). — [3] Quasiconformal mappings and Teichmüller's theorem. Analytic functions, 89—119. Princeton, N.J.: Princeton Univ. Press 1960.

Beurling, A., and L. Ahlfors: [1] The boundary correspondence under quasiconformal mappings. Acta Math. **96**, 125—142 (1956).

Bojarski, B. V. (Б. В. Боярский): [1] Гомеоморфные решения систем Бельтрами. Докл. Акад. Наук СССР **102**, 661—664 (1955). — [2] Обобщенные решения системы дифференциальных уравнений первого порядка эллиптического типа с разрывными коэффицентами. Мат. Сб. **43** (85), 451—503 (1957).

Calderón, A. P., and A. Zygmund: [1] On the existence of certain singular integrals. Acta Math. **88**, 85—139 (1952).

Carleson, L.: [1] On mappings, conformal at the boundary. J. Analyse Math. **19**, 1—13 (1967).

Dunford, N., and J. T. Schwartz: [1] Linear operators. Part I. New York: Interscience 1958.

Fuglede, B.: [1] Extremal length and functional completion. Acta Math. **98**, 171—219 (1957).

Gehring, F. W.: [1] The definitions and exceptional sets for quasiconformal mappings. Ann. Acad. Sci. Fenn. A I **281** (1960).

Gehring, F. W., and O. Lehto: [1] On the total differentiability of functions of a complex variable. Ann. Acad. Sci. Fenn. A I **272** (1959).

Gehring, F. W., and E. Reich: [1] Area distortion under quasiconformal mappings. Ann. Acad. Sci. Fenn. A I **388** (1966).

Gehring, F. W., and J. Väisälä: [1] On the geometric definition for quasiconformal mappings. Comment. Math. Helv. **36**, 19—32 (1962). — [2] Hausdorff dimension and quasiconformal mappings. To appear in J. London Math. Soc.

Golusin, G. M.: [1] Geometrische Funktionentheorie. Berlin: Deutscher Verlag der Wissenschaften 1957.

Gross, W.: [1] Über das Flächenmaß von Punktmengen. Monatsh. Math. **29**, 145—176 (1918).

Grötzsch, H.: [1] Über einige Extremalprobleme der konformen Abbildung. Ber. Verh. Sächs. Akad. Wiss. Leipzig **80**, 367—376 (1928). — [2] Über die Verzerrung bei schlichten nichtkonformen Abbildungen und über eine damit zusammenhängende Erweiterung des Picardschen Satzes. Ber. Verh. Sächs. Akad. Wiss. Leipzig **80**, 503—507 (1928).

Hersch, J.: [1] Longueurs extrémales et théorie des fonctions. Comment. Math. Helv. **29**, 301—337 (1955).

Hersch, J., et A. Pfluger: [1] Généralisation du lemme de Schwarz et du principe de la mesure harmonique pour les fonctions pseudo-analytiques. C. R. Acad. Sci. Paris. **234**, 43—45 (1952).

Juve, Y.: [1] Über gewisse Verzerrungseigenschaften konformer und quasikonformer Abbildungen. Ann. Acad. Sci. Fenn. A I **174** (1954).

Kelingos, J. A.: [1] Contributions to the theory of quasiconformal mappings. University of Michigan dissertation 1963.

Koebe, P.: [1] Zur konformen Abbildung unendlich-vielfach zusammenhängender schlichter Bereiche auf Schlitzbereiche. Nachr. Ges. Wiss. Göttingen, 60—71 (1918).

Künzi, H. P.: [1] Quasikonforme Abbildungen. Berlin-Göttingen-Heidelberg: Springer 1960.

Kuusalo, T.: [1] Verallgemeinerter Riemannscher Abbildungssatz und quasikonforme Mannigfaltigkeiten. Ann. Acad. Sci. Fenn. A I **409** (1967).

Lavrentieff, M. A.: [1] Sur une classe de représentations continues. Rec. Math. **48**, 407—423 (1935). — (М. А. Лаврентьев): [2] Основная теорема теории квази-конформных отображений плоских областей. Изв. Акад. Наук СССР. Сер. Мат. **12**, 513—554 (1948).

Lehto, O.: [1] On the differentiability of quasiconformal mappings with prescribed complex dilatation. Ann. Acad. Sci. Fenn. A I **275** (1960). — [2] Remarks on the integrability of the derivatives of quasiconformal mappings. Ann. Acad. Sci. Fenn. A I **371** (1965).

Lehto, O., and K. I. Virtanen: [1] On the existence of quasiconformal mappings with prescribed complex dilatation. Ann. Acad. Sci. Fenn. A I **274** (1960).

Lehto, O., K. I. Virtanen and J. Väisälä: [1] Contributions to the distortion theory of quasiconformal mappings. Ann. Acad. Sci. Fenn. A I **273** (1959).

Marstrand, J. M.: [1] The dimension of Cartesian product sets. Proc. Cambridge Philos. Soc. **50**, 198—202 (1954).

Martio, O.: [1] Boundary values and injectiveness of the solutions of Beltrami equations. Ann. Acad. Sci. Fenn. A I **402** (1967).

Mori, A.: [1] On an absolute constant in the theory of quasiconformal mappings. J. Math. Soc. Japan **8**, 156—166 (1956). — [2] On quasi-conformality and pseudo-analyticity. Trans. Amer. Math. Soc. **84**, 56—77 (1957).

Morrey, C. B.: [1] On the solution of quasilinear elliptic partial differential equations. Trans. Amer. Math. Soc. **43**, 126—166 (1938).

Munroe, M. E.: [1] Introduction to measure and integration. Cambridge, Mass.: Addison-Wesley 1953.

Näätänen, M.: [1] Maps with continuous characteristics as a subclass of quasiconformal maps. Ann. Acad. Sci. Fenn. A I **410** (1967).

Newman, M. A.: [1] Elements of the topology of plane sets of points. Cambridge University Press 1961.

Pesin, I. N. (И. Н. Песин): [1] Метрические свойства Q-квазиконформных отображений. Мат. Сб. **40** (82), 281—295 (1956).

Pfluger, A.: [1] Quasikonforme Abbildungen und logarithmische Kapazität. Ann. Inst. Fourier Grenoble **2**, 69—80 (1951). — [2] Über die Äquivalenz der geometrischen und der analytischen Definition quasikonformer Abbildungen. Comment. Math. Helv. **33**, 23—33 (1959). — [3] Über die Konstruktion Riemannscher Flächen durch Verheftung. J. Indian Math. Soc. **24**, 401—412 (1961).

Reed, T. J.: [1] Quasiconformal mappings with given boundary values. Duke Math. J. **33**, 459—464 (1966).

Reich, E.: [1] On a characterization of quasiconformal mappings. Comment. Math. Helv. **37**, 44—48 (1962).

Reich, E., and H. Walczak: [1] On the behavior of quasiconformal mappings at a point. Trans. Amer. Math. Soc. **117**, 338—351 (1965).

Renggli, H.: [1] Quasiconformal mappings and extremal lengths. Amer. J. Math. **86**, 63—69 (1964).

Rickman, S.: [1] Characterization of quasiconformal arcs. Ann. Acad. Sci. Fenn. A I **395** (1966).

Šabat, B. V. (Б. В. Шабат): [1] Об обобщенных решениях одной системы уравнений в частных производных. Мат. Сб. **17** (59), 193—210 (1945).

Saks, S.: [1] Theory of the integral. Warsaw 1937.

Sario, L.: [1] Über Riemannsche Flächen mit hebbarem Rand. Ann. Acad. Sci. Fenn. A I **50** (1948).

Schiffer, M.: [1] On the modulus of doubly-connected domains. Quart. J. Math. Oxford Ser. **17**, 197—213 (1946).

Springer, G.: [1] Fredholm eigenvalues and quasiconformal mapping. Acta Math. **111**, 121—142 (1964).

Stoïlow, S.: [1] Leçons sur les principes topologiques de la théorie des fonctions analytiques. Paris: Gauthier-Villars 1938.

Strebel, K.: [1] On the maximal dilation of quasiconformal mappings. Proc. Amer. Math. Soc. **6**, 903—909 (1955).

Taari, O.: [1] Charakterisierung der Quasikonformität mit Hilfe der Winkelverzerrung. Ann. Acad. Sci. Fenn. A I **390** (1966).

Teichmüller, O.: [1] Untersuchungen über konforme und quasikonforme Abbildung. Deutsche Mathematik **3**, 621—678 (1938). — [2] Extremale quasikonforme Abbildungen und quadratische Differentiale. Abh. Preuß. Akad. Wiss. **22**, 1—197 (1940).

Tienari. M.: [1] Fortsetzung einer quasikonformen Abbildung über einen Jordanbogen. Ann. Acad. Sci. Fenn. A I **321** (1962).

Väisälä, J.: [1] On quasiconformal mappings in space. Ann. Acad. Sci. Fenn. A I **298** (1961). — [2] Remarks on a paper of Tienari concerning quasiconformal continuation. Ann. Acad. Sci. Fenn. A I 324 (1962).

Vekua, I. N. :[1] Verallgemeinerte analytische Funktionen. Berlin: Akademie-Verlag 1963.

Volkovyskij, L. I.: (Л. И. Волковыский): [1] Квазиконформные отображения. Львов: Издательство Львовского Университета 1954.

Wittich, H.: [1] Zum Beweis eines Satzes über quasikonforme Abbildungen. Math. Z. 51, 278–288 (1948).

Yûjôbô, Z.: [1] On absolutely continuous functions of two or more variables in the Tonelli sense and quasiconformal mappings in the A. Mori sense. Comment. Math. Univ. St. Paul 4, 67–92 (1955).

Zygmund, A.: [1] Trigonometric series. Vol. II. Cambridge University Press 1959.

Index

Absolutely continuous
—, additive set function 117
—, function on an arc 126
—, function on an interval 125
—, homeomorphism 118, 162
— in the sense of Tonelli 143
—, locally 117
— on lines 127, 162
Agard 181
Ahlfors 2, 16, 21, 28, 44, 66, 79, 82, 85, 98, 100, 132, 159, 166, 195, 206, 213, 214, 219
Analytic
—, arc 20
—, characterization of a quasiconformal function 246, 247
—, curve 20
—, definition of quasiconformality 168
—, quadrilateral 27
—, ring domain 36
Angular distortion 181
Annulus 31
Apostol 149
Approximation
—, good 185
— in the L^pmetric 141
— of a function with L^p-derivatives 144, 147
— of a measurable function 136, 138
— of a quadrilateral from inside 26
— of a quasiconformal mapping 207
— of a quasisymmetric function 91
— of a ring domain from inside 36
— theorem of Weierstrass 138
Arc
—, analytic 20
—, Jordan 6
— length 122
—, locally rectifiable 122
—, of bounded turning 102
—, quasiconformal 97
—, rectifiable 122
Area measure 110
Areal derivative 120, 130

Behnke 13, 56
Belinskij 227
Beltrami differential equation 184
—, generalized solution 184
—, generalized L^p-solution 184
Bers 2, 165, 170, 187, 195, 207, 219
Beurling 21, 79, 82, 85, 132, 159, 166, 206
Bojarski 195, 214, 215, 219, 236
Borel function 112
Borel set 111
Boundary
— behaviour
— — of conformal mappings 13
— — of quasiconformal mappings 42, 79
— — of topological mappings 11
— correspondence theorem (for conformal mappings) 14
—, positively oriented 8
— sets, corresponding to each other 11
— value problem
— — for a Jordan domain 79
— — for a multiply connected domain 85
— — for the half-plane 80, 82
Bounded turning 100, 102
Bounded variation 123, 126
—, on lines 127

Calderón 159, 213, 214
Canonical annulus 31
Canonical mapping
— of a quadrilateral 15
— of a ring domain 31
Canonical rectangle 15

Cantor set 125
Carleson 219
Characteristic function 112
Characterization of a (homeomorphic) quasiconformal mapping 16, 30, 41, 48, 168, 171, 173, 174, 176, 177, 185, 211
Characterization of a quasiconformal function 239, 240, 245, 247
Circular dilatation 105, 177
Class
— C^n 139
— C^∞ 139
— C_0^∞ 139
— L^p 135
— O_{AD} 206
Closed polygon 6
Completely additive class 111
Completeness of L^p 158
Complex derivatives 49
Complex dilatation 182
Conformal equivalence classes of simply connected domains 13
Conformal module, see Module
Conformality at a point 219
Conjugate quadrilaterals 99
Connectivity 10
Continuation, see Extension
Continuity
—, absolute 117, 118, 125, 126, 127, 143, 162
—, Hölder 70
— of the module 26, 27, 36
Convergence
— in the sense of Fréchet 27
— of quadrilaterals 26, 27
— of ring domains 35
—, weak 186
Convexity theorem of Riesz 214
Curve
—, analytic 20
—, Jordan 6
— of bounded turning 100
—, oriented 8
—, quasiconformal 97
—, rectifiable 122

Degenerate quadrilateral 25
Degenerating ring domains 34
Density, see Point of density
Derivative
—, areal 120, 130
—, complex 49
— in the direction α 17
—, L^p- 143, 162
— of a set function 119
Differentiability 9, 49
— of a homeomorphism 128
— of a quasiconformal mapping 164, 233, 235
Dilatation
—, circular 105, 177
—, complex 182
— condition 50, 164
—, maximal 16
—, —, at a point 47
—, —, of quadrilaterals 16
—, —, of ring domains 38
— quotient 17, 49
Dimension of a point set 116
Distance
— between a-sides (b-sides) 22
—, hyperbolic 66
—, spherical 5
Distortion function
— λ 81, 82
— φ_K 63, 64
Distortion theorem of Koebe 56
Dunford 214

Egoroff's theorem 113
Equicontinuity 68
Existence theorem 194
— for the local maximal dilatation 197
Extension
— of quasisymmetric functions 89
— over a free boundary arc 47, 98
— to an isolated point 41
— to free boundary arcs 42
— to the exterior of a compact set 96
Extremal domain
— of Grötzsch 53
— of Mory 58
— of Teichmüller 55
Extremal length 22, 133

Finite plane 5
Free boundary arc 11
Fubini's theorem 113
Fuglede 132, 134, 135
Function
—, absolutely continuous

—, — on an arc 126
—, — on an interval 125
—, — on lines 127, 162
—, —, set- 117
—, admissible for a curve family 132
—, Borel 112
—, Hölder continuous 70
—, integrable 113
—, measurable 112
— of bounded variation
— — on an arc 126
— — on an interval 124
— — on lines 127
—, quasiconformal 239
—, quasisymmetric 88
—, singular 125
— with L^p-derivatives 143

Gauss 195
Gehring 39, 105, 128, 170, 174, 177, 178, 181, 198, 216
General module condition 171
Generalized Cauchy integral formula 154
Generalized solution of a Beltrami equation 184
Geometric characterization of a quasiconformal function 240, 245
Geometric definition of quasiconformality 16
Golusin 27
Good approximation 185
Green's formula 148
Gross 116
Grötzsch 1, 17, 18, 53
—, extremal domain 53
—, inequality 18
—, module theorem 54

Hausdorff dimension 116
Hausdorff measure 116
Hersch 59, 63
Hilbert transform 157
—, differentiation 157
—, integral representation 156, 159
—, L^2-norm 157
Hilbert transformation 157
—, extension to L^2 159
—, L^p-norm 214
Hölder condition 70
Hölder continuous 70
Hölder's inequality 141
Homeomorphism 6
—, absolutely continuous 118
—, differentiability 128
—, locally absolutely continuous 118
—, sense-preserving 9
Horizontal rectangle 164
Hyperbolic metric 66

Inequality
— of Grötzsch 18
— of Hölder 141
— of Rengel 22
Inner measure 112
Integrable 113
—, locally 139
Integral 113
—, Lebesgue 113
—, Legendre normal 60
—, over an oriented arc 126
—, Stieltjes 126
Interior mapping 241
Interval 6
Invariance
— of domains 6
— of open sets 6
Isolated boundary point 41

Jacobian 9, 49, 130
Jordan arc 6
—, oriented 8
Jordan curve 6
—, oriented 8
Jordan curve theorem 7
Jordan domain 8
Juve 233

Kelingos 88
Kernel of a sequence of sets 76
Koebe 205
—, distortion theorem 56
—, one-quarter theorem 56, 233
K-quasiconformal function 239
K-quasiconformal mapping 16
k-quasisymmetric function 88
Künzi 3
Kuusalo 195

Landau's symbols O and o 49
Lavrentieff 2, 66, 194

Lebesgue
— area measure 110
— convergence theorem 114
—, extended measure 132
— integral 113
— linear measure 110
— theorem 120
— volume measure 111
Legendre normal integral 60
Lehto 92, 106, 108, 128, 170, 216, 236
Length of an arc 122, 123
Length preserving parametrization 123
Light mapping 241
Limit exponent $p(K)$ 215
Limit function of quasiconformal mappings 29, 73, 74, 210
Line 6
—, extended 6
—, finite 6
— segment 6
Linear measure 116
—, Hausdorff 116
—, Lebesgue 110
Local maximal dilatation 48
Locally absolutely continuous 117
Locally of bounded variation 124
Locally rectifiable 122
L^p-derivatives 143, 162
L^p metric 135
L^p-norm 135
— of Hilbert transformation 214
Lusin's theorem 137

Mapping 5
—, affine 19
—, antiquasiconformal 16
—, canonical
—, —, of a quadrilateral 15
—, —, of a ring domain 31
—, conformal 13
—, homeomorphic 6
—, interior 241
—, light 241
—, open 241
—, quasiconformal 16
—, —, K- 16
—, —, 1- 28
—, —, regular 17
Mapping theorem 194
—, Riemann 13
Marstrand 116
Martio 195
Maximal dilatation
— at a point 47
— of a homeomorphism 16
— of quadrilaterals 16
— of ring domains 38
Measurable function 112
Measurable set 111
Measure
—, area 110
—, Hausdorff 116
—, inner 112
—, Lebesgue 110
—, linear 110, 116
—, outer 110
Metric
—, hyperbolic 66
—, L^p 135
—, spherical 5, 34
Module
—, conformal invariance 173
— of a family of arcs or curves 133
—, —, monotonicity 133
—, —, subadditivity 133
— of a quadrilateral 15, 22, 173
— —, characterization without conformal mapping 21
— —, continuity 26, 27
— —, monotonicity 25
— —, superadditivity 25
— of a ring domain 31, 32, 173
— —, characterization without conformal mapping 32
— —, continuity 35
— —, monotonicity 35
— —, superadditivity 35
Module condition 171
Module theorem
— of Grötzsch 54
— of Mori 59
— of Teichmüller 56
Mori 2, 58, 66, 106, 162, 165, 169
— extremal domain 58
— module theorem 59
Morrey 2, 165, 195
Munroe 109, 111, 112, 113, 114, 115, 117, 118, 158

Näätänen 182
Net 138
Newman 4, 6, 7, 8, 9, 10, 11, 12, 242
Nikodym 118
Norm

—, L^p 135
—, —, of Hilbert transformation 214
Normal family 72
—, closed 72
— of quasiconformal mappings 72
n-tuply connected 10
Null set 111

One-quarter theorem of Koebe 56, 233
Open mapping 241
Orientation 8
Orientation theorem 9
Oriented arc 8
Oriented curve 8
Outer measure 110
—, Hausdorff 116
—, —, α-dimensional 116
—, —, linear 116
— in the sense of Carathéodory 110
—, Lebesgue 110
—, —, area 110
—, —, linear 110
—, —, volume 111
—, regular 110

Pesin 205
Pfluger 2, 16, 63, 92, 99, 104, 162, 170, 233
Piranian 105
Plane 5
—, finite 5
Point of density 115
—, linear 115
—, xy- 115
Polygonal arc 6
Polynomial approximation 138, 210
Positively oriented boundary 8

Quadrilateral 14
—, analytic 27
—, approximation from inside 26
—, a-side (b-side) 14
—, canonical mapping 15
—, canonical rectangle 15
—, conformal mapping 15
—, conjugate pair 99
—, convergence 26, 27
—, convergence from inside 26
—, degenerate 25
—, dilatation under a homeomorphism 16
—, homeomorphism 15
—, module 15, 22, 173
—, quasiconformal mapping 30
—, side 14
—, vertex 14
Quasiconformal arc 97
Quasiconformal curve 97
Quasiconformal extension 96, see also Extension
Quasiconformal function
—, analytic characterization 247
—, definition 239
—, geometric characterization 240, 245
Quasiconformal mapping (quasiconformal homeomorphism) 16
—, absolute continuity on arcs 170
—, absolute continuity on lines 162
—, absolute continuity with respect to area measure 165
—, analytic definition 168
—, approximation 185, 207
—, boundary behaviour 42, 79
—, characterization 16, 30, 41, 48, 168, 171, 173, 174, 176, 177, 185, 211
—, convergence of derivatives 186, 216
—, degree of continuity 71
—, differentiability 164, 233, 235
—, dilatation condition 50, 164
—, geometric definition 16
—, good approximation 185, 207
—, K- 16
—, L^p-derivatives 165, 215
—, limit mapping 29, 73, 74, 210
—, of a quadrilateral 30
—, of a ring domain 38
—, 1- 28
—, preservation of null sets 165, 216
—, regular 17, 210, 235
—, regular points 166, 199, 233
—, representation 218
Quasiconformal reflection 98
Quasisymmetric function 88

Rademacher-Stepanoff theorem 165
Radó 27
Radon-Nikodym theorem 118
Real-analytic 208
Rectangle 15
—, horizontal 164
Rectifiable 122
Reed 85
Reflection
— in a circular arc 46
—, quasiconformal 98

Reflection principle 47
Regular point 10
Regular quasiconformal mapping 17, 210, 235
Reich 27, 39, 216, 237
Removability
— of an analytic arc 44
— of an isolated point 41
Removability problems 199
Removable 41, 199
Rengel's inequality 22
Renggli 177, 199
Rickman 103
Riemann mapping theorem 13
Riemann sphere 5
Riesz convexity theorem 214
Ring domain 30
—, analytic 36
—, canonical mapping 31
—, convergence from inside 35
—, degenerating 34
—, module 31, 32, 173
—, quasiconformal mapping 38

Šabat 236
Saks 109, 115, 120, 123, 127, 130, 165
Sario 206, 207
Schiffer 55
Schwartz 214
Sense-preserving homeomorphism 9
Separation theorems 6
Set function 110, 132
Sewing theorem 92
σ-finite 112
Simply connected 10
Singular function 125
Sommer 13, 56
Spherical metric 5, 34
Springer 98
Step function 138
Stepanoff 130, 165
Stoïlow 241
Strebel 162, 202
Support 139

Taari 181
Teichmüller 2, 54, 56, 59, 224
—, extremal domain 55
—, module theorem 56
Tienari 99, 104
Turning 100, 102

Uniform Hölder continuity 70
Uniformization theorem 195
Uniqueness theorem 183

Väisälä 39, 86, 104, 105, 106, 108, 153, 171, 174, 177
Variation
—, bounded 123, 126
—, —, on lines 127
—, on an arc 126
—, on an interval 124
Vekua 213, 219
Vertex of a quadrilateral 14
Virtanen 92, 106, 108
Volkovyskij 2, 181, 236

Walczak 237
Weak convergence 186
Weierstrass's approximation theorem 138
Wittich 224

Yûjôbô 2, 170, 178

Zygmund 159, 213, 214

721/10/73 V/12/6

Die Grundlehren der mathematischen Wissenschaften in Einzeldarstellungen mit besonderer Berücksichtigung der Anwendungsgebiete

Eine Auswahl

---

1. Blaschke/Leichtweiß: Elementare Differentialgeometrie. 5. Aufl.
27. Hilbert/Ackermann: Grundzüge der theoretischen Logik. 6. Aufl.
51. van der Waerden: Einführung in die algebraische Geometrie. 2. Aufl.
63. Eichler: Quadratische Formen und orthogonale Gruppen.
65. Fejes Tóth: Lagerungen in der Ebene, auf der Kugel und im Raum. 2. Aufl.
67. Byrd/Friedman: Handbook of Elliptic Integrals for Engineers and Scientists. 2nd edition
77. Behnke/Sommer: Theorie der analytischen Funktionen einer komplexen Veränderlichen. Nachdruck der 3. Aufl.
87. van der Waerden: Mathematische Statistik. 3. Aufl.
96. Bachmann: Aufbau der Geometrie aus dem Spiegelungsbegriff
99. Cassels: An Introduction to the Geometry of Numbers. 2nd printing
102. Nevanlinna/Nevanlinna: Absolute Analysis
121./122. Dynkin: Markov Processes
123. Yosida: Functional Analysis. 3rd edition
124. Morgenstern: Einführung in die Wahrscheinlichkeitsrechnung und mathematische Statistik. 2. Aufl.
125. Itô/McKean jr.: Diffusion Processes and Their sample Paths
126. Lehto/Virtanen: Quasiconformal Mappings in the plane
127. Hermes: Enumerability, Decidability, Computability. 2nd edition
128. Braun/Koechler: Jordan-Algebren
129. Nikodym: The Mathematical Apparatus for Quantum Theories
130. Morrey jr.: Multiple Integrals in the Calculus of Variations
131. Hirzebruch: Topological Methods in Algebraic Geometry. 3rd edition
132. Kato: Perturbation Theory for Linear Operators
133. Haupt/Künneth: Geometrische Ordnungen
134. Huppert: Endliche Gruppen I
135. Handbook for Automatic Computation. Vol. 1/Part a: Rutishauer: Description of ALGOL 60
136. Greub: Multilinear Algebra
137. Handbook for Automatic Computation. Vol. 1/Part b: Grau/Hill/Langmaack: Translation of ALGOL 60
138. Hahn: Stability of Motion
139. Doetsch/Schäfke/Tietz: Mathematische Hilfsmittel des Ingenieurs. 1. Teil
140. Collatz/Nicolovius/Törnig: Mathematische Hilfsmittel des Ingenieurs. 2. Teil
141. Mathematische Hilfsmittel des Ingenieurs. 3. Teil
142. Mathematische Hilfsmittel des Ingenieurs. 4. Teil
143. Schur: Vorlesungen über Invariantentheorie
144. Weil: Basic Number Theory
152. Hewitt/Ross: Abstract Harmonic Analysis. Vol. 2: Structure and Analysis for Compact Groups, Analysis on Locally Compact Abelian Groups
154. Singer: Bases in Banach Spaces I
155. Müller: Foundations of the Mathematical Theory of Electromagnetic Waves
156. van der Waerden: Mathematical Statistics
157. Prohorov/Rozanov: Probability Theory
158. Constantinescu/Cornea: Potential Theory on Harmonic Spaces

159. Köthe: Topological Vector Spaces I
160. Agrest/Maksimov: Theory of Incomplete Cylindrical Functions and Their Applications
161. Bhatia/Szegö: Stability Theory of Dynamical Systems
162. Nevanlinna: Analytic Functions
163. Stoer/Witzgall: Convexity and Optimization in Finite Dimensions I
164. Sario/Nakai: Classification Theory of Riemann Surfaces
165. Mitrinović: Analytic Inequalities
166. Grothendieck/Dieudonné: Eléments de Géométric Algebriqué I
167. Chandrasekharan: Arithmetical Functions
168. Palamodov: Linear Differential Operators with Constant Coefficients
169. Rademacher: Topics in Analytic Number Theory
170. Lions: Optimal Control of Systems Governed by Partial Differential Equations
171. Singer: Best Approximation on Normed Linear Spaces by Elements of Linear Subspaces
172. Bühlmann: Mathematical Methods in Risk Theory
173. F. Maeda/S. Maeda: Theory of Symmetric Lattices
174. Stiefel/Scheifele: Linear and Regular Celestial Mechanics. Perturbed two-body Motion — Numerical Methods — Canonical Theory
175. Larsen: An Introduction of the Theory of Multipliers
176. Grauert/Remmert: Analytische Stellenalgebren
177. Flügge: Practical Quantum Mechanics I
178. Flügge: Practical Quantum Mechanics II
179. Giraud: Cohomologie non abélienne
180. Landkof: Foundations of Modern Potential Theory
181. Lions/Magenes: Non-Homogeneous Boundary Value Problems and Applications I
182. Lions/Magenes: Non-Homogeneous Boundary Value Problems and Applications II
183. Lions/Magenes: Non-Homogeneous Boundary Value Problems and Applications III
184. Rosenblatt: Markov Processes. Structure and Asymptotic Behavior
185. Rubinowicz: Sommerfeldsche Polynommethode
186. Wilkinson/Reinsch: Handbook for Automatic Computation II, Linear Algebra
187. Siegel/Moser: Lectures on Celestial Mechanics
188. Warner: Harmonic Analysis on Semi-Simple Lie Groups I
189. Warner: Harmonic Analysis on Semi-Simple Lie Groups II
190. Faith: Algebra: Rings, Modules, and Categories I
191. Faith: Algebra: Rings, Modules, and Categories II
192. Mal'cev: Algebraic Systems
193. Pólya/Szegö: Problems and Theorems in Analysis. I.
194. Igusa: Theta Functions
195. Berberian: Baer*-Rings
196. Athreya: Branching Processes
197. Benz: Vorlesungen über Geometrie der Algebren
198. Gaal: Linear Analysis and Representation Theory
199. Nitsche: Vorlesungen über Minimalflächen
200. Dold: Lectures on Algebraic Topology
201. Beck: Continuous Flows in the Plane
202. Schmetterer: Introduction to Mathematical Statistics
203. Schoeneberg: Elliptic Modular Functions
204. Popov: Hyperstability of Control Systems